T0139757

Society of Earth Scientists Series

Series Editor

Satish C. Tripathi, Lucknow, India

The Society of Earth Scientists Series aims to publish selected conference proceedings, monographs, edited topical books/text books by leading scientists and experts in the field of geophysics, geology, atmospheric and environmental science, meteorology and oceanography as Special Publications of The Society of Earth Scientists. The objective is to highlight recent multidisciplinary scientific research and to strengthen the scientific literature related to Earth Sciences. Quality scientific contributions from all across the Globe are invited for publication under this series. Series Editor: Dr. Satish C. Tripathi

More information about this series at http://www.springer.com/series/8785

Manvendra Singh Chauhan ·
Chandra Shekhar Prasad Ojha
Editors

The Ganga River Basin: A Hydrometeorological Approach

Editors
Manvendra Singh Chauhan
Department of Civil Engineering
Holy Mary Institute of Technology
and Science
Hyderabad, Telangana, India

Chandra Shekhar Prasad Ojha
Department of Civil Engineering
Indian Institute of Technology Roorkee
Roorkee, Uttarakhand, India

ISSN 2194-9204 ISSN 2194-9212 (electronic)
Society of Earth Scientists Series
ISBN 978-3-030-60871-2 ISBN 978-3-030-60869-9 (eBook)
https://doi.org/10.1007/978-3-030-60869-9

This Springer imprint is published by the registered company Springer Nature Switzerland AG
The registered company address is: Gewerbestrasse 11, 6330 Cham, Switzerland

Series Editor's Foreword

Rivers are considered as 'Mother' in Hindu literature because they upkeep the people thriving on them by all means: food, water, livelihood, environment, etc. But unfortunately mankind was not kind enough to protect them. As such, ecological health of rivers is deteriorating significantly. The Ganga river basin is one of the largest basins of the world and is home of nearly half the population of India. Many ancient cities (e.g., Varanasi) are situated at its bank. Ganga basin provides over one-third of the available surface water in India and contributes to more than half of the national water use. As such, its importance in the social, spiritual, and economic growth of the large population is unaccountable. However, while people overutilized its resources, they returned nothing which led to ill health of mighty Ganga and its tributaries. Of late, government started attempts to regain its health by strategic planning and execution. As such, even a small piece of scientific work has importance in achieving this goal.

The hydrometeorological studies of Ganga basin are very important because large population depends upon Ganga water for drinking, irrigation, industrialization, etc. Social and industrial effluents have contaminated the surface and groundwater of the Ganga basin on the one hand, and climate change scenario impacted the water supply on the other. The proper river water management of the basin is need of the hour. I am thankful to Prof. Manvendra Singh Chauhan and Prof. Chandra Shekhar Prasad Ojha for coming up with this topical volume on Ganga basin. I hope contributions on hydrometeorological aspects of Ganga basin will help planners in river basin management.

Lucknow, India

Satish C. Tripathi
Series Editor

Introduction

Ancient Indian literature highlights the importance of various river systems and states that these rivers are lifeline of human civilization. Thus, if we want to be well, these river systems should be in a healthy state. River health does not limit itself to good water quality but also its abundance in terms of water quantity. Among many river systems, the one which has been the voice of the Indian subcontinent is River Ganga. Its journey of approximately 2500 km from Gangotri Glacier, higher Himalaya to its confluence with the Bay of Bengal, has economic, social, religious, and cultural bearing on the huge population it hosts in five states of India, viz. Uttarakhand, Uttar Pradesh, Bihar, Jharkhand, and West Bengal. Rainfall, subsurface water flows, and snowmelt from glaciers are the main sources of water in the River Ganga. This linkage between human well-being and river health has motivated political thinkers to pay attention to rejuvenate this river system for past several decades.

At this stage, two slogans (Aviral Dhara and Nirmal Dhara) related to continuous flow in the flow of the pollution-free water have been very popular. Continuous flow is particularly important as recently, the Ganges River has witnessed the launch of navigation transport between Varanasi and Haldia. Similarly, quality has always been important as many human activities including bathing, drinking, and river water use for agricultural activities have been prevalent. For the last few decades, there is a growing emphasis on the sustainable use of water resources. Toward this, the assessment of water resources in the river basin has been acquiring regular attention.

As the River Ganges originates from Himalaya, there is a concern about whether glacier retreats and the extinction of glacier will adversely affect the flow in the upper stretches of River Ganges. It is also equally important that whether the precipitation patterns in the Ganges basin will lead to further attenuation of the river flow. Similarly, if there is an increase in temperature over the basin, will it lead to more depletion of the soil moisture leading to drought conditions in many parts of the river basin. Thus, variation in the hydrometeorological parameters may significantly impact the water yield, and their future projections may be of considerable interest.

Global spatial patterns of standard deviation, signal-to-noise ratio, skewness, and kurtosis of eight hydrometeorological variables, i.e., potential evapotranspiration, average temperature, precipitation, maximum temperature, minimum temperature, wet day frequency, diurnal temperature range, and vapor pressure, are studied by Sharma and Ojha (2020). Hydrometeorological and hydrological variables influence river system's water yield and quality. These variables may influence many hydrological processes, including contaminant transport through porous media, aquifers, river systems, and its interface. The world is mostly witnessing the onset of climate change which may lead to uncertainty in projection of water resources (Sharma et al. 2019). It is vital to identify the sources of hydrometeorological variables to assess how climate change has taken place in different parts of the basin (Sharma and Ojha 2019).

These issues have motivated us to bring out this special publication on River Ganga. The book contains twenty chapters covering topics related to water quality, water yield, the variability of hydrological variables, floods, droughts, water supply, and contaminant transport. Some of the modeling tools discussed in this book may help practitioners and researchers.

Shukla et al. deal with water quality challenges in Ganga river basin. For assessment of water quality indicators, such as biochemical oxygen demand (BOD), chemical oxygen demand (COD), dissolved oxygen (DO), and fecal coliform are assessed during 2010–2013. Analysis of river sediments indicates the presence of many heavy metals. Even if steps are taken to reduce influent quality to zero levels in the river, it will still take a considerable time for the river to attain its very high-quality levels because of the interaction with underlying sediments and flowing water. This study is based on analysis of data collected till 2015. Singh and Datta deal with the sequential characterization of contaminant plumes using feedback information. The proposed methodology is based on an optimization model that utilizes feedback information obtained from sequentially designed contaminant monitoring sites to characterize the contaminant plume when adequate initial concentration measurements are not available, and the contaminant sources are unknown. Zahra et al. deal with the evaluation of groundwater quality using multivariate analysis in the Raebareli district of Uttar Pradesh, India. This chapter reflects results from a large sample of groundwater quality data. The analysis is done using multivariate statistical techniques, namely principal component analysis (PCA)/factor analysis (FA), cluster analysis, and multiple linear regression (MLR) to compute the variability of the water quality data and to identify the sources that presently affect the groundwater. Gangwar and Singh explained optimal groundwater monitoring network models developed to determine the mass estimation error of contaminant concentration over different management periods in groundwater aquifers. The mass estimation error of contamination concentration over time is determined using the various computer software such as method of characteristics (MOC, USGS), Surfer 7.0, and simulated annealing (SA). Omar et al. deal with a case study of the Varanasi district in India. In this, the impact of leachate parameters on the river water (surface water) from a few landfill sites was observed. Interaction of groundwater with surface water was quantified using the groundwater flow modeling program. Thakur et al. deal with the augmentation of drinking water supply by riverbank filtration (RBF) in the Indo-Gangetic Basin. Here, the water quality data of different RBF sites are studied to assess the efficacy of the riverbank filtration system. Shekhar et al. give an overview of a new monitoring system using acoustic Doppler current profiler (ADCP) measurements and river flow simulation at the eight sections of the River Ganga at Varanasi. This ADCP was used to attain accurate and continuous monitoring for river discharge at a low cost. Kumar and Ojha focus on the review for the field performances of the application of different permeable and impermeable-type structures in several river basins to emphasize the effectiveness and non-effectiveness of these structures. The outcomes may help river engineers in utilizing the information to select a better alternative as per their field requirement.

Sreedevi et al. present the performance of a physically based spatially distributed model SHETRAN which is compared with two simple lumped conceptual rainfall–runoff models, namely the Australian Water Balance Model (AWBM) and GR4J in India. The results showed comparable performance of SHETRAN and AWBM. The study concludes that conceptual models are best suited for data-scarce regions, and the choice of distributed and lumped models for hydrologic studies is dependent on data availability and output requirements. Ojha explained the regression relationships are developed to estimate average minimum, average, and average maximum near-surface air temperature lapse rates (at monthly timescales) for the Ganga basin. Normal daily air temperature data from 178 stations and the latitude, longitude, and elevation of the stations were used for developing the regression equations. Pal and Ojha deal with the identification of the relationship between precipitation and atmospheric oscillations in the upper Ganga basin. The study outcomes are particularly beneficial for hydrometeorological analyses and climate impact assessment-based studies in the region. Medhi and Tripathi include the application of regional flood frequency analysis (RFFA) applied to 53 sub-basins in the Ganga basin, India. This study shows the usefulness of stream network information for RFFA and suggests the need for future research to incorporate stream network information in flood quantile estimation. In the chapter by Swetapadma and Ojha, plotting positions for the generalized extreme value (GEV) distribution are described and modified for better accuracy. The necessity of modifying plotting positions is to address the effect of sample size and skewness coefficient. The outcome of this chapter may be useful to estimate flood or precipitation or other hydrometeorological variables corresponding to a return period. Kumar and Khan demonstrate the application of machine learning like clustering over the 20 different stations across the Ganga basin using the average monthly rainfall of 102 years. Clusters are sensitive to the average and seasonal rainfall. Pathak et al. demonstrate the performance of the integrated valuation of ecosystem services and trade-off (InVEST) model in estimating water yield in one of the most diverse and undulated topographic basins of India, i.e., upper Ganga basin. The estimated water yield values are compared with the observed, to understand the variability of the yield models in predicting water yield in the upper Ganga basin. Das and Goyal review the literature regarding the climate change impact analysis over the Ganga river basin. Also, the precipitation and temperature extreme indices are analyzed using the trend analysis. Sharma and Ojha analyze changes in twentieth-century seasonal precipitation in the Ganga river basin. Overall, it was observed that the precipitation started decreasing significantly after the year 1960 in the basin. The another chapter by Sharma and Ojha deals with the analysis of twenty-first-century projections of precipitation and temperature in the Ganga river basin (GRB). The statistical downscaling method is used to downscale coarse-scale GCM data at the local station. The twenty-first-century future projections of temperature, and annual and seasonal precipitation were estimated considering RCP4.5 and RCP8.5. Pal and Ojha deal with trends in rainfall, mean temperature, and soil moisture over the Ganga river basin. Long-term trends in hydrologic and climatic annual time series are analyzed using the Mann-Kendall test for the period 1948–2015.

The last chapter by Patil et al. presents the strategic analysis of water resources in the Ganga basin using an integrated tool developed by Deltares and its partners AECOM India and FutureWater, in cooperation with the Government of India. The Ganga water information system (GangaWIS), which combines the models with a database and tools, presents the input data, and the simulation results in graphical and map format. The system is operational for the impact assessment of socioeconomic and climate change scenarios and management strategies.

In recent decades, the water crisis has received attention of the world community. There is a need to protect existing water resources in terms of quality and quantity. To understand the impact of anthropogenic activities and climate change, there is upsurge of studies. Also, geospatial technology (Chawla et al. 2020) coupled with machine learning tools (Goyal et al. 2017) has gained momentum in water resources arena. Many hydrometeorological aspects are not necessary feature in the book. Considering the uncertainty of future and the way changes will shape our planet, it is expected that this brief compilation of articles will motivate future generations to achieve the mission of Aviral Dhara (continuous flow) and Ujjawal Dhara (pollution-free flow), not only for Mother River Ganges but also for all other blessed rivers of the planet earth.

Manvendra Singh Chauhan
Chandra Shekhar Prasad Ojha
Editors

References

Chawla, Ila, Karthikeyan, L. and Mishra Ashok K. (2020) A review of remote sensing applications for water security: Quantity, quality, and extremes. Journal of Hydrology, v. 585, DOI: https://doi.org/10.1016/j.jhydrol.2020.124826.

Goyal, M.K, Ojha, C.S.P and Burn D. (2017) Machine Learning Algorithms and Their Application in Water Resources Management" In: Sustainable Water Resources Management, Edited by C.S.P. Ojha, Rao Y Surampalli, Bardossy, A., Tian C. Zhang, and Kao, C.M. ASCE, https://doi.org/10.1061/9780784414767.ch06

Sharma Chetan and Ojha, C.S.P. (2019) Changes of Annual Precipitation and Probability Distributions for Different Climate Types of the World. Water, 11(10), pp 1-21, https://doi.org/10.3390/w11102092

Sharma C. and Ojha C.S.P. (2020) "Statistical Parameters of Hydrometeorological Variables: Standard Deviation, SNR, Skewness and Kurtosis", In: Advances in Water Resources Engineering and Management, DOI: https://doi.org/10.1007/978-981-13-8181-2_5

Contents

About the Editors

Dr. Manvendra Singh Chauhan has completed his M.Tech. and Ph.D. degrees from Indian Institute of Technology, (Banaras Hindu University), Department of Civil Engineering, with specialization in Hydraulics and Water Resources Engineering. He received fellowships from MHRD as JRF and SRF. Currently, he is Professor of Civil Engineering Department, Holy Mary Institute of Technology and Science, Hyderabad, India. He also obtained field experience while working with Reliance Refinery Jamnagar, Gujarat. He is also Member of Institution of Engineers (India) and Life Member of many other reputed societies. His research interest is in the field of water resources, environmental engineering and geohydrological engineering. He has supervised three M.Tech. students and more than ten students of B.Tech. He published one book and fifteen research papers in national and international journals. He has also recently organized International Conference on 'Recent Trends in Civil Engineering and Water Resources Engineering'.

Prof. Chandra Shekhar Prasad Ojha (F.ASCE), Civil Engineering, IIT Roorkee, India, has over 36 years of professional experience. He is Institute Chair Professor, IIT Roorkee, since June 2018, and Adjunct Professor, Civil and Environmental Engineering Department, University of Missouri, Columbia, USA, since September 2018. His research interest is in the area of water resources and environmental engineering. He has published eight books, over 350 research publications including 175 research articles in peer-reviewed international journals. He stared his professional career from IIT BHU, Varanasi, in 1984 where he was involved in Environmental Impact Assessment of River Ganges. He has supervised more than 50 Ph.D. students. He has been Commonwealth Research Scholar at Imperial College of Science, Technology and Medicine, London, UK (1990–1993); Visiting Scholar at Louisiana State University, USA (2001); Alexander Von Humboldt Fellow at Water Technology Centre, Karlsruhe, Germany, and Guest AvH Fellow, Institute for Hydromechanics, University of Karlsruhe, Germany (2002–2003); Visiting Professor, Civil Engineering, AIT Bangkok (August 2004–November 2004); and Curtis Visiting Professor at Purdue University (August 2012–June 2013). He has been awarded Distinguished Visiting Fellowship of Royal Academy of Engineering and Visiting Fellowships of ASCE, JSPS, and STINT. He has received Young Engineer Award from Central Board of Irrigation and Power and AICTE Young Teacher Career Award. His research work has received several awards from national and international organizations including Institution of Engineers (India) (1989, 1992, 2005); Indian Water Works Association (2007); Indian Society of Hydraulics (2007); and ASCE (2001, 2009, 2010, 2013, and 2018). He also received ASCE State-of-the-Art of Civil Engineering Award in 2014.

Water Quality Challenges in Ganga River Basin, India

Anoop Kumar Shukla, C. S. P. Ojha, Satyavati Shukla, and R. D. Garg

Abstract Ganga is considered to be the most important and holiest river all over the world, having its own economic, environmental, and cultural value in India. During the past few decades, fast-developing industrialization and urbanization have led to an alarming threat to groundwater and surface water quality. With a highly non-uniform pattern of precipitation in the Ganga basin, it is facing extreme water shortage in several sections. Due to the extensive use of groundwater in the river basin, the water table has decreased at an alarming rate, which has resulted in the reduction of flows in the majority of the streams across this river basin. Increasing prosperity and quickly developing urbanization has resulted in increased generation of wastewater. River Ganga and its tributaries are receiving a considerable quantity of treated and untreated sewage generated from industrial operations and municipal discharges. Hence, serious water quality issues are posed in the river basin due to the combined effect of increased waste loads and decreased water flows. The physicochemical parameters such as Biochemical Oxygen Demand (BOD), Chemical Oxygen Demand (COD), and Dissolved Oxygen (DO) in the river water were explored at various sampling sites using data from Central Water Commission (CWC) for years 2010–2013. The water quality data is also supplemented with new pollution levels reported in the literature. The analysis of available data in research reveals that the Ganga River water quality remains declined because of the presence of industrial waste, domestic waste, heavy metals, sewage, animal, and human skeletons. The presence of abundant lethal chemicals such as Cr, Zn, Cu, Pb, Fe, Ni, and Cd also show the unhygienic state of the water quality of river Ganga. Water quality analysis of Ganga water revealed the presence of various microbial species such as Faecal Coliform and Total Coliform. The efforts from the Government to clean river Ganga are worth appreciation. However, to ensure the success of such programs, an accurate

A. K. Shukla (✉) · C. S. P. Ojha · R. D. Garg
Civil Engineering Department, Indian Institute of Technology Roorkee, Roorkee, India
e-mail: anoopgeomatics@gmail.com

S. Shukla
Centre of Studies in Resources Engineering (CSRE), Indian Institute of Technology Bombay, Mumbai, Maharashtra 400076, India

© The Author(s), under exclusive license to Springer Nature Switzerland AG 2021
M. S. Chauhan and C. S. P. Ojha (eds.), *The Ganga River Basin: A Hydrometeorological Approach*, Society of Earth Scientists Series,
https://doi.org/10.1007/978-3-030-60869-9_1

1

inventory of streams and drains joining river Ganga along with the pollution potential of each of these still need to be done. Also, the use of eco-friendly technologies
needs to be promoted wherever feasible in the Ganga basin.

Keywords Ganga basin · Biochemical oxygen demand (BOD) · Urbanization ·
Faecal coliform

1   Introduction

Ganga River is one of the very significant rivers of India. It flows from the Himalayan
Mountain to the Indian Ocean. The origin place of river Ganga is Gaumukh. The river
has an immense religious significance as a large number of cities are located nearby
the Ganga Riverbank. The Ganga is the only river, which serves the world's highest
population densities and covers the drainage area approximately 861,404 km^2. As
Ganga has its extreme importance for all living beings living in its vicinity, several
studies were done in the Ganga basin related to the assessment of quality and quantity of water. For example, Singh (2010) studied the physicochemical properties (i.e.,
acidity, alkalinity, temperature, pH, BOD, COD, DO, Cl$^-$, Electrical conductance,
NO$_3$$^-$, PO$_4$$^{3-}$) of water in river Ganga and impact of pollution on river Ganga at
different sites in Varanasi. Singh et al. (2012) measured the amount concentrations
of heavy metals such as Cd, Zn, Cu, Cr, Ni in river water and sediments of river
Ganga in the middle Ganga plain for a period of two years 2007–08. The concentrations observed in the river water were Zn (up to 0.87 mg/L); Ni (up to 0.12 mg/L);
Cr (up to 1.09 mg/L); Cu (up to 0.12 mg/L) and in the sediments the range were
Zn (up to 0.87 mg/g); Ni (up to 0.09 mg/g); Cr (up to 0.14 mg/g); and Cu (up
to 0.09 mg/g). Beg and Ali (2008) assessed the quality of sediment in the Ganga
River at Kanpur city, which is infamous for contamination of river water because of
discharges from tannery industries. The observations were shocking as Chromium
(Cr) concentrations in downstream river reach were 30 folds higher than that of in
upstream river sediment. A similar study was done by Bhatnagar et al. (2013), but
some heavy metals (Ni, Cd, Pb, Zn, Mn, Cu, Fe, Co, As, and Cr) were assessed
and were found in the significant amount. Many of these are health hazards. Few
studies are also reported on the contamination of water due to Mercury (Hg). Sinha
et al. (2006) have attempted to assess the extent of pollution caused by mercury
in the Ganga River at Varanasi. The observations included the concentrations and
amount of deposited mercury in river water. The aspects, which were covered for this
study, were water, sediment, benthic macro-invertebrates, fish, aquatic macrophytes
of the Ganga River, and soil and vegetation present in the associated floodplain. The
yearly mean concentration of mercury in the river water was 0.00023 ppm, in sediment was 0.067 ppm, in benthic-invertebrate biota was 0.118 ppm, in fish of Ganga
River was up to 91.679 ppm. In flood, plain soil was reported up to 0.269 ppm. Pal
et al. (2012) found high levels of methyl mercury (MeHg) in samples of freshwater

fishes collected from the Ganga River in West Bengal. The high levels of accumulated mercury, which was found in some fish muscles, contained about 50–84% of organic mercury. The highest amount of organic mercury found was around 0.93 ± 0.61 μg Hg/g, which was surprisingly very high and toxically unacceptable. The consumption may result in early nervous dysfunction. That is why the proper monitoring of such kind of pollution, which is hazardous, must be promptly taken care of. Some studies related to Arsenic (As) poisoning in the Ganga basin are also reported in the literature. Nickson et al. (1998), Acharyya et al. (1999) have studied it for alluvial Ganga aquifer in Bangladesh and West Bengal. Chakraborti et al. (2003) have studied it for Middle Ganga Plain, Bihar. In these cases, the river was observed polluted due to naturally occurring Arsenic in alluvial Ganga aquifers. Ahamed et al. (2006) observed the contamination of groundwater due to Arsenic in the upper and middle Ganga plain, mainly in the three districts of Uttar Pradesh such as Gazipur, Ballia, and Varanasi.

Ahamed et al. (2006) have done a preliminary medical check-up in eleven affected villages, i.e., one from Gazipur district and ten from Balliadistrict. Typical arsenical skin lesions, Arsenical neuropathy, and adverse obstetric outcome were observed, which indicated the severity of Arsenic exposure in the population. These studies give an idea about the severe effects of Arsenic, which are to be handled with higher priority. Thus, based on some of these studies on water quality, it can be realized that Ganga sediment, the water within it, and the water in the vicinity of its course, along with living organisms within the river, have been affected. With this in view, the chapter is aimed and organized as follows: (i) To investigate the effects on surface water quality of the Ganga River basin (ii) To introduce some preventive measures to improve as well as control the Ganga River water quality.

2 Study Area and Its Water Resources

The Ganga River system is composed of the Ganga-Brahmaputra-Meghna river basin. The Ganga River is extended over India, Nepal, Bangladesh, and China, and the total drains area is 1,086,000 km^2. In the east, it is surrounded by Brahmaputra range and in the west by Aravalli hills, which separates the Ganga basin from the Indus basin. The Himalayas are located in the north and the south, it is surrounded by Vindhyas and Chhotanagpur Plateau. This basin comprises the complete states of Uttar Pradesh (240,798 km^2), Uttarakhand (53,566 km^2), West Bengal (71,485 km^2), Bihar (143,961 km^2) and Delhi (1484 km^2), and some part of Rajasthan (112,490 km^2), Madhya Pradesh (198,962 km^2), Himachal Pradesh (4317 km^2) and Haryana (34,341 km^2) by Tripathi and Tripathi (2012). Figure 1 shows the extent and monitoring stations of the Ganga River basin. The climate in the basin varies from temperate in the north to tropical monsoon in the south. Ganga River is essential to its inhabitants, particularly concerning water resources. According to the Indian Meteorological Department, the normal annual rainfall across the river basin is approximately 1051 mm, and almost three fourth part of the rainfall is received during monsoon

Fig. 1 Study area map showing the monitoring stations of the Ganga River basin

season (Jain and Kumar 2012). A considerable amount of water is lost in the evaporation and transpiration process, and the remaining water is utilized for land practices. However, rainfall varies in the basin from <500 mm in Hissar (Haryana), 2209 mm in the upper Himalayan region to about 1600 mm in Kolkata, Alipore (West Bengal). During the monsoon period, the excess water available from rainfall is beyond the holding capacity of natural and human-made structures across the basin, which results in flash floods.

Conversely, during the summer period, the excessive demands of water coupled with extreme evaporation rates in the basin cause drought conditions, and hence, the import of water is demanded. Flood and droughts are indicators of severity in the hydrological cycle of the Ganga River basin. The circumstances are worsened because of exhausting forest cover and a huge increasing population in the basin. The Central Water Commission (2005) has evaluated the catchment-wise average yearly flow in Indian river systems, and it is found that the average annual flow for the Ganga basin is 525 km^3.

Because of geographical, hydrological, and different limitations, it is not possible to utilize all the available surface water. Using conventional development methods, about 250 km^3 of surface water of the basin can be put to valuable use for various activities. Ministry of Water Resources conducted a study on catchment wise groundwater potential of Ganga River, and it was observed that extreme groundwater potential of about 171 km^3 exists in these catchments compared to other watersheds of the country. In excess, a total groundwater potential of 39% is observed in the Ganga

basin of the country. Groundwater is highly accessible in these catchments; therefore, groundwater use is high compared to other watersheds of the country. All the
main streams of the river Ganga are not perennial. Some important sub-tributaries
like Sone, Betwa, Chambal, Khan, Yamuna, and Kshipra were perennial. Still, from
the last few decades, it has become seasonal due to overexploitation of groundwater
in the catchments, which results in a decline of the groundwater table at an alarming
rate and also reduces the baseflow in the non-monsoon period.

3 Water Quality Variations

For assessment of water quality indicators, such as BOD (Biochemical Oxygen
Demand), COD (Chemical Oxygen Demand), DO (Dissolved Oxygen), Faecal
Coliform, etc. are in use. A sample of observed data (courtesy, CWC) for these
parameters at four sampling stations is shown in Fig. 2 over a period of 4 years.
This gives a comparison of four different places for four different years. As per the
plots of maximum BOD, the most vulnerable site is Kanpur. The reason behind this
may have a large influence on the tannery industries lying in that area. However,
the government has put restrictions on it and do a regular check on the sewage and
industrial treatment, but even then, there may be many loopholes, which cause the
untreated water, are drained directly into the river and results in pollution of river
water. Again, the DO was found to be minimum for sites located at Kanpur city.
Figure 2 is provided to give an idea about the levels of water quality indicators in the
Ganga River at different sites. It can be seen from Fig. 2 that the water quality of the
Ganga River does indicate some sign of improvement in the last two years. Cumulative level of contamination from urban and industrial areas has resulted in severe
water quality deterioration in the Ganga River. Cascading medium and large cities
located along the course of river Ganga and its tributaries release a large amount of
treated and untreated urban municipal wastes as well as industrial effluents into the
river water, which further aggravates the water pollution problem in the basin.

It is observed that about three-fourths of the Ganga water pollution is contributed
from untreated domestic sewage from urban households. This river receives approximately 900 million liters of sewage consistently. Middle reaches of the basin
consisting of Kanpur and Buxar are the most industrialized and urbanized, hence the
most polluted part of the basin. The introduction of industrial and urban hazardous
wastes into the watercourse of this part of the basin is a grave threat to the health of
society. In the hilly northern reaches of the basin, particularly up to Rishikesh, except
for the sediments, the Ganga River water is immaculate. After Rishikesh, the large
amounts of pollutants start introducing into the river water. In addition to municipal
sewage, partially treated industrial effluents are also released into the river water from
Haridwar and Rishikesh. The total population of Haridwar district is 1.5 lakh, and as
it is a religious site, on an average of about 60,000 people visit this town daily. This
number further increases to lakhs (may go up to 15 lakhs) on critical religious days
during the Kumbha fair. The large sewer lines in this area get obstructed from the

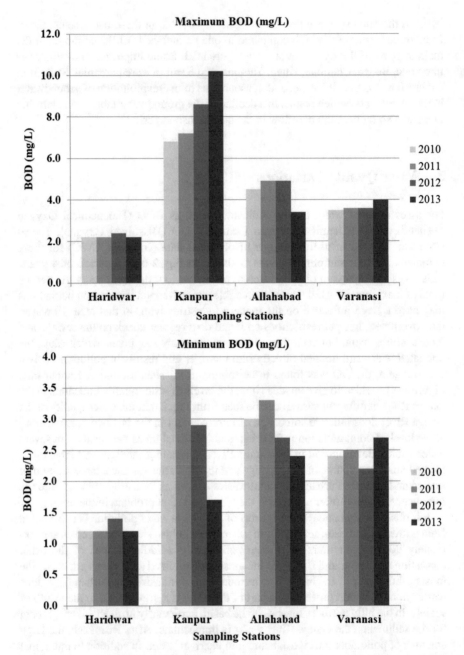

Fig. 2 BOD, COD and DO at Haridwar, Kanpur, Allahabad and Varanasi sites for the year 2010–2013

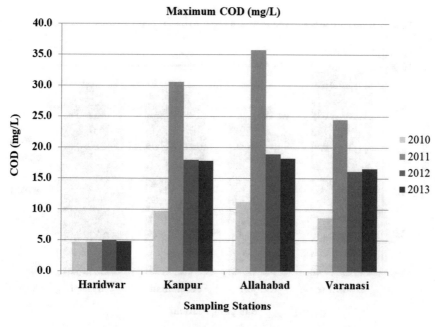

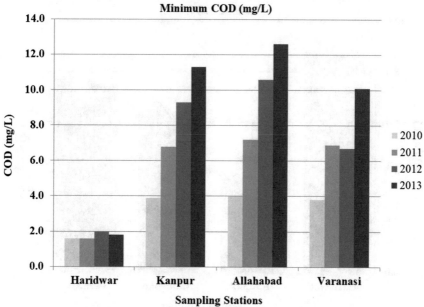

Fig. 2 (continued)

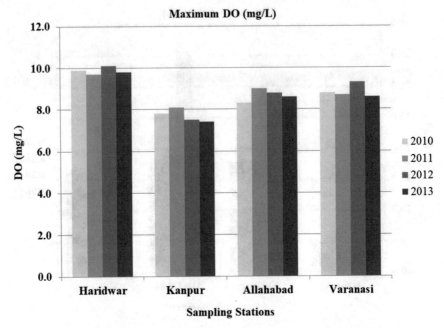

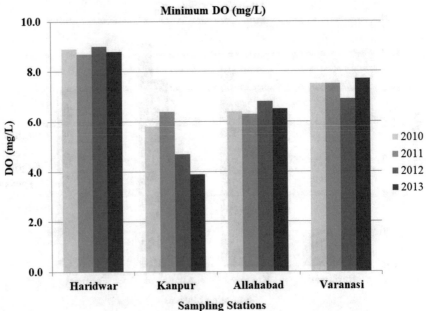

Fig. 2 (continued)

sediments that flow in from the adjoining mountainous regions. After Haridwar, river Ganga flows downstream through towns such as Bijnor, Garhmukteshwar, Narora, and Kannauj. Due to the absence of large industries in this region, the river water is not much polluted. Further downstream, when the river reaches Kanpur, the river water gets severely polluted. Municipal sewage from Kanpur city (2.7 million population) and untreated toxic industrial effluents released into the river water from about 150 industrial units result in severe water quality degradation. At the Jajmau urban-industrial zone of Kanpur, an average DO value of 3 mg/L is observed, which is a result of organic pollution of river Ganga from about 80 tanneries and other industrial units. Downstream when the river passes through the holy city of Allahabad (approximate population of >1 million), the urban sewage is the main contributor to water pollution. The Yamuna, which is one of the most polluted rivers in India, joins river Ganga in Allahabad at Sangam. Hence, the extremely polluted river water of Yamuna further pollutes Ganga waters. Further, when the river reaches the religious city of Varanasi, river Varuna consisting of several drains introduces a large number of wastes into the river water. With about 1.2 million population, it is identified as one of the most densely populated cities in India. There is a religious belief that if people die and get cremated in Varanasi, they will go to heaven for sure. Hence, on an average, >40,000 dead bodies are cremated and remains dumped into river Ganga in Varanasi. Additionally, a massive amount of industrial and urban wastes generated from Varanasi city are released into the Ganga, which further degrades its water quality. Downstream when the Ganga River enters Bihar state, Patna is the most populated city which contributes a huge amount of untreated domestic urban sewage into the river water. In addition, the water quality is degraded from the various industries, mainly fertilizer industries and oil refineries located along the banks of the river Ganga. Similarly, downstream of river Ganga, Kolkata (West Bengal) is the most populated city in the Ganga (Hooghly) river basin. The wastewater from municipal sewage and various industries contaminate Ganga water. Similar observations were made in a study done by Shukla et al. (2018a, b).

Water quality deterioration issues generally come about because of the release of unprocessed or incompletely processed domestic/household wastewater from city sprawl and urban slums came about because of expanding inhabitants across the Ganga River and its sub-tributaries. Chaudhary et al. (2017) monitored pre and post-monsoon month's water quality data at nine sampling locations from Haridwar to Garhmukteswar. In the year 2014–2015, the results indicated that the water quality of the Ganga River was appropriate for drinking, and the authors also found that river water was acutely polluted due to heavy metals, the 9 sample locations used by the author included Bhimgoda and downstream of Garhmukteswar. The authors also monitored water quality in different major drains such as Haridwar and Laksar drain and three river tributaries Solani, Malin, and Chhoiya River. The contribution from Chhoiya River was significant because this river water had a BOD of 60 mg/L, COD of 791 mg/L, and turbidity of 354 NTU. As a result of this river, the water quality of Ganga, which was found having BOD 2.65 mg/L, COD 38 mg/L and turbidity 7.78 NTU at Bhimgoda (Haridwar) changed to BOD 5.18 mg/L, COD of 112 mg/L and turbidity 22.7 NTU at Garhmukteswar in a stretch of 160 km only. The Ganga water quality was deteriorated simply because of the pollutant addition by drains, rivers, and tributaries of rivers. Therefore, to protect the Ganga River quality, proper measures must be adopted to reduce the polluting potential of intermediate drains and tributaries; unless it is done the effort of the Government to clean river Ganga will not be realized in the near future. Paul (2017) again reported the concentration of the heavy and toxic metals in the Ganga River and riverbed sediments. Using the data reported in Paul (2017), Fig. 3 has been plotted between the concentration (mg/L) and Min.-Max. values of the heavy metal for 5 sites of Ganga river viz. Rishikesh, Haridwar, Kanpur, Allahabad, and Varanasi. The results indicated that the concentration of heavy metals Pb, Zn, and Cr had crossed the permissible limit (Fig. 3). Recent investigations by Thakur et al. (2017) revealed that Arsenic was not present in the river waters. However, the adjacent aquifers to the Ganges River had undesirable Arsenic concentrations. In such situations, where aquifer quality is adversely affected by Arsenic, the use of riverbank filtration technology was found to be useful to consume the Ganges water for drinking purposes. In a series of investigations (Ojha and Thakur 2010; Thakur et al. 2012) showed the advantage of using Ganga water through adaptation of infiltration wells close to river Ganga bank at Haridwar.

4   Major Causes of Water Quality Degradation

There is a considerable literature highlighting significant causes for water quality degradation in Indian rivers. We prefer to include this here through the material in this section is borrowed from different sources including those published by the Central Pollution Control Board (2013).

4.1   Wastewater from Metropolitan Cities

On Ganga River banks, about 29 urban areas (approximate population >100,000), 23 municipalities (approximate population >50,000) and 50 small towns (approximate population >20,000) are present. About 3000 MLD of domestic wastewater is generated from these urban regions and municipalities. In a study done by Trivedi and Trivedi (2012), out of 150 major industries located close to the banks of river Ganga, 85 are identified as highly polluting. Besides major industries, various small scale and large-scale industries are also situated along the Ganga Riverbank.

The majority of the small scale and large-scale industries in the region don't have proper waste treatment facilities. Therefore, treated and untreated wastes generated from these industries along with the municipal discharge, are introduced directly or indirectly into the river system. Results of MPN of total coliform reveals the bacterial contamination of river water. Table 1 illustrates the physicochemical properties of

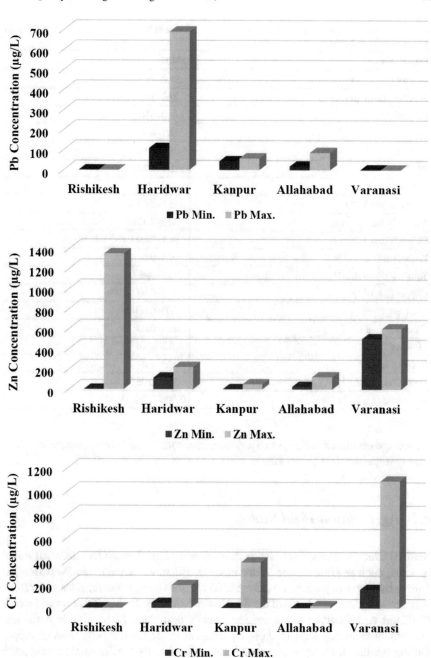

Fig. 3 The concentration of heavy metals Pb, Zn and Cr in Rishikesh, Haridwar, Kanpur, Allahabad and Varanasi

Table 1 The range of water quality parameters in river Ganga at Kanpur (Mean of 32 samples from the years 2015 to 2017)

Parameters	Value
Acidity (mg L^{-1})	58.2–78.9
Total alkalinity (mg L^{-1})	260–332
Ec (μmhos cm^{-1})	260–570
pH	7.7–8.7
COD (mg L^{-1})	3.8–142.4
BOD (mg L^{-1})	1.7–10.4
DO (mg L^{-1})	2.4–10.9
Sulphate (mg L^{-1})	5.3–22.1
Nitrate-N (mg L^{-1})	0.70–2.17
Chloride (mg L^{-1})	18.1–28.0
Potassium (mg L^{-1})	6.6–9.8
Total Coliform (MPN)	5–398
Copper (μg L^{-1})	0.6–0.97
Lead (μg L^{-1})	0.05–0.09
Cadmium (μg L^{-1})	0.04–0.09
Zinc (μg L^{-1})	0.1–49.4
Iron (μg L^{-1})	0.25–3.6
Chromium (μg L^{-1})	0.01–0.06
Phosphate (mg L^{-1})	0.004–2.16

the sewage generated at Kanpur city, which are the main factors responsible for water quality impairment in river Ganga.

4.2 Agricultural Field Runoff

Starting from Gomukh, river Ganga flows downstream through five different states of India, such as Uttarakhand, Uttar Pradesh, Bihar, Jharkhand, and West Bengal. During its course of journey, i.e., along 2525 km, the Ganga River receives water from very fertile agricultural lands of the river basin. In a study done by Trivedi and Trivedi (2012), it was observed that a huge amount of pesticides, herbicides, insecticides, and rodenticides are used in these agricultural fields for the protection of crops. During rainfall, these agro-chemicals containing mainly fertilizers are transported to the river water through surface runoff, causing severe water pollution problems (Shukla et al. 2018a, b).

4.3 Discarding of Bio-medical and Solid Wastes

In urban agglomerations and small towns located along the banks of Ganga River, a huge amount of solid waste materials are generated. Bio-medical wastes are composed of various toxic and harmful materials. These wastes (both bio-medical and solid wastes) are generally dumped into landfill sites located close to the river banks. Enormous toxic chemical leachates from these sites are introduced into the river water directly or indirectly. Sometimes these wastes are openly thrown into the river water causing severe water quality impairment.

4.4 Washing of Clothes and Animal Cleaning

While passing through urban areas and small towns, Ganga River receives a huge amount of inorganic and organic wastes from various anthropogenic activities such as the washing of clothes and animal bathing. It is observed that washing of clothes introduces a high amount of phosphate, alkaline, and carbonate materials that are directly introduced into the river system. Due to such activities, the pH of the river water has increased from 8.5 to 9.5 in the washing Ghats near Kanpur, Allahabad, and Varanasi. Similarly, due to lack of sufficient water and its conveyance facilities, the majority of the bathing Ghats are being used for bathing pets such as cows, dogs, buffalos, etc. Therefore, these events also contribute towards the lowering of water quality in river Ganga.

4.5 Temple Waste Materials

Ganga River has great importance to Hinduism and having its own spiritual values. Ganga is sacred, holistic, worshipped, and revered by the Hindu community. Hence it is referred to as "Mother Ganga". Almost more than 1000 tons of flowers, as well as garlands, are introduced into the river every day. These flowers, once thrown into the river water, further decompose, providing favorable conditions for the expansion of bacterial growth and lead to a decrease in the dissolved oxygen (DO) content of the river water. Many plastics bags are simply thrown into the river, which was used to bring the worship materials. This leads to physical contamination of the water.

4.6 Use of Lift Canals for the Extraction of Water

Various dams and lift canals are built over Ganga River to extract a large amount of river water and to supply it for various purposes such as domestic, industrial and agricultural (irrigation). Tehri dam is constructed over Bhagirathi River in Uttarakhand state, which is utilized for generating electricity. At Haridwar, barrages alter the flow of huge amount of water to the upper Ganga canal. The course of Ganga water changes into the Madhya Ganga channel due to another barrage at Bijnor. Further deviation of water is observed into the lower Ganga Channel at Narora. Few other lift canals are additionally developed on river Ganga to extract the water and supply it to the vast agricultural fields of Gangetic plains. Because of substantial water extraction from the river Ganga, the flow regime of the river is also altered. A significant reduction of water quantity in the river has reduced the dilution capacity and hence the self-purification capacity of the river.

4.7 Deforestation in Himalayan Region and Development of Reservoirs

Various multipurpose dams/reservoirs are constructed on river Ganga and its tributaries in the Himalayan region, mainly for hydropower production. Due to the construction of these hydropower projects, lots of forests are removed in this region. Heavy blasting in the region results in enormous soil erosion. The silts/sediments generated from the erosion reach the Ganga River through runoff, causing heavy siltation in the riverbed of this region. It results in the reduction of water holding capacity of the river.

5 Cleaning Efforts from Government of India

Without the intervention of Government, things should have been worse. In this respect, authors personally appreciate the efforts of the Government from time to time. It is important that major initiatives adopted by the Government must be highlighted. Some such programs are worth mentioning here.

5.1 NamamiGange Programme (NGP)

An integrated Ganga development project titled "NamamiGange," which means "Obeisance to the Ganga River" was declared on 10 July 2014 by Union Finance Minister of Indian Government. Spending cost of Rs. 20,000 crores are declared for this program for the next 5 years, and about 48 industrial units operating near river Ganga are shut down on the strict orders from the Government of India as a part of this program. Under this plan, the Central Government is undertaking 100% funding for various projects and activities. Learning lessons from the unacceptable consequences of previous Ganga Action Plans (GAP), the government has planned to facilitate the maintenance and operation of the assets for a period of 10 years. For pollution hotspots, "Special Purpose Vehicle/Public-Private Partnership" (SPV/PPP) approach is planned to be adopted. "NamamiGange" program is focused on pollution abatement interventions such as diversion, interception and treatment of wastewater from open drains; proper in-situ treatment of wastes in effluent treatment plant (ETPs)/sewage treatment plants (STPs); expansion and restoration of existing STPs; bio-remediation or utilization of creative innovations; immediate on the spot, short term pollution abatement measures by capturing and treating pollution at exit points or prevention of untreated sewage inflow into the river water, etc.

5.2 Ganga Manthan (GM)

Ganga Manthan is another plan inaugurated on 7 July 2014, where all different kind of problems along with the solution to clean Ganga water was discussed for restoring the water quality. The National Conference was conducted to get feedback from stakeholders and prepare an action plan to clean river Ganga. This program was composed of the National Mission for Clean Ganga at Vigyan Bhawan New Delhi.

5.3 National Mission for Clean Ganga (NMCG)

Under the society registration Act of 1860, the NMCG plan was registered as a society on 12 August 2011. National Ganga River Basin Authority (NGRBA) was constituted by the Government of India under the provisions of the Environment Protection Act (EPA) of 1986. NMCG plan acted as a task force of NGRBA. However, on 7 October 2016, the NGRBA plan was abandoned after the constitution of the National Ganga Council (NGC), also referred to as the National Council for Rejuvenate, Protection, and Management of River Ganga. Under this act, a five-tier structure is indicated at national, state, and district level's for control and prevention of environmental pollution in Ganga River while ensuring sufficient and continuous river flow regime. The five-tier structures are as follows:

1. National Ganga Council under the administration of Hon'ble Prime Minister of India
2. Empowered Task Force (ETF) on river Ganga under the administration of Hon'ble Union Minister of Water Resources, River Development and Ganga Rejuvenation
3. District Ganga Committees in each district along Ganga River and its tributaries in the respective states
4. State Ganga Committees
5. National Mission for Clean Ganga (NMCG).

Thus, the newly formed 5-tier structure is aimed to bring all stakeholders on a single platform to adopt a comprehensive strategy to clean Ganga, water quality enhancement and rejuvenation.

5.4 National Ganga River Basin Authority (NGRBA)

National Ganga River Basin Authority is one of the plan to reduce pollution and maintain the water quality. The authority of the plan can take all important measures to carry the proper planning and execution of the program. This can be achieved through four sectors viz.

1. Solid waste management
2. Wastewater management
3. Riverfront development and management
4. Industrial pollution management.

On 20 February 2009, the NGRBA plan was established at New Delhi aiming (a) conservation of Ganga River using a river basin approach which promotes inter-sectoral co-ordination for comprehensive management as well as planning, and guaranteeing actual reduction of Ganga water pollution; and (b) for guaranteeing water quality and environmentally sustainable development in the Ganga basin keeping of environmental flows.

For comprehensive management and planning in the Ganga basin, a river basin approach is required for conserving river Ganga and for the actual reduction of water pollution. The NGRBA is accounted to take up developmental and monitoring functions to meet the objectives while ensuring sustainability needs. The nodal Ministry for the NGRBA plan is the River Development and Ganga Rejuvenation, Ministry of Water Resources (RD and GR, MoWR). NGRBA committee is led by honorable Prime Minister of India. Its members are the concerned Union ministers and the Chief Ministers of the states (Uttarakhand, Uttar Pradesh, Bihar, Jharkhand, and West Bengal) through which Ganga River flows, among many others. Various activities to be implemented under this plan are aimed to restore combined endeavors of state and central government for cleaning river Ganga.

The role of the NGRBA is to develop a River Basin Management Plan for river Ganga; to prepare guideline of activities aimed at prevention; to conserve the water quality of river Ganga; control and reduction of pollution; and to take significant measures for restoration of a river ecology in the states of Ganga basin. For sustainability, minimum ecological flows are required to be maintained in the Ganga River, ensuring reduction in the water pollution by financing, planning, and executing programs which include:

1. Enhancement of Sewage Structure
2. Management of Watershed Area
3. Creating Public Awareness
4. Protection of Flood Plains.

5.5 Ganga River Basin Management Plan (GRBMP)

The consortium of seven Indian Institutes of Technology (IITs) (Kanpur, Delhi, Madras, Bombay, Kharagpur, Guwahati, and Roorkee) is preparing a comprehensive River Basin Management Plan for Ganga. The objective of the Ganga River basin management plan is to restore and rejuvenate the wholesomeness of the river ecosystem and the betterment of the environment. The wholesomeness of the river can be achieved in terms of four different approaches: "AviralDhara" (Continuous Flow"), "Nirmal Dhara" ("Unpolluted Flow"), Ecological Entity and Geologic Entity. The vision of the GRBMP are as follows:

1. The flow of the river should be continuous and smooth
2. The river should have lateral and longitudinal connectivity
3. The river should provide adequate space for its various functions
4. The river should not carry the waste load.

5.6 Other Initiatives

In addition, IIT Roorkee students are working in close collaboration with NIH, Roorkee and MoWR, India to identify all the drains and tributaries, which are likely to degrade the quality of river Ganga between a reach covering Uttarkashi to Narora. Similar efforts are being made at other places also. As Government is of the people and for the people, so should be the exercise of the river cleaning.

6 Summary

It is a general perception that in view of reduced flow in the river along with depletion of its discharge carrying capacity, the self-cleaning ability of the river is on the decline. Under these situations, river Ganga may be conserved with the involvement of stakeholders. Prevention is desirable and must be taken seriously. For this, there has to be a proper policy and planning. Analysis of river sediments also indicates the presence of many heavy metals, and even if, steps are taken to reduce influent quality to zero levels in the river, it will still take a considerable time for the river to attain its very high-quality levels because of the interaction with underlying sediments and flowing water. Riverbank filtration can be a viable technology for adaptation to provide clean drinking water to the residents dependent on river Ganga water. Efforts from the Government of India are in the right direction, and with the involvement of all stakeholders, there lies a better future for Clean Ganga.

References

Acharyya SK, Chakraborty P, Lahiri S, Raymahashay BC, Guha S, Bhowmik A (1999) Arsenic poisoning in the Ganges delta. Nature 401(6753):54

Ahamed S, Kumar SM, Mukherjee A, Hossain AM, Das B, Nayak B, Pal A, Chandra Mukherjee S, Pati S, Nath Dutta R, Chatterjee G, Mukherjee A, Srivastava R, Chakraborti D (2006) Arsenic groundwater contamination and its health effects in the state of Uttar Pradesh (UP) in upper and middle Ganga plain, India: a severe danger. Sci Total Environ 370(2–3):310–322

Beg KR, Ali S (2008) Chemical contaminants and toxicity of Ganga river sediment from up and down stream area at Kanpur. Am J Environ Sci 4(4):362–366

Bhatnagar MK, Singh R, Gupta S, Bhatnagar P (2013) Study of tannery effluents and its effects on sediments of river Ganga in special reference to heavy metals at Jajmau, Kanpur, India. J Environ Res Dev 8(1):56–59

Chakraborti D, Mukherjee SC, Pati S, Sengupta MK, Rahman MM, Chowdhury UK, Lodh D, Chanda CR, Chakraborti AK, Basu GK (2003) Arsenic groundwater contamination in Middle Ganga Plain, Bihar, India: a future danger? Environ Health Perspect 111(9):1194–1201

Chaudhary M, Mishra S, Kumar A (2017) Estimation of water pollution and probability of health risk due to imbalanced nutrients in River Ganga, India. Int J River Basin Manage 15(1):53–60

CPCB (2013) Pollution assessment: River Ganga. Central Pollution Control Board (Ministry of Environment and Forests, Govt. of India)

CWC (2005) Water sector at a glance. Central Water Commission, Ministry of Water Resources, Govt. of India, New Delhi

Jain SK, Kumar V (2012) Trend analysis of rainfall and temperature data for India. Curr Sci 102(1):37–49

Nickson R, McArthur J, Burgess W, Ahmed KM, Ravenscroft P, Rahmanñ M (1998) Arsenic poisoning of Bangladesh groundwater. Nature 395(6700):338

Ojha CSP, Thakur AK (2010) Turbidity removal during a subsurface movement of source water: case study from Haridwar, India. J Hydrol Eng 16(1):64–70

Pal M, Ghosh S, Mukhopadhyay M, Ghosh M (2012) Methyl mercury in fish-a case study on various samples collected from Ganges River at West Bengal. Environ Monit Assess 184(6):3407–3414

Paul D (2017) Research on heavy metal pollution of river Ganga: a review. Ann Agrar Sci 15(2):278–286

Shukla AK, Ojha CSP, Mijic A, Buytaert W, Pathak S, Garg RD, Shukla S (2018a) Population growth, land use and land cover transformations, and water quality nexus in the Upper Ganga River basin. Hydrol Earth Syst Sci 22(9):4745–4770

Shukla S, Gedam S, Khire MV (2018b) Implications of demographic changes and land transformations on surface water quality of rural and urban subbasins of Upper Bhima River basin, Maharashtra, India. Environ Dev Sustain 1–43

Singh N (2010) Physicochemical properties of polluted water of river Ganga at Varanasi. Int J Energy Environ 1(5):823–832

Singh L, Choudhary SK, Singh PK (2012) Status of heavy metal concentration in water and sediment of River Ganga at selected sites in the middle Ganga plain. Int J Res Chem Environ 2(4):236–243

Sinha RK, Sinha SK, Kedia DK, Kumari A, Rani N, Sharma G, Prasad K (2006) A holistic study on mercury pollution in the Ganga river system at Varanasi, India. Curr Sci 92(9):1223–1228

Thakur AK, Singh VP, Ojha CSP (2012) Evaluation of a probabilistic approach to simulation of alkalinity and electrical conductivity at a river bank filtration site. Hydrol Process 26(22):3362–3368

Thakur AK, Ojha CSP, Singh VP, Chaudhur BB (2017) Potential for river bank filtration in arsenic-affected region in India: case study. J Hazard Toxic Radioact Waste 21(4):04017015

Tripathi BD, Tripathi S (2012) Issues and challenges of river Ganga. In: Sanghi R (ed) Our national river Ganga. Springer, Cham, pp 211–211. https://doi.org/10.1007/978-3-319-00530-0_8

Trivedi RC, Trivedi RC (2012) Water quality challenges in Ganga Basin, India. In: Sanghi R (ed) Our national river Ganga. Springer, Cham, pp 189–210. https://doi.org/10.1007/978-3-319-005 30-0_7

Sequential Characterization of Contaminant Plumes Using Feedback Information

Deepesh Singh and Bithin Datta

Abstract In many practical field problems, it may not be possible to identify the actual characteristics (location, magnitude and duration of contamination) of the groundwater contaminant sources in a contaminated aquifer. Also, most of the time, very sparse information regarding spatiotemporal contaminant concentration is available initially, which is inadequate for reliable identification and simulation of the contaminant plume. Simulation of the contaminant plume movement is necessary to predict the future distribution of the contaminant in the groundwater aquifer. Reliable simulation and prediction are also essential for developing an efficient contamination monitoring strategy. To address this practical problem of data inadequacy, an interactive methodology is proposed, incorporating the sequential design of optimal monitoring networks. These sequentially developed and implemented monitoring networks provide feedback information on measured concentrations. This measurement information helps in progressively improving the prediction of the contaminant plume, starting with very sparse initial information about the contaminant sources and spatial distribution of concentration. The proposed methodology is based on an optimization model that utilizes feedback information obtained from sequentially designed contaminant monitoring sites to sequentially characterize the contaminant plume when adequate initial concentration measurements are not available, and the contaminant sources are unknown.

Keywords Monitoring network design · Simulated annealing · Kriging · Feedback based network design

D. Singh (✉)
Department of Civil Engineering, Harcourt Butler Technical University, Kanpur, Uttar Pradesh 208002, India
e-mail: dr.deepeshsingh@gmail.com

B. Datta
Discipline of Civil and Environmental Engineering, School of Engineering and Physical Sciences, James Cook University, Townsville, QLD 4811, Australia
e-mail: bithin.datta@jcu.edu.au

1   Introduction

Monitoring networks for the detection of contaminants in groundwater systems can serve two purposes. It can be used to detect the contaminant plume movement. Also, a designed monitoring network can be used to identify unknown contamination sources. In many real-life cases, due to various reasons, it may not be possible to identify the contaminant sources accurately. However, it may be possible to design an optimal monitoring network that can help in detecting the contaminant plumes without having to identify the contaminant sources. This can be accomplished by sequentially designing a monitoring network which is dynamic in nature, and by obtaining the concentration observations for the installed monitoring network (Datta et al. 2002). A methodology for simulation of contaminant plumes in an aquifer based on the sequentially designed and monitored network is proposed. The problem of groundwater monitoring network design has been formulated by various researchers. Some of the earlier attempts using integer and mixed-integer programming (MIP) are reported in Meyer and Brill (1988), Loaiciga (1989), Meyer et al. (1994), Loaiciga and Hudak (1992, 1993), Datta and Dhiman (1996). In recent years Simulated Annealing (SA) has been used in the groundwater problems as an optimization algorithm. SA is based on heuristic search techniques like a genetic algorithm (GA) and tabu search (TS). In this study, SA is used as an optimization algorithm. Simulated annealing is analogous to the annealing of solids in metallurgical parlance where the initial energy of a system is raised to allow molecules to be mobile; later, the system is cooled to a lower energy crystalline form at a slow rate. A search objective such as the cost is mapped onto the energy of the system and the feasible solutions onto the state of the system (Rogers et al. 1998). SA was first introduced by Kirkpatrick et al. (1983). This method of optimization has been used successfully in large scale applications in groundwater. Kuo et al. (1992) used it for pump and treated the problem, Lee and Ellis (1996) presented an empirical comparison of eight heuristic algorithms for solving a typical nonlinear integer optimization problem of groundwater monitoring network design. They found simulated annealing performs better among the other algorithms used. Wang and Zheng (1998) coupled two global search optimization techniques, GA and SA, with a flow simulation model. SA was found to outperform GA. Rao et al. (2006) discussed the problem of pumping from the existing series of skimming wells to maximize the pumpages. Geostatistical kriging in the groundwater has been used by various authors as an estimator for the unknown parameter values. This, being a linear interpolation procedure, provides a best linear unbiased estimator (BLUE) for quantities that vary spatially. Many researchers have considered geostatistical approaches for designing groundwater monitoring network (Rouhani 1985; Rouhani and Hall 1988; Loaiciga 1989; Loaiciga and Hudak 1993; Benjema et al. 1994; McKinney and Loucks 1992). Other authors have used kriging for variance reduction analysis and estimation of spatial maps (Loaiciga et al. 1995; Prakash and Singh 2000; Lin and Rouhani 2001; Lin et al. 2001; Passarella et al. 2003; Reed and Minsker 2004; Yeh et al. 2006; Chadalavada and Datta 2007). Singh and Singh (2013) presented a methodology using the Method of Characteristics (MOC) to model groundwater for non-biodegradable contaminant transport for varying transmissivity of the aquifer. The methodology developed here is essentially based on the kriging linked optimization approach. Simulated annealing (SA) is utilized as the optimization algorithm. The flow and contaminant transport processes are numerically simulated using MODFLOW-96 (Harbaugh and McDonald 1996) and MT3DMS (Zheng and Wang 1999) simulation, respectively. The spatial extrapolation of measured and simulated concentration is performed by using geostatistical estimation based on GSLIB (Deutsch and Journel 1998).

2 Methodology

The proposed kriging linked SA based methodology has three main components: (1) groundwater flow and transport simulation, (2) global mass estimation, and (3) optimization using SA.

2.1 Groundwater Flow and Transport Simulation

In the study area for the assumed conditions, it is essential to have a calibrated flow and transport model. The simulation model is used to simulate the contaminant scenario of the site from specified initial time t_0, to time t_n.

The equation describing the transient, two-dimensional areal flow of groundwater through a non-homogeneous, anisotropic, saturated aquifer can be written in Cartesian tensor notation (Pinder and Bredehoeft 1968) as:

$$\frac{\partial}{\partial x_i}\left(T_{ij}\frac{\partial h}{\partial x_j}\right) = S\frac{\partial h}{\partial t} + W; \quad i, j = 1, 2 \tag{1}$$

where K_{ij} = hydraulic conductivity tensor; T_{ij} = transmissivity tensor $K_{ij}b$; h = hydraulic head; W = volume flux per unit area b = saturated thickness of aquifer; and x_i, x_j = cartesian coordinates.

The partial differential equation describing the fate and transport of the kth contaminant species (Zheng and Wang 1999):

$$\frac{\partial(\theta C^k)}{\partial t} = \frac{\partial}{\partial x_j}\left(\theta D_{ij}\frac{\partial C^k}{\partial x_j}\right) - \frac{\partial}{\partial x_i}(\theta v_i C^k) + q_s C_s^k + \Sigma R_n \tag{2}$$

where θ = porosity of the subsurface medium, dimensionless; $x_{i,j}$ = distance along the respective Cartesian coordinate axis; t = time; C^k = dissolved concentration of species k; v_i = seepage or linear pore water velocity; q_s = volumetric flow rate per unit volume of aquifer representing fluid sources (positive) and sinks (negative); C_s^k = concentration of the source or sink flux for species k; D_{ij} = hydrodynamic dispersion coefficient tensor; ΣR_n = chemical reaction term.

In this study, a single species ($k = 1$) and conservative contaminant ($\Sigma R_n =$ 0) is considered. The flow simulation model MODFLOW is linked to the fate and transport simulation model MT3DMS.

2.2 Global Contaminant Mass Estimation

Cooper and Istok (1988) suggest that the global contaminant plume estimates can be obtained by combining the local estimates obtained from kriging. The output from MT3DMS as concentration values at the potential monitoring locations and their coordinates are used to randomly generate different design set of monitoring wells using SA. Each design set is used to compute contaminant mass based on geostatistical kriging estimation of the contaminant concentration at all the nodes of unknown concentration over the entire area. The global mass estimate for a particular design set is calculated as by summing up the concentration values at all the nodes and multiplying this value by porosity and effective volume of the contaminant plume. A larger number of the potential monitoring wells provides a better approximation of the actual contaminant mass present in the aquifer.

Kriging is a collection of generalized linear regression techniques for minimizing an estimation variance defined from a prior model for a covariance (Deutsch and Journel 1998). Kriging provides the best linear unbiased estimate for unsampled value. Kriging possesses some unique properties, including conditional unbiasedness, smoothing effect, additivity, and exact interpolation (Zheng et al. 2005). Kriging has been applied extensively to estimate hydrogeologic variables at unknown locations from scattered data points, such as transmissivity and hydraulic conductivity (Lavenue and Pickens 1992; McKinney and Loucks 1992; Eggleston et al. 1996; Fabbri 1997) and contaminant concentrations (Zhu et al. 1997; Reed et al. 2000; Chadalavada and Datta 2007). The kriging subroutine used in this study is based on GSLIB (Deutsch and Journel 1998).

2.3 Optimization Using SA

The optimization algorithm used for the design of the monitoring network is based on Simulated Annealing. Annealing is the cooling process of molten metals. At the high-temperature atoms with high energy moves freely and when the temperature is reduced, get ordered, and finally form crystals having minimum possible energy. This crystalline stage will not be achieved if the temperature is reduced at a fast rate (Deb 2002; Singh and Datta 2014).

2.4 Optimization Model Formulation

The contaminant plume characterization problem is solved by utilizing the optimal groundwater monitoring network design in each management period. The sequential design of a monitoring network for each management period is used to collect field contamination data (Singh and Datta 2016). The specified objective determines the optimal set of monitoring locations for which the normalized mass estimation error is minimum while limiting the total number of monitoring wells to a specified maximum number. The objective function is defined as:

$$OF1 = \underset{k}{Minimize} : ABS\left(\frac{M_{ac} - M_{est}^k}{M_{ac}}\right) \tag{3}$$

$$M_{est}^k = F\left\{\underset{k}{K}R(C_{ij}^s)\right\} \quad \forall i, j \in k \tag{4}$$

$$C_{ij}^s = f(I, BC, S) \quad \forall i, j \in k \tag{5}$$

Subject to:

$$\sum_{n=1}^{N} W_n \leq W_N \quad \forall n \in N \tag{6}$$

where M_{ac} = Actual mass based on simulated concentrations at all potential monitoring locations; M_{est}^k = total estimated contaminant mass for kth set of chosen monitoring locations; C_{ij}^s = simulated contaminant concentration at spatial location i, j belongs to the kth subset of the potential monitoring locations N at the end of the management period; $KR_k\left(C_{ij}^s\right)$ = spatially kriged concentrations at all nodes of the study area using the simulated concentration data at the kth chosen set of the potential monitoring locations; $F\left\{KR_k\left(C_{ij}^s\right)\right\}$ = estimated contaminant mass as function of the kriged concentration values at all the nodes of the study area; $f(I, BC, S)$ = simulated concentration obtained by solution of the simulation model as function of different initial conditions, boundary conditions, and source characteristics; I = initial conditions, at the beginning of the management period; $B.C.$ = boundary conditions as applicable; S = contaminant source characteristics, if any; W_n = a binary decision variable, 1 indicates a monitoring well is selected and 0 indicating otherwise; W_N = maximum number of installed monitoring wells; n = allowable number of potential locations in a design set.

The optimal monitoring network design based methodology is developed for the identification of contaminant plumes in a contaminated aquifer with very sparse initial contamination data and without identification of the contaminant sources. The steps in the application of this methodology are as follows.

In the first step of the proposed methodology, the field concentration measurements for the arbitrary existing locations are obtained. The study area is discretized into finite difference grids. These spatially sparse data of concentration measurements at arbitrary existing wells are then extrapolated to all finite-difference grid locations using geostatistical kriging. These spatially extrapolated concentration data are utilized as initial conditions for simulation of spatial concentrations at the end of a specified management period (assumed as 1 year in this study for illustration purpose). The boundary condition and hydrogeologic parameter values are specified based on available information. These simulated/predicted concentrations are utilized to design an optimal monitoring network to be installed at the end of the management periods. The objective function and the constraints (Eqs. 3–6) are specified in the above section. Once the network is designed, it is implemented, and actual field measurement data would be collected at these selected monitoring locations. These concentration measurements may also include previously existing observation locations. These collected observations are again extrapolated over the entire study domain using geostatistical kriging (Singh 2015). These spatial estimates are now utilized as initial concentration for simulating concentration at the end of the next management period ($f(I, BC, S)$).

It is expected that by using feedback information on actual field values of concentration at designed locations without any specific knowledge of the contaminant sources, it is possible to sequentially improve the identification and prediction of contaminant plumes. It is also expected that this sequential information would be very useful for the prediction of contaminant measurement in a contaminated aquifer, where it is difficult to identify the unknown sources of contaminant.

2.5 Model Application

The developed methodologies are applied to an illustrative study area. This study area forms a portion of a hypothetical aquifer with varying head boundary conditions on all sides. The illustrative study area is shown in Fig. 2. There are three contaminant sources $S1$, $S2$, and $S3$. It is assumed that the contaminant sources are conservative in nature and is in a solute form. The flow and transport model for the study area is based on the MODFLOW-96 (Harbaugh and McDonald 1996) and MT3DMS (Zheng and Wang 1999). The model covers an area of 7.56 km^2. The study area is discretized into finite difference grids of size 100×100 m. The study area is considered homogeneous and isotropic. The input parameter values used in the flow and transport simulation model are listed in Table 1.

The sources are assumed to be continuous in each management period (specified to be one year). The management period is divided into two management time intervals of 90 and 275 days. Three management periods are considered for the performance evaluation of the methodology. The recharge rate, pumping rate, and boundary conditions change with management time intervals. A total of 108 potential monitoring locations are assumed in this study (Fig. 1).

Table 1 Primary data used in flow and transport model

Parameters	Values
Porosity of the aquifer	0.29
Hydraulic conductivity (same in x and y-direction)	15 m/day
Number of nodes in x-direction	28
Number of nodes in y-direction	27
Management period	365 days
Management time interval 1	90 days
Management time interval 2	275 days
Number of pumping wells	756
Mass flux at the injection wells (only for evaluation purpose)	27.4 g/s
Injection rate	45 m^3/day
Storage coefficient	0.2
Longitudinal dispersivity	40 m
Ratio of the horizontal to the longitudinal dispersivity	0.1

2.6 Performance Evaluation

The performance of the methodology is evaluated for the illustrative study area with different scenarios of existing arbitrary initial monitoring wells and potential monitoring well locations. The aquifer is considered as unconfined, homogeneous, and isotropic in nature. Three management periods each of one-year duration and consisting of two-time intervals within a management period are considered. The actual contaminant plume at the end of the three management periods is shown in Fig. 3. This plume is used as a reference to evaluate the simulated plume as obtained using the proposed methodology. For a Scenario, ten arbitrary concentration monitoring locations are specified to be initially existing for measuring the contaminant concentration. These initial arbitrary locations are different from the potential monitoring locations specified. The exact locations and number of contaminant sources are assumed to be. In some cases, contaminant source data may be available and can be incorporated. Figure 2 shows the location of initial existing arbitrary monitoring locations, and the designed monitoring network locations for the three management periods.

Initial concentration measurements taken at these ten initially existing arbitrary locations are spatially kriged over the entire study area. Concentration measurement data from these arbitrary monitoring locations are utilized to design the optimal monitoring network at the end of the first management period. These new monitoring locations, as well as remaining arbitrary initial monitoring locations, are utilized to collect the concentration measurement data at the end of the second management period. The process is repeated for the third management period. Only for performance evaluation purposes for a hypothetical study area, the concentration

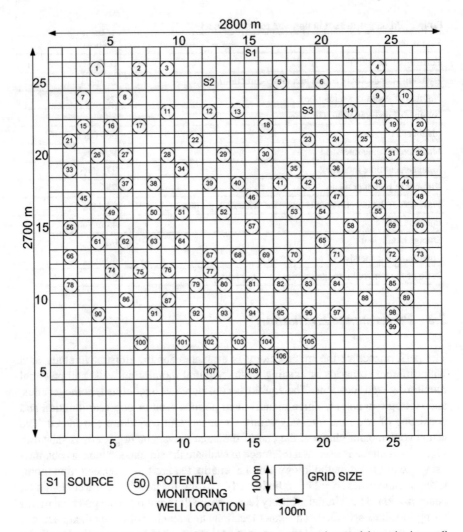

Fig. 1 Illustrative study area showing its physical parameters and potential monitoring well locations

measurement data at the monitoring locations are also simulated using all the specified known parameter values and boundary conditions, as no field measurements are available in such an illustrative problem. In real life, application field measurements will have to be collected over time. In real-life cases, the concentrations at monitoring locations are to be measured in the field. After three management periods, there would be forty monitoring locations, including ten arbitrary initially existing monitoring locations. In each management period, the monitoring network design model is solved for a maximum permissible ten new monitoring locations based on the budgetary constraint. In this illustrative problem, J is equal to 10, and K_i is equal to

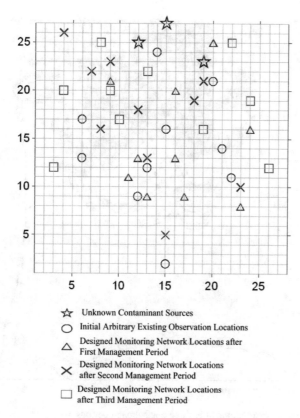

Fig. 2 Network design results

10 for $i = 1, 2, 3$. The selection of K_i optimal observation locations at the ith management iteration/period does not overlap existing arbitrary locations and designed monitoring locations in previous management periods. Figure 4 shows the concentration contours at the end of the third management period based on simulated concentration measurement data from the designed network covering three management periods and also from the initial arbitrary observation locations. These contaminant plume contours are compared with actual contaminant plume contours (Fig. 4) for evaluation purposes (simulated with all known sources, boundary conditions, initial conditions, and all parameter values). It is evident that the simulated plumes, as obtained based on monitored concentrations and actual plumes (also obtained by simulation using the extensive monitoring network consisting of 108 locations); do not show a close visual match.

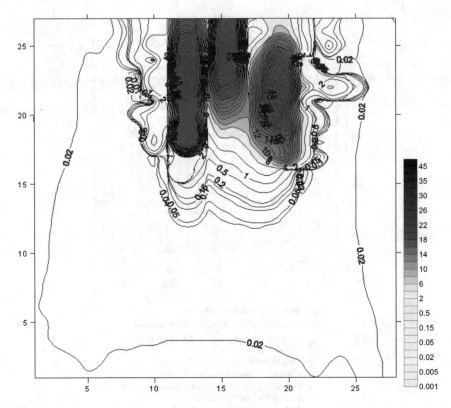

Fig. 3 Actual contaminant plume after three management periods

The performance of the plume characterizing methodology is also evaluated using the Root Mean Square Error (*RMSE*). The *RMSE* is defined as:

$$RMSE = \sqrt{\frac{1}{N_P} \sum_P \left(\frac{C_{ij}^s - C'}{C}\right)^2} \tag{7}$$

where C_{ij}^s = concentration value as actual

C' = concentration value as estimated

N_P = number of grid node points included.

This estimation is based on spatial estimation using kriging. The kriged or estimated values depend upon the choice of the variogram and its parameters. Kriging estimates depend upon the number of known data points. If the data measurement points are small in number, the spatial estimates will be poor at unknown locations. The *RMSE* value depends upon the squared difference of the simulated and estimated concentration values.

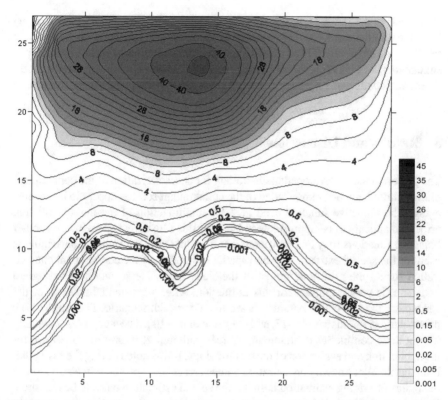

Fig. 4 Characterized contaminant plume contours after three management periods

Figure 5 shows the plot between the *RMSE* value and the number of grid node points used in the computation of the *RMSE*. Grid points starting from the lower southern boundary were first included in the computation. More and more numbers of grid points were added by including more grid points towards the northern boundary. It is seen from Fig. 5 that with nearly 90% of all points included, the *RMSE* error is

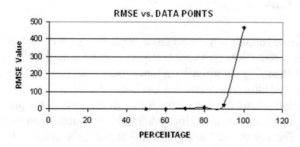

Fig. 5 Plot between RMSE value and number of data points selected after three management periods

very small in the range of 10.00. As more grid points towards the northern (upper) boundary are included, the *RMSE* suddenly increases and reaches a value of 468.39 for 100% of the points. These results show that the contaminant plume boundary is simulated quite well using the proposed procedure. However, it is not able to correctly detect the location of very large concentrations close to the contaminant sources.

3 Results and Discussion

The performance of developed methodology for characterization of the contaminant plume without any prior knowledge of the contaminant sources appear to be encouraging. The objective function used here for minimization of the mass estimation error also appears to be appropriate at least for this illustrative site. However, other objective functions may be incorporated and evaluated in the future. The kriging is used for the spatial extrapolation and spatial estimation of the contaminant concentration for all the grid node points of the study area. The selection of variogram parameter is very important for this estimation. After a substantial analysis of the variogram, the associated parameters are found for each scenario. The range of the parameter c is found to be 11–17, and the value of a is 100. The study area is actually having three continuous contaminant sources, although it is assumed unknown for the application and evaluation of the methodology. It was noted during the extrapolation, that smaller known concentration values create some problem in estimation. If the radius of kriging estimation is small, the spatial estimation tends to be clustered. There is no doubt that these performance evaluation results are also dependent on the kriging parameters chosen. It is evident that the performance of the methodology is dependent on the initial arbitrary monitoring network. It can be inferred that the contaminant plume characterization does improve with the increase in the design iterations. If the radius is increased for kriging estimation, the nearby values will have the same order of magnitude as the known values.

4 Conclusions

New kriging linked simulation-optimization methodology is developed in this study that combines a numerical flow and transport simulator, a global mass estimator, and simulated annealing algorithm to design a groundwater monitoring network and simulate contaminant plume using feedback monitoring information. The performance evaluation results presented here establish the potential applicability of the proposed kriging linked SA model for spatial and temporal estimation of the contaminant plume. The use of the SA algorithm is found quite suitable for solving the optimization problem. The choice of the objective function as the minimization of the contaminant mass estimation error appears to be adequate. These evaluations of the developed methodology also establish its potential applicability to actual field problems. Further improvement in the characterization of the contaminant plume is possible if a larger number of monitoring network design iterations or management periods are utilized.

However, it can also be concluded that the performance of the methodology is somewhat dependent on the initially existing arbitrary monitoring locations, and to a certain extent, on the correct choice of the kriging parameters. It is often possible to have some knowledge of the possible sites of potential contaminant sources. Even with very sketchy information, it may be possible to have a rough idea of possible contaminant sources. In this case, the first monitoring locations can be close to these possible source locations. In such cases, this proposed methodology is expected to perform better. More rigorous evaluations are, however, necessary, considering various real-life scenarios before this methodology can be applied to different study areas. The proposed method provides a practical approach for predicting future contaminant plume scenarios in an unmonitored groundwater aquifer.

References

Benjema F, Mario MA, Loaiciga HA (1994) Multivariate geostatistical design of groundwater monitoring networks. Water Resour Plan Manage ASCE 120(4):505–522

Chadalavada S, Datta B (2007) Dynamic optimal monitoring network design for transient transport of pollutants in groundwater aquifers. Water Resour Manage 22:651–670

Cooper RM, Istok JD (1988) Geostatistics applied to groundwater contamination. III global estimates. J Environ Eng 114(2):287–299

Datta B, Dhiman SD (1996) Chance-constrained optimal monitoring network design for pollutants in ground water. J Water Resour Plan Manage ASCE 122(3):180–188

Datta B, Brantigan J, Rosbjerg D, Nilsson B (2002) Hydrological information content and tracking of contaminant plume movement at an existing aquifer cleanup in Denmark. In: 4th international model care 2002 conference, Prague, Czech Republic, 17–20 June, 2002. IAHS Proceedings, vol 2, pp 377–381

Deb K (2002) Optimization for engineering design, Prentice Hall of India. In: Delhi N

Deutsch CV, Journel AG (1998) GSLIB: geostatistical software library and user's guide. 2nd ed. Oxford University Press, New York

Eggleston JR, Rojstaczer SA, Peirce JJ (1996) Identification of hydraulic conductivity structure in sand and gravel aquifer: Cape Cod data set. Water Resour Res 32(5):1209–1222

Fabbri P (1997) Transmissivity in the geothermal Euganean basin: a geostatistical analysis. Groundwater 35(5):881–887

Harbaugh AW, McDonald MG (1996) User's documentation for MODFLOW-96, an update to the U.S. Geological Survey modular finite-difference ground-water flow model: U.S. Geological Survey Open-File Report 96-486, p 220

Kirkpatrick S, Gelatt CD, Vecchi MP (1983) Optimization by simulated annealing. Science 220(4598):671–680

Kuo CH, Michel AN, Gray WG (1992) Design of optimal pump-and-treat strategies for contaminated groundwater remediation using the simulated annealing algorithm. Adv Water Resour 15:95–105

Lavenue AM, Pickens JF (1992) Application of a coupled adjoin sensitivity and kriging approach to calibrate a groundwater flow model. Water Resour Res 28(6):1543–1569

Lee YM, Ellis JH (1996) Comparison of algorithms for nonlinear integer optimization: application to monitoring network design. J Environ Eng ASCE 122(6):524–531

Lin YP, Rouhani S (2001) Multiple-point variance analysis for optimal adjustment of a monitoring network. Environ Monit Assess 69(3):239–266

Lin YP, Tan YC, Rouhani S (2001) Identifying spatial characteristics of transmissivity using simulated annealing and Kriging methods. Environ Geol 41:200–208

Loaiciga HA (1989) An Optimization approach for groundwater quality monitoring network design. Water Resour Res 25(8):1771–1782

Loaiciga HA, Hudak PF (1992) A location modeling approach for groundwater monitoring network augmentation. Water Resour Res 28(3):643–649

Loaiciga HA, Hudak PF (1993) An optimization method for monitoring network design in multilayered groundwater flow systems. Water Resour Res 29(8):2835–2845

Loaiciga HA, Hudak PF, Marino MA (1995) Regional-scale ground water quality monitoring via integer programming. J Hydrol 164:153–170

McKinney DC, Loucks DP (1992) Network design for predicting groundwater contamination. Water Resour Res 28(1):133–147

Meyer PD, Brill ED (1988) A method for locating wells in a groundwater monitoring network under conditions of uncertainty. Water Resour Res 24(8):1277–1282

Meyer PD, Valocchi AJ, Eheart JW (1994) Monitoring network design to provide initial detection of groundwater contamination. Water Resour Res 30(9):2647–2659

Passarella G, Vurro M, D'Agostino V, Barcelona MJ (2003) Co. kriging optimization of monitoring network configuration based on fuzzy and non-fuzzy variogram evaluation. Environ Monit Assess 82:1–21

Pinder GF, Bredehoeft JD (1968) Application of the digital computer for aquifer evaluations. Water Resour Res 4(5):1069–1093

Prakash MR, Singh VS (2000) Network design for groundwater monitoring-a case study. Environ Geol 39(6):628–632

Rao SVN, Kumar S, Shekhar S, Chakraborty D (2006) Optimal pumping from skimming wells. J Hydrol Eng ASCE 11(5):464–471

Reed PM, Minsker BS (2004) Striking the balance: long-term groundwater monitoring design for conflicting objective. J Water Resour Plan Manage ASCE 130(2):140–149

Reed P, Minsker B, Valocchi AJ (2000) Cost-effective long term groundwater monitoring design using a genetic algorithm and global mass interpolation. Water Resour Res 36(12):3731–3741

Rogers LL, Johnson VM, Knapp RB (1998) Remediation tradeoffs addressed with simulated annealing optimization. In: XII international conference on computational methods in water resources, the Chersonese, Crete, Greece, Report no. UCRL-JC-129850

Rouhani S (1985) Variance reduction analysis. Water Resour Res 21(6):837–846

Rouhani S, Hall TJ (1988) Geostatistical schemes for groundwater sampling. J Hydrol 103:85–102

Singh D (2015) Groundwater monitoring network design: an optimal approach. Lambert Academic Publishing, Deutschland. ISBN 978-3-659-78092-9

Singh D, Datta B (2014) Optimal groundwater monitoring network design for pollution plume estimation with active sources. Int J Geomate 6(2):864–869

Singh D, Datta B (2016) Linked optimization model for groundwater monitoring network design. In: Urban hydrology, watershed management and socio-economic aspects. Springer International Publishing, pp 107–125. ISBN 978-3-319-40194-2

Singh D, Singh RK (2013) Non-biodegradable contaminants transport modeling with varying transmissivity for aquifer at West Campus HBTI Kanpur. Int J Innov Res Sci Eng Technol 2(10):5731–5740

Wang M, Zheng C (1998) Ground water management optimization using genetic algorithms and simulated annealing: formulation and comparison. J Am Water Resour Assoc 34(3):519–530

Yeh MS, Lin YP, Chang LC (2006) Designing an optimal multivariate geostatistical groundwater quality monitoring network using factorial kriging and genetic algorithms. J Environ Geol 50:101–121

Zheng C, Wang PP (1999) MT3DMS: a modular three-dimensional multispecies transport model for simulation of advection, dispersion, and chemical reactions of contaminants in groundwater systems; documentation and user's guide total 169 pages, Contract Report SERDP-99-1, US Army Engineer Research and Development Center, Vicksburg, Mississippi

Zheng C, Wu J, Chien CC (2005) Cost-effective sampling network design for contaminant plume monitoring under general hydrogeological conditions. J Contam Hydrol 77:41–65

Zhu XY, Xu SH, Zhu JJ, Zhou NQ (1997) Study on the contamination of fracture-krastwater in Boshan District, China. Ground Water 35(3):538–545

Evaluation of Groundwater Quality Using Multivariate Analysis: Rae Bareli District, Ganga Basin, Uttar Pradesh

Tahzeeb Zahra, Anand Kumar Tiwari, Manvendra Singh Chauhan, and Deepesh Singh

Abstract Ground Water is the principal source for drinking and other activities in Rae Bareli district, Uttar Pradesh, India, which makes its monitoring relatively important. Despite the importance of groundwater in Rae Bareli, Uttar Pradesh water quality assessment has mustered very minute attention; thus, efforts to utilize water quality data to compute particular problems are lesser. This following paper, reflects results from a large sample of groundwater quality data examined using multivariate statistical techniques namely, Principal Component Analysis (PCA)/Factor Analysis (FA), Cluster Analysis and Multiple Linear Regression (MLR) with the aim of computing the variability of the water quality data and to distinguish the spring of pollution that presently affect the groundwater (Ismail et al. 2014). The parameters analyzed were pH, electric conductivity, total dissolved solids, total hardness, chlorides, calcium and magnesium, sodium, potassium, bicarbonates, and sulphates for the same. Methodology for characterizing groundwater quality in regard to the use of PCA to qualify and sort the data available for groundwater (Hooper et al. 2008). The aim of this work is to analyze this data in order to explore the groundwater samples and use the mathematical method and modeling. Thus, with the help of MATLAB R2010a and NCSS11 for all the techniques, a regression equation is explored for the sampled groundwater. The results obtained from the same depicts variability in the data wherein variables are indicators of different properties of the data. PCA reduced

T. Zahra · D. Singh (✉)
Department of Civil Engineering, H. B. Technical University, Kanpur,
Uttar Pradesh 208002, India
e-mail: dr.deepeshsingh@gmail.com

T. Zahra
e-mail: tahzeeb31193@gmail.com

A. K. Tiwari
Department of Civil Engineering, Babu Banarsi Das University, Lucknow,
Uttar Pradesh 226020, India
e-mail: tiwari465anand@gmail.com

M. S. Chauhan
Department of Civil Engineering, Holy Mary Institute of Technology and Science, Hyderabad,
Telangana 501301, India

© The Author(s), under exclusive license to Springer Nature Switzerland AG 2021   37
M. S. Chauhan and C. S. P. Ojha (eds.), *The Ganga River Basin: A Hydrometeorological Approach*, Society of Earth Scientists Series,
https://doi.org/10.1007/978-3-030-60869-9_3

the data dimensionality from 12 original physio-chemical and micro-organic parameters known in drinking water samples to certain specific principal components about wherein, on the other hand, CA displayed dendrograms which categorized blocks with similar water parameters in the data. The MLR performance resulted in a trend equation showing the R^2 value to be as close as 0.99, explaining 99% variability in the data-keeping TDS as the dependant variable. All three techniques were helpful in elaborating on different aspects of the water quality of the area. Also, the Piper diagram (a graphical representation of a water sample chemistry) was used to explain the groundwater type and facies either of Calcium bicarbonate or sodium bicarbonate in the site samples using AqQA software.

Keywords PCA/FA · Cluster analysis · Multiple linear regressions · AqQA

1 Introduction

With ever-increasing demand for water for agriculture, industrial and domestic needs, efforts are being directed towards this development of water resources and their scientific management. Due to optimum utilization of surface water resources by way of the canal irrigation system, their high initial cost input, their dependence upon rainfall pattern and side effects like waterlogging and impediment in the natural drainage pattern of the area, attention of the planners have been diverted towards groundwater resource utilization and their management. The development of groundwater resources for domestic and irrigation needs have brought into focus the presence of variable quality of groundwater with depth span in Rae Bareli district. This important resource is useful for many purposes like irrigation, industrial chores, and human use which, directly or indirectly pollutes the endangered resource. Basically, groundwater contamination is first, the result of human activities followed by other activities. In areas where population density is high, and intensively land is used by a human, groundwater is especially vulnerable. Intentionally or accidentally, various activities hold the potential to pollute the groundwater and the cleaning of which is an expensive cum difficult process (Tyagi and Singh 2017).

The following study includes numerous techniques viz. Cluster Analysis (CA), Multiple Linear Regression Analysis (MLR) and Principal Component Analysis (PCA)/Factor Analysis (FA), which are primarily used in groundwater data analysis. The above-mentioned techniques were used to obtain a relationship between the water quality parameters and assessment sites, to recognize the factors and sources that influence groundwater quality (Singh and Vishwakarma 2017). These studies also bring into picture various tools for managing the resources of water and supervise the monitoring of groundwater quality (Singh and Datta 2014, 2016). Cluster Analysis was deployed to examine the bunch groupings of the various sampling sites, making it helpful in classifying the variables into various clusters (Massart and Kaufman 1983). CA and FA are basically followed by MLR to serve as a confirmation for the previous two techniques and are recognized as pattern recognition techniques.

This use of different pattern recognition techniques for analyzing the complexity of large data set has proved to provide an outcome with improved interpretation and understanding of water quality data. Along with all these techniques to examine the complex groundwater data in different ways, a graphical representation called the piper diagram is also applied to the data by the use of the software AqQa. This AqQa software gives matrix transformation of the anion/cation graph to categorize the facies in the samples available. These all the multivariate statistical tools helped to study complex groundwater data and to quantify the pollution sources and to understand the water quality for effective water quality management (Singh and Singh 2013) .

Therefore, the aim of the following study is to determine the variability of groundwater data and to recognize the cause of the pollution, affecting groundwater.

2 Materials and Methods

2.1 Study Area

The city or district of Rae Bareli covering an area of about 4609 km^2. Rae Bareli District is amongst the 71 Districts of Uttar Pradesh State, India, and is located 88 km north towards the capital city, Lucknow. Rae Bareli District holds a large population of about 3,404,014, making it the twenty-seventh largest district by population. Rae Bareli is at latitude 26.2N and longitude 81.2E. The district is divided into five tehsils and 18 blocks. The total population of the district is 3,404,004 souls (Source: census 2011). The entire district is almost a flat country with a gentle slope towards the south-west. The availability of groundwater is 98,303.63 hams.

2.2 Data Collection and Analysis

This study includes data collected from secondary sources. To evaluate the quality of drinking water of groundwater total of 751 water samples were collected from Ground Water Monitoring Stations (GWMS) during the session 2014–15. The monitoring sites are open dug wells drawing perched aquifer (Bamousa and Maghraby 2016). The samples were collected from hand pumps where dug wells are not used. The water quality data obtained from the Central Ground Water Board was registered, which was processed in MS Excel 2010 for all the parameters. The data from monitoring sites were converted into a normalized data set for easier analysis.

2.2.1 Data Pre-processing

As the groundwater quality is an important issue, and it can easily be analyzed by understanding the status of all the essential parameters, which reflect the main pollution sources according to water pollution level standards. As all parameters are measured on a different scale, then to analyze these parameters simultaneously, we need to comprise all these parameters in one scale. Thus, the collected data from all the sampling sites were pre-processed and normalized between 0 and 1. Normalization was done to the data because all the parameters were measured on different scales Barrett (2007). Thus, for accurate results, normalization is applied to the dataset to settle down each of the parameters in a specific range. The equation used for normalizing was

$$Xnor = \frac{(X - X\min)}{(X\max - X\min)} \tag{1}$$

where Xnor is the normalized value, and X is the original parameter, and Xmax and Xmin are the maximum and minimum values of the data set.

2.3 Analytical Methods

Environmetric Statistical Method is chosen as one of the efficient interpretation techniques of huge complex monitoring data (Simeonov et al. 2003). The most common of this data is used to analyze the imbibed variability and, also recognize the various pollution sources which are again subjected to the statistical procedures.

2.3.1 Principal Component Analysis

The methodology used in PCA is unique in its own way, where factors are formed out of the process. The statistical technique acts as an extractor of factors for which an estimate of the variation is computed, variation in the individual groundwater parameters (Chenini and Khemiri 2009). For computing the variability in factors, eigenvalues are calculated with conditions wherein an eigenvalue, more than one, describes notable variation in the parameters of groundwater data, whereas the same for a value less than one indicates less variation in totality. Thus, factors holding eigenvalues more than one were selected and used for varimax rotation (VR) (Kaiser 1960), where VR is a perpendicular rotation method that is useful in minimizing the number of variables that depict high loading on each factor. Also, this is notable that the VR coefficient with a correlation greater than 0.8 are assumed to be stable as strong and indicates high proportionality in its variance as described by the factors, which lie between 0.5 and 0.8 and are considered moderate in terms of loading but the moderate loading is said to be a week, if lies between the range 0.30–0.52 (Raghunath et al. 2001).

2.3.2 Multiple Linear Regression Analysis (MLR)

The regression analysis is used to compute the relationship between two variables, where some acts as an independent variable and one dependent variable where these variables are treated as linear or non-linear, which in return makes predictions for the output value. Also, a linear relationship is justified when the dependent variable shows uniform change with respect to change in the independent variable

Linear regression analysis is conducted to predict values corresponding to the dependent variable with respect to one or more independent variables. In simple linear regression analysis, there is only one independent variable and is expressed as (Landau and Everitt 2004)

$$Y = \beta_0 + \beta_1 X_1 + \varepsilon_i. \tag{2}$$

Y = dependent variable as the observed values, n = sample size, X_i = explanatory or independent variables, are the observed values, i = 1, ..., n; ε_i = residual or error for individual i, β_0 = constant; β_1 = multiple regression coefficients (Landau and Everitt 2004).

MLR creates a trend equation for the model development, which includes regression coefficient for each independent variable.

2.3.3 Cluster Analysis (CA)

Cluster analysis is used to form clusters wherein the similarities and dissimilarities within the clusters is categorized. The set of data was normalized to form clusters, by the use of Ward's Method.

The aim of the cluster formation is to categorize the data into values with homogenous data and heterogeneous so that blocks can be grouped likewise. The same is illustrated using dendrograms, which can group the clusters.

2.3.4 Piper Analysis

This piper diagram is a graphical representation of the water sample that comprises anions and cations percentage in triangle fields and their combined condition in rhombus Square in three different fields. The ternary plots are used to show the cations and anions on the graph. This representation defines the concentration of anions as well as cations which helped to determine the type of water according to the major anion and type of facies according to major cations at each location.

3   Results and Discussion

3.1   *Principal Component Analysis/Factor Analysis*

The results for PCA is shown in Table 1. The elaboration of the result shows the variation in pH for 7.9–8.1, pointing alkalinity in water. The nature of water being alkaline is because of the dissolution of carbonate minerals; the effluent from the wastewater treatment plant adds alkalinity to groundwater.

Total hardness ranges between 140 and 495 mg/l, which is caused by calcium, magnesium. Calcium values were recorded within the range of 20–70 mg/l where higher values might end up creating health issues, scaling in pipelines. The alkalinity was found to be between 232 and 891 mg/l where the higher value of the same indicates mixing in the sewerage. Here, TDS ranges from 288 to 1394 mg/l corresponding to dissolved solids in the water consisting of salts and organic materials where higher values affect dissolved oxygen. Here, the electronic conductivity varies from 430 to 2080 µS/cm, where a high value of EC is an indication of sewerage water mixing (Mustapha et al. 2012). The correlation coefficient of the parameters is presented in Table 2, where there is a strong correlation, which is positive between TDS and EC where r = 0.97. In the matrix there is strong correlation which is positive between total dissolved solids and EC (r = 0.97), between HCO_3 and Na (r = 0.90), similarly positive correlation has been found between Mg and TH (r = 0.93) and Na and TDS (r = 0.97) also between total hardness and chloride (r = 0.84).

Kaiser-Mayer-Olkin (KMO) and Barlett's test of sphericity when performed, suitability check for principal component analysis, KMO was capable of measuring the sampling adequacy, which in return determine the number of variables

Table 1 Data statistic summary

	Min	Max	Mean	SD
pH	7.9	8.1	7.9825	0.0795
EC	430	2080	847.3	427.25
HCO_3	232	891	412.65	148.56
Cl	7	351	54.9	79.31
SO_4	2.1	130	25.4	37.31
TH	140	495	272.7	93.472
Ca	20	72	33.3	13.657
Mg	19	107	46.7	23.305
Na	20	299	80.5	72.891
K	2.9	9.5	5.03	1.5169
TDS	288	1394	549.35	293.5
F	0.04	1.52	0.4945	0.358

Table 2 Correlation matrix

Parameters	pH	EC	HCO$_3$	Cl	SO$_4$	TH	Ca	Mg	Na	K	TDS	F
pH	1											
EC	0.015686	1										
HCO$_3$	0.140472	0.860846	1									
Cl	0.150173	0.811714	0.481381	1								
SO$_4$	0.001517	0.804509	0.771401	0.514983	1							
TH	0.083141	0.875874	0.707404	0.845534	0.7056	1						
Ca	−0.31066	−0.17054	−0.38048	0.149133	0.23036	0.082951	1					
Mg	0.001215	0.901899	0.818683	0.758573	0.76266	0.930737	0.27778	1				
Na	0.076155	0.955948	0.904669	0.745715	0.81234	0.772174	0.26747	0.838319	1			
K	0.234073	0.718031	0.486314	0.741826	0.59122	0.710545	−0.0294	0.697673	0.62598	1		
TDS	0.01205	0.977884	0.87896	0.822599	0.81524	0.900749	0.15721	0.923845	0.96674	0.68247	1	
F	0.022709	0.049103	0.053334	−0.02629	0.17453	−0.14521	0.20369	−0.05	0.14375	−0.2102	0.0627	1

Table 3 KMO and Barlett's test

Kaiser-Meyer-Olkin measure of sampling adequacy		0.59483
Barlett'test	Approx. Chi-square	362.65
	DF	66
	Prob.	0

having common variance. Also, KMO values, which are higher than 0.5, were considered satisfactory for the PCA technique. In the present study, this value is 0.59483 (Table 3), depicting a set of fit data for PCA.

On closer consideration, Barlett's test of sphericity shows a significant correlation where the correlation matrix is the identity matrix. This should be noted that the correlation matrix now obtained is not an identity matrix because the significance level is 0.00 in the analysis, which clearly indicates that variables are significantly related.

The principal component analysis was applied to the data using NCSS11.0 software. The correlation matrix was obtained using PCA, and factors were extracted by using the centroid method, which is rotated by varimax rotation.

As shown in (Fig. 1), this can be said a larger part of the variance in the original data is computed by the first two factors.

These eigen values are actually the magnitude of eigen vectors this can also consider as the measurement of covariance of the data. And we get Principal component by arranging the eigen vectors according to the eigen values, highet to lowest. This Eigen Values are basically the core of the principal component analysis because its value explains the variance in the data. For Example, if we get a zero value eigen value the it is actually showing no variance in the data. This is how it actually narrates the variance according to its values.

The results showed that PCA computed three principal components which describes for 89.71% of the total variance within the 12 parameters. Also

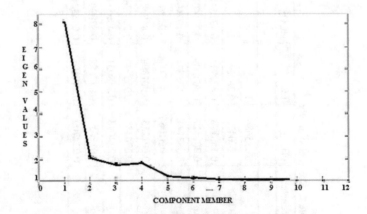

Fig. 1 Plot B/W eigen values and component (output of software NCSS11.0)

Table 4 Eigen values

No.	Eigen value
1	7.131774
2	1.081783
3	0.770193
4	0.832967
5	0.241012
6	0.161268
7	0.064979

from Table 4. From the respective analysis this is explicable that first component This can be explained that the first component (PC1) reports about 71.2% of total computed variance and have positive loading of hardness and TDS while there is negative loading of Calcium. The second component (PC2) is then responsible for 10.80% of total variance and indicated towards strong positive loading of TH, EC, Na, Mg, Ca and TDS. TDS and EC are indicative of high positive loading showing huge presence of ions.

3.2 Multiple Linear Regression (MLR)

The measurement of TDS is one of the characteristics, which decides the quality of drinking water. According to Indian standard Drinking Water-Specification IS 10500: 2012, acceptable limit for total dissolved solids in drinking water is 500 mg/L. Table 1 shows that in the present study, TDS is ranging from 288 to 1394 mg/l. The acceptable limit of total hardness of drinking water is 200 mg/L according to IS 10,500: 2012.

In the present study, TH is ranging from 140 to 495 mg/L. So we conclude that the quality of water is acceptable drinking and agricultural purposes. A linear relation of the form TDS = k * EC between TDS and EC (Hem 1985) has been explored for the data in Table 1, which revealed k = 0.8965 with an R^2 value 0.9563. Mostly, k lies between 0.55 and 0.9, and lesser values suggest a lower presence of ions (Ali 2010). This is again confirmed from the higher k obtained for the area, which is more vulnerable to different kinds of pollution when compared to rural areas. However, notice that the k value 0.8965 for the Rae Bareli district is lying in its range, suggesting a higher presence of ions.

MATLAB R2010a was used to develop an MLR model for predicting TDS; we performed a PCA of the normalized data. The percentages of variance explained by the first two principal components (PC) were 71.22, 10.84, respectively, which accounts for around 82.02% of variance. Table 4 gives the component loadings of the first two PCs. It shows the variable pH has low loadings, especially for the first two components, which explains almost 82% of variance. The significance of TDS and insignificance of pH can be easily viewed in Fig. 2, which is a Biplot of the PCA.

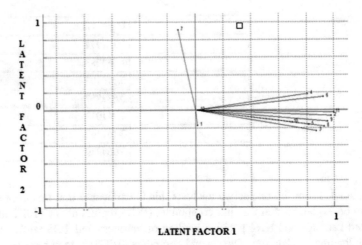

Fig. 2 Coefficients of first two Pcs' (black vectors) with component numbers (output of NCSS11.0)

Thereby we opted to form a regression model for predicting TDS, excluding pH. As can be viewed in Fig. 2, the variable Ca^{2+} is not much significant than pH; however, we decided to exclude Ca^{2+} in the initial regression model. We made an initial regression model for predicting TDS in terms of Cl^-, SO_4^{2-}, Na^+, HCO_3^-, Mg^{2+}, NO_3^- and K, the results of which are given in Table 5. It shows that the p-values of all the variables except Cl^-, HCO_3^- and SO_4^{-2} are greater than 0.05, suggesting

Table 5 Factor loadings after varimax rotation

Variables	Factors	
	Factor1	Factor2
pH	0.012958	−0.17390
EC	**0.978856**	−0.06484
HCO₃	0.868862	−0.23450
Cl	0.804231	0.185179
SO₄	0.814249	−0.16328
TH	**0.921158**	0.151433
Ca	−0.124183	**0.910927**
Mg	**0.927168**	−0.18496
Na	**0.952052**	−0.12758
K	0.693766	−0.13760
TDS	**0.998443**	−0.02677
F	0.023182	−0.09236
Eigen value	7.13	1.08
Cumulative	71.22	82.02

Table 6 Result for first MLR analysis

Regression	Coefficients	t-value	p-value
Constant	−0.021133346	−0.955	0.358
HCO₃	0.542297501	3.275	**0.006**
Cl	0.567485824	4.766	**0.000**
SO₄	0.156658244	3.101	**0.009**
Mg	0.065565989	0.743	0.471
Na	0.0349117601	0.205	0.840
K	−0.0872346493	−1.795	0.097
F	0.0037862789	0.1178	0.908
R-square	0.993	MSE	0.00

that Cl^-, HCO_3^- and SO_4^{-2} are the only variables that are statistically significant for the regression model. This made us consider a second regression model for predicting TDS, which involved only Cl^-, HCO_3^- and SO_4^{-2}. Results from this study are given in Table 6, which shows a slight downfall in the R-square value, indicating a probably inferior model with respect to the data. Also, the p-value indicates a better fit of the data by the second model. Later in the second model, only the p-value associated with the constant term is just above 0.05, and the other p-values are less than 0.000. Considering all the point, this was concluded that the second regression model better fits the data when compared to the first model. We thus concluded the MLR model for predicting TDS as

$$TDS = 0.60740 * HCO_3^- + 0.56292 * Cl^- + 0.15040 * SO_4^{2-}$$
$$- 0.03598. \tag{3}$$

3.3 Cluster Analysis

Cluster analysis is used to form clusters wherein the similarities and dissimilarities within the clusters is categorized. The set of data was normalized to form clusters, by the use of Ward's Method. Figure 3. represents Dendrograms formed from the cluster analysis. (Including 4 cluster) Cluster 1 holds up Bachharawa, and Dalmauwhere in Cluster 2 consists of Sareni and Lalganj. Cluster 1 and cluster 2 points out relative variation in pollution, high and low, respectively.

Also, it was noted that Cluster 3 consists of no blocks showing similarity in the same manner and showing moderate variation, explaining fair water quality of water in Rae Bareli district as a whole (Table 7).

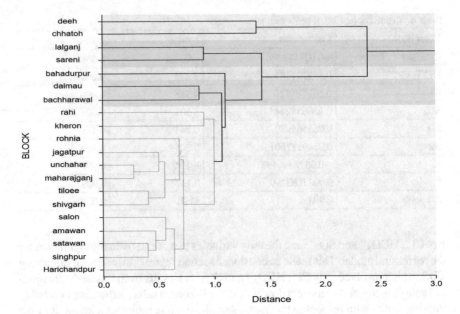

Fig. 3 Dendrogram (tree diagram)

Table 7 Results for second MLR analysis

Regression	Coefficients	t-value	p-value
Constant	−0.0359	−3.382	0.0037
HCO$_3$	0.6070	13.405	**0.000**
Cl	0.5629	17.11	**0.000**
SO$_4$	0.1504	4.200	**0.000**
R-square	0.9906	MSE	0.0044

3.4 Piper Analysis

The piper diagram for the data analysis for 20 wells explained the types and facies of water for 15 wells as Magnic Bicarbonate and for three wells as sodic Bicarbonate and other two wells as Calcic Bicarbonate facies. The dominant anions were Cl > HCO$_3$ > SO$_4$ (Figs. 4, 5 and Table 8).

Name	Unit	Lalganj	Dalmau	Rahi	Jagatpur	Unchahar	Rohnia	Salon	Chhatoh		Mix
Sample ID		Lalganj	Dalmau	Rahi	Jagatpur	Unchahar	Rohnia	Salon	Chhatoh		
pH		8	7.9	8	8	8.1	8.1	8	7.9		
Conductivity	µmho/cm	1290	525	920	923	636	710	534	1837		
Bicarbonate	mg/L	537	287	329	317	390	439	317	537		403
Chloride	mg/L	71	28	121	14	14	14	11	351		14
Sulfate	mg/L	130	9.9	4.7	2.1	4.6	2.5	10	56		6.2
Hardness	mg CaCO₃/L	350	210	380	190	220	290	200	495		240
Calcium	mg/L	24	36	56	28	24	36	46	44		32
Magnesium	mg/L	71	29	58	29	39	49	27	94		39
Sodium	mg/L	156	35	55	43	57	46	40	235		54
Potassium	mg/L	6.6	3	4.1	5.3	4.5	3.1	4.1	9.5		5
Dissolved Solids	mg/L	864	352	616	350	426	476	358	1231		441
Fluoride	mg/L	1.03	1.52	0.16	0.55	0.29	0.61	0.67	0.54		0.34

Fig. 4 Blockwise characteristics data (output of software AqQA)

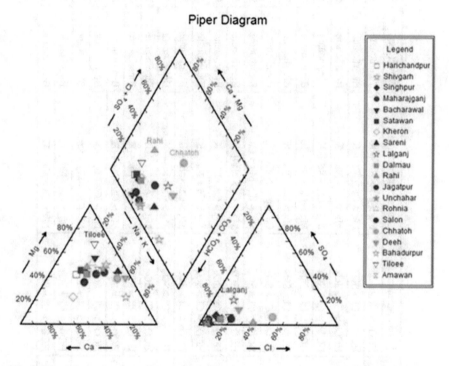

Fig. 5 Piper diagram (output of software AqQA)

4 Conclusion

The multivariate statistical methods such as PCA, CA sorted data and explained the variability within data of 12 parameters and also stated the blocks(sites) sharing similarities and dissimilarities applied for the groundwater quality assessment of Rae Bareli using one-year data. In addition, MLR created a trend equation for predicting TDS while TDS/EC ratio was analyzed using linear regression. Piper Diagram helped

Table 8 Results for type and facies

S. No.	Sampling	Anion concentration	Cation concentration	Water type	Facies	Conclusion
1	Harichandpur	$HCO_3 > Cl > SO_4$	$Mg > Ca > Na + K$	Bicarbonate	Magnic	Magnic Bicarbonate
2	Shivgarh	$HCO_3 > Cl > SO_4$	$Mg > Ca > Na + K$	Bicarbonate	Magnic	Magnic Bicarbonate
3	Singhpur	$HCO_3 > Cl > SO_4$	$Mg > Ca > Na + K$	Bicarbonate	Magnic	Magnic Bicarbonate
4	Maharajganj	$HCO_3 > Cl > SO_4$	$Mg > Ca > Na + K$	Bicarbonate	Magnic	Magnic Bicarbonate
5	Bacharawan	$HCO_3 > Cl > SO_4$	$Mg > Ca > Na + K$	Bicarbonate	Magnic	Magnic Bicarbonate
6	Satawan	$HCO_3 > Cl > SO_4$	$Mg > Ca > Na + K$	Bicarbonate	Magnic	Magnic Bicarbonate
7	Kheron	$HCO_3 > Cl > SO_4$	$Ca\ Mg > \ > Na + K$	Bicarbonate	Calcic	Calcic Bicarbonate
8	Sareni	$HCO_3 > Cl > SO_4$	$Na + K > Mg > Ca$	Bicarbonate	Sodiac	Sodiac Bicarbonate
9	Lalganj	$HCO_3 > Cl > SO_4$	$Mg > Ca > Na + K$	Bicarbonate	Magnic	Magnic Bicarbonate
10	Dalmau	$HCO_3 > Cl > SO_4$	$Mg > Ca > Na + K$	Bicarbonate	Magnic	Magnic Bicarbonate
11	Rahi	$HCO_3 > Cl > SO_4$	$Mg > Ca > Na + K$	Bicarbonate	Magnic	Magnic Bicarbonate
12	Jagatpur	$HCO_3 > Cl > SO_4$	$Mg > Ca > Na + K$	Bicarbonate	Magnic	Magnic Bicarbonate
13	Unchahar	$HCO_3 > Cl > SO_4$	$Mg > Ca > Na + K\ ara>$	Bicarbonate	Magnic	Magnic Bicarbonate
14	Rohnia	$HCO_3 > Cl > SO_4$	$Mg > Ca > Na + K$	Bicarbonate	Magnic	Magnic Bicarbonate
15	Salon	$HCO_3 > Cl > SO_4$	$Ca > Mg > Na + K$	Bicarbonate	Calcic	Calcic Bicarbonate
16	Chhatoh	$HCO_3 > Cl > SO_4$	$Mg > Ca > Na + K$	Bicarbonate	Magnic	Magnic Bicarbonate
17	Deeh	$HCO_3 > Cl > SO_4$	$Na + K > Mg > Ca$	Bicarbonate	Sodiac	Sodiac Bicarbonate
18	Bahadurpur	$HCO_3 > Cl > SO_4$	$Na + K > Mg > Ca >$	Bicarbonate	Sodiac	Sodiac Bicarbonate
19	Tiloee	$HCO_3 > Cl > SO_4$	$Mg > Ca > Na + K$	Bicarbonate	Magnic	Magnic Bicarbonate
20	Amawan	$HCO_3 > Cl > SO_4$	$Mg > Ca > Na + K$	Bicarbonate	Magnic	

in calculating the type and facies of water so that the direct safeguard can be provided to keep the water standards up to the mark. Thus, these results can be helpful for the management and control of pollution to the authorities.

Also, this multivariate statistical approach will be helpful for the analysis of future groundwater of Rae Bareli district by adding more techniques along with the no. of years of data this not only helps in understanding complex data but also determine the primary pollution sources and contaminants responsible for effecting it on a different level.

References

Ali MH (2010) Fundamentals of irrigation and on-farm water management, vol 1. Springer, Berlin. ISBN 978-1-4419-6335--2

Bamousa AO, El Maghraby M (2016) Groundwater characterization and quality assessment, and sources of pollution in Madinah, Saudi Arabia. Arab J Geosci 9:536

Barrett P (2007) Structural equation modeling: adjudging model fit. Pers Individ Differ 42(5):815–824

Chenini I, Khemiri S (2009) Evaluation of ground water quality using multiple linear regression and structural equation modeling. Int J Environ Sci Technol 6(3):509–519

Hem JD (1985) Study and interpretation of the chemical characteristics of natural water. U.S. Geological Survey, Water Supply Paper 2254, 3rd edn.

Hooper D, Coughlan J, Mullen MR (2008) Structural equation modeling: guidelines for determining model fit. Electron J Bus Res Methods 6(1):53–60

Ismail H, Basim SA, Shahla AQ (2014) Application of multivariate statistical techniques in the surface water quality assessment of Tigris River at Baghdad stretch. Iraq J Babylon Univ/Eng Sci 22(2):450–463

Kaiser HF (1960) The application of electronic computers to factor analysis. Educ Psychol Meas 20:141–151

Landau S, Everitt BS (2004) A handbook of statistical analyses using SPSS. Chapman & Hall/CRC Press LLC, Washington, DC

Massart D Kaufman L (1983) The interpretation of analytical chemical data by the use of cluster analysis. Wiley, New York.

Mustapha A, Aris Z, Ramli MF, Juahir H (2012) Spatial-temporal variation of surface water quality in the downstream region of the Jakara River, north-western Nigeria: a statistical approach. J Environ Sci Health 47:1551–1560

Raghunath R, Sreedhara Murthy TR, Raghavan BR (2001) Spatial distribution of pH, EC and total dissolved solids of Nethravathi river basin, Karnataka state, India. Pollut Res 20(3):413–418

Singh D, Datta B (2014) Optimal groundwater monitoring network design for pollution plume estimation with active sources. J Geomate 6(2):864–869

Singh D, Datta B (2016) Linked optimization model for groundwater monitoring network design. In: Urban hydrology, watershed management and socio-economic aspects. Springer International Publishing, pp 107–125. ISBN 978-3-319-40194-2

Singh D, Singh RK (2013) Non-biodegradable contaminants transport modeling with varying transmissivity for aquifer at West Campus HBTI Kanpur. Int J Innov Res Sci Eng Technol 2(10):5731–5740

Singh D, Vishwakarma RD (2017) Modelling of heat and solute transport in groundwater. In: 33rd national convention of environmental engineers on status of technological advancement to meet the environmental norms for Indian mining and allied industries. The Institution of Engineers (India), IIT (ISM) Dhanbad, 24–25 Augu 2017, pp 111–123

Simeonov V, Stratis JA, Samara C, Zachariadis G, Voutsac D, Anthemidis A, Sofoniou M, Kouimtzis
    Th (2003) Assessment of the surface water quality in Northern Greece J Water Res 37(1):4119–
    4124
Tyagi M, Singh D (2017) Study of biodegradable contaminants transport in groundwater. In:
    Proceedings of international conference on modelling of environmental and water resources
    systems, ICMEWRS 2017, at HBTU Kanpur

A Supervising Grid Model
for Identification of Groundwater Pollute

Sumit Gangwar, Manvendra Singh Chauhan, and Deepesh Singh

Abstract Groundwater asset is the most critical freshwater asset. A large portion of the number of inhabitants in our nation relies upon this asset for their essential needs of water. Groundwater has assumed a noteworthy part in expanding nourishment generation and accomplishing sustenance security. Groundwater, an inexhaustible wellspring of water, has an exceedingly reliable water supply for horticulture, household, business, and mechanical needs. Groundwater sullying is a significant issue in our reality. Optimal groundwater monitoring network design models are developed to determine the mass estimation error of contaminant concentration over different management time periods in groundwater aquifers. The objective of the paper is to determine the mass estimation error of contamination concentration at 2.5 years and 7.5 years. The mass estimation error of contamination concentration over time is determined by using the various computer software such as Method of Characteristics (MOC, USGS), Surfer 7.0, and Simulated Annealing (SA). The Method of Characteristics (MOC) is used in this model to solve the solute-transport equation. Simulated annealing is a worldwide improvement strategy that is utilized to locate optimal monitoring well locations. The error of the estimated concentration at potential well locations is extrapolated over the entire study area by geostatistical instrument, kriging. The outlined observing system is dynamic in nature, as it gives time-shifting system plans to various management periods. The optimal monitoring wells design incorporates budgetary constraints in the form of limits on the number of monitoring wells installed in any particular management period. The solution results are evaluated for an illustrative study area comprising of a hypothetical aquifer. The performance evaluation results establish the potential applicability of the

S. Gangwar · D. Singh (✉)
Department of Civil Engineering, H. B. Technical University, Kanpur, Uttar Pradesh 208002, India
e-mail: dr.deepeshsingh@gmail.com

S. Gangwar
e-mail: sumit13aug@gmail.com

M. S. Chauhan
Department of Civil Engineering, Holy Mary Institute of Technology and Science, Hyderabad, Telangana 501301, India

© The Author(s), under exclusive license to Springer Nature Switzerland AG 2021   53
M. S. Chauhan and C. S. P. Ojha (eds.), *The Ganga River Basin: A Hydrometeorological Approach*, Society of Earth Scientists Series,
https://doi.org/10.1007/978-3-030-60869-9_4

proposed methodology for the optimal design of the dynamic monitoring networks for determining the mass estimation error of contamination concentration.

Keywords Dynamic monitoring network design · Groundwater contamination · Contamination detection · MOC · Optimization · Simulated annealing · Kriging · Mass estimation error

1 Introduction

The motivation behind streamlining a groundwater monitoring system is to get adequate hydrogeological data for the required accuracy utilizing the least wells. Continuous observation of groundwater contamination is essential for the effective management of aquifer systems. The need for designing an ideal monitoring network is required due to the uncertainties associated with the prediction contaminant concentration movement and due to economic constraints, which limits the number of monitoring wells installation. The establishment of the groundwater observing system in an ideal way to satisfy the diverse goals of checking is turning into an essential issue. There are some imperative explanations behind building up a monitoring system are: (1) status of groundwater quality and amount in a specific report zone can be determined, (2) checking gives the standard information to the groundwater administration programs, (3) observing data encourages reactions to essential circumstances, (4) it gives a significant human effect on groundwater quality, (5) it might prompt productive land to utilize the technique in tainted zones. An approach is created for ideal groundwater pollution observing system configuration considering the transient stream and transport process in the aquifer. The composed checking system is dynamic as it gives time fluctuating system outlines to various administration periods, to represent the transient contaminant crest. The ideal framework fuses budgetary requirements through observing wells introduced in a specific administration period. Execution of the proposed technique is assessed for illustrative investigation involving speculative aquifer. Numerous numerical models were created for the ideal checking system plan, which upgrades diverse destinations of the outline. Meyer and Brill (1988) recommended the whole number program advancement demonstrate, which augments the likelihood of discovery of contaminant tufts that surpass a predetermined standard focus. Loaiciga (1989) give a survey of various systems for outlining a groundwater quality observing systems. McKinney and Loucks (1992) introduced a strategy for an ideal observing system plan, which limits the vulnerability in groundwater reenactment show forecasts by the decision of new areas at which aquifer properties to be estimated. Meyer et al. (1994) built up an ideal observing system configuration model to limit the quantity of checking wells, boost the likelihood of distinguishing a contaminant spill, and limits the normal zone of sullying at the season of recognition. Reproduced tempering has been utilized as a part of this investigation as streamlining calculations. The hereditary calculation is utilized as an enhancement device. The utilization of the Genetic Algorithm in taking care of the groundwater checking issue has been exhibited by Cieniawski et al. (1995). They have considered the Multi-target detailing of the observing outline models. Reed et al. (2000) built up a financially savvy long haul groundwater ideal observing systems. The checking system brings about limiting the worldwide mass estimation mistake. Hereditary Algorithm has been utilized to comprehend streamlining model. The long haul checking procedure consolidates obliged multi-target issue detailing, tuft introduction utilizing kriging and advancement utilizing NSGA-2 (Deb 2002). Singh (2015) introduced a model in view of the location of contaminant fixation in checking wells. They built up philosophy to discover an ideal observing system in light of minimization of summation of fixation deviations utilizing mimicked strengthening. The vast majority of the related works considered cost minimization as one of the fundamental targets of the ideal checking system outline. This procedure is created for an ideal outline of a dynamic observing system. The goal of this model is to decide the mass estimation mistake of contaminant focus at various time interims. The created enhancement models are unraveled utilizing Simulated Annealing. The mass estimation blunder of contaminant focus after some time at potential ideal checking areas is processed utilizing geostatistical kriging.

2 Methodology

2.1 Groundwater Flow Equation

The equation describing the transient, two-dimensional areal flow of groundwater through a non-homogeneous, anisotropic, saturated aquifer can be written in Cartesian tensor notation (Pinder and Bredehoeft 1968) as:

$$\frac{\partial}{\partial x_i}\left(T_{ij}\frac{\partial h}{\partial x_i}\right) = S\frac{\partial h}{\partial t} + W; \quad i, j = 1, 2 \tag{1}$$

where T_{ij} = transmissivity tensor ($L^2\,T^{-1}$) = $K_{ij}b$;

K_{ij} = hydraulic conductivity tensor (LT^{-1});

b = saturated thickness of aquifer (L);

h = hydraulic head (L);

W = volume flux per unit area (LT^{-1}); and

x_i, x_j = Cartesian coordinates (L).

In this investigation the stream demonstrate is simulated utilizing MOC.

The partial differential equation describing the fate and transport of contaminants of species k in 3 – D, transient groundwater flow systems can be written as follows (Zheng and Wang 1999):

$$\frac{\partial(\theta c^k)}{\partial t} = \frac{\partial}{\partial x_i}\left(\theta D_{ij}\frac{\partial C^k}{\partial x_i}\right) - \frac{\partial}{\partial x_i}\left(\theta v_i^{C^k}\right) + q_s C_s^k + \Sigma R_n \tag{2}$$

where,

θ = porosity of the subsurface medium, dimensionless;

C^k = dissolved concentration of species k, ML^{-3};

t = time, T;

$x_{i,j}$ = distance along the respective Cartesian coordinate axis, L;

D_{ij} = hydrodynamic dispersion coefficient tensor, L^2T^{-1};

v_i = seepage or linear pore water velocity, LT^{-1}; it is related to the specific discharge or Darcy flux through the relationship;

$v_i = q_i/\theta$; q_s = volumetric flow rate per unit volume of aquifer representing fluid sources (positive) and sinks (negative), T^{-1};

C_s^k = concentration of the source or sink flux for species k, ML^{-3};

ΣR_n = chemical reaction term, $ML^{-3}\ T^{-1}$.

2.1.1 Method of Characteristics (MOC)

The Method of Characteristics (USGS) is used in this model to solve the solute-transport equation. This method was developed to solve hyperbolic differential equations. If solute transport is dominated by convective transport, as common in many field problems, then Eq. (2) may closely approximate a hyperbolic partial differential equation and be highly compatible with the method of characteristics.

2.1.2 Optimization Models

The monitoring network design model is developed for the optimal design of the dynamic monitoring networks. In this model, the number of monitoring wells installed in a given management period is limited as imposed budgetary constraints.

2.1.3 Monitoring Network Design Model

The objective function for the monitoring network design Model is the minimization of the summation of all positive deviations between the simulated contaminant concentration and a specified low threshold value of contaminant concentration for all realization, at all potential locations where monitoring well is not to be installed as per the design solution (Singh 2008).

This model is mathematically represented as

$$\text{Minimize} : \sum_{i=1}^{M} \sum_{j=1}^{N} (C_{ij} - C^*) P_{ij} X_i \tag{3}$$

Subjected to:

$$P_{ij} = 0, \text{For}(C_{ij} - C^*) < 0 \tag{4}$$

$$P_{ij} = 1, \text{For}(C_{ij} - C^*) \geq 0 \tag{5}$$

Constraints sets (4) and (5) ensure that only positive concentration deviations are added in the objective function value 1.

$$1 - W_i = X_i \text{ for all } i \in S \tag{6}$$

Constraints set (6) ensure that X_i would we zero if monitoring well is installed as per design at location i, so that objective function is properly defined.

$$\sum_{i=1}^{M} W_i \leq W \tag{7}$$

Constraints (7) ensure that the number of monitoring wells installed in any management period does not exceed the number permitted by the budgetary considerations.

$$X_i = (0, 1) \tag{8}$$

$$W_i = (0, 1) \tag{9}$$

where,

$i = $ Index number for potential monitoring location.

$S = $ the set of potential monitoring well locations.

$M = $ Total number of potential monitoring locations.

$N = $ Total number of contaminant concentration realizations.

$j = $ Index number for representing a particular realization.

$C_{ij} = $ Simulated at ith potential location for jth realization.

W_i = A binary decision variable, one indicates monitoring well is to be installed at potential location i, and 0 indicates otherwise.

W = Most extreme number of wells allowed to be introduced for the given time frame.

X_i = Compliment of W_i.

$\in$ = Belongs to.

C^* = Threshold value of contaminant concentration.

2.2 Kriging

Kriging is a geostatistical estimation method that has wide applications in many engineering fields. It is a widely used spatial estimation technique. The contaminant mass estimates were made by utilizing the spatial estimation technique, Kriging. It is a widely used spatial estimation technique. Kriging is applied in this study for the estimation of the contaminant mass from a limited amount of spatial concentration data. Kriging is applied in this work for performance evaluation of the generated optimal monitoring network design. In this study, Surfer (vr.7.0) is used for extrapolation of concentration data over the study area (Singh and Datta 2014, 2016).

Almost no endeavors have been made to build up philosophy for the ideal plan of the observing system using improvement models connected to the models for geostatistical estimations, i.e., kriging. Reed et al. (2000) built up an approach that consolidates a destiny and transport show, tuft insertion, and a hereditary calculation to recognize financially savvy testing designs that precisely measure the aggregate mass of broke down contaminant for crest introduction. Reverse separation weighting, Ordinary kriging, and a half and half technique are utilized.

Kriging is a collection of generalized linear regression techniques for minimizing an estimation variance defined from a prior model for a covariance.

Consider the estimate of an un-sampled value z(u) from neighboring data values $z(u_\alpha), \alpha = 1 \ldots n$. The RF model z(u) is stationary with mean m and covariance C(h) in its simplest form, also known as Simple Kriging (SK), the algorithm considers the following linear estimator:

$$Z_{SK}^*(u) = \sum_{\alpha=1}^{n} \lambda_\alpha(u) Z(u_\alpha) + \left(1 - \sum_{\alpha=1}^{n} \lambda_\alpha(u)\right) \tag{10}$$

The weights $\lambda_\alpha(u)$ are determined to minimize the error variance, also called the "estimation variance". That minimizes results in a set of normal equations:

$$\sum_{\beta=1}^{n} \lambda_\beta(u) C(u_\beta - u_\alpha) = C(u - u_\alpha), \quad \forall \alpha = 1 \ldots n \tag{11}$$

The corresponding minimized estimation variance, or Kriging variance, is:

$$\sigma_{SK}^2(\mathbf{u}) = C(0) - \sum_{\alpha=1}^{n} \lambda_\alpha(\mathbf{u})C(u - u_\alpha) \geq 0 \tag{12}$$

2.3 Simulated Annealing

Simulated annealing is introduced and connected to the improvement of groundwater management issues. No continuity requirements are imposed on objective (cost) function. The key is to formulate the problem in terms of a discrete number of decision variable, each with a discrete number of possible values. Simulated annealing is then applied to the resulting combinatorial problem. The combinatorial problem has a finite number of possible solutions; a solution that results in the "best" or globally optimal value of the objective function is non-unique, in general. In theory, simulated annealing can be guaranteed to locate a global optimum; however, in practice, computational limitations lead to the search for nearly optimum solutions (i.e., local optima with objective function values near the optimal one). Constraints may be added to the cost function via penalties, imposed by the designation of the solution domain, or embedded in sub-models used to evaluate the cost. The application of simulated annealing to water resources issue is new, and its evolution is immature, so further performance enhances can be expected (Deb 2002).

3 Performance Evaluation

3.1 Study Area

The area under the study is hypothetical. As required by the model, the whole area was enclosed in a rectangular area. The rectangle in itself 20×20 grid is 4000 feet in both x and y directions, with each cell having dimensions of 200 feet in both x and y directions. The upper left corner was taken as the point of origin as required by the model. The area concerned was the confined aquifer with variable thickness, variable permeability, and variable pumping well rate. The boundaries of the hypothetical study area have been shown in Fig. 1. The aquifer was confined with variable thickness, variable permeability, and variable pumping well rate. The study area is divided into six zones namely, Z_1, Z_2, Z_3, Z_4, Z_5 and Z_6, respectively. Different zone of the aquifer has a different thickness, water table heads, and permeability values and areas, which are shown in Table 1. The total number of the cells thus falling inside the boundary was numbered as 324, and thus the total area to be studied came out to be $324 \times 61 \times 61$ m^2 ($324 \times 200 \times 200$ ft^2) which is equal 1,205,604 m^2

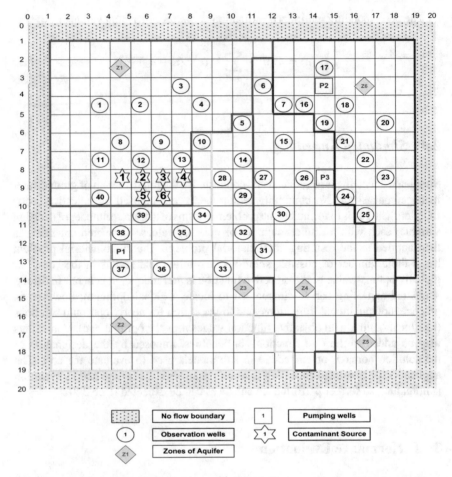

Fig. 1 Study area showing specified observation well locations, pumping well locations, contaminant source and various zones of aquifer

Table 1 Different Zones of the aquifer and their respected thicknesses, water table heads, permeability, and areas

Zones	Thickness (m)	Water table heads (m)	Permeability (m/d)	Area (m²)
Z_1	7.32	45.14	9.15	290,238
Z_2	6.71	44.53	9.98	267,912
Z_3	6.10	43.31	10.98	137,677
Z_4	5.49	42.09	12.19	234,423
Z_5	6.10	42.09	10.98	55,815
Z_6	5.19	41.18	12.97	219,539

Table 2 Hydro geological inputs for the model

Parameters	Value
Longitudinal Dispersivity	100
Leakance	1.0
Effective Porosity	0.4
Storage coefficient	0.0
The ratio of T-YY to T-XX execution parameter	1.0

$(12,960,000 \text{ ft}^2)$. As the model requires, the boundary was so chosen that the outermost cells of the whole grid be designated as a noflow boundary, essential care was taken to fit the grid in such a manner that the outermost cells fall under a noflow region while the concerned area would fall in the aquifer region. This condition was sufficed by putting the value of transmissivity of all cells falling outside the boundary equal to 0.0 while all the rest 324 cells were allocated a value in accordance with their respective aquifer zones.

3.2 Hydro-geological Parameters

Apart from the above stated inputs, several other inputs regarding the hydro geological characteristics of the aquifer were also needed by the program. The following inputs were given to the program which was listed in Table 2.

3.3 Pumping Well Locations

Based on the data collected during the survey of the area, three pumping wells with variable rate were found to exist in the area. The coordinates of the pumping wells and pumping rates have been tabulated in Table 3.

Table 3 Co-ordinates of the pumping wells

Pumping well no.	Co-ordinates	Pumping rates (m³/s)
1	(5, 13)	0.142
2	(15, 4)	0.567
3	(15, 8)	0.142

3.4   Observation Wells

Open or dug wells could have been used as observation wells. In the study area, 40 observation wells have been located on the grid. Their locations on the grid have been tabulated in Table 4.

4   Results and Discussion

The developed observing system configuration Model is solved for two different management time period that is for 2.5 years and 7.5 years. The pollution source, pumping well, and monitoring well locations in the study area shown in Fig. 1.

Table 4 Co-ordinates of the observation wells

Observation Well No.	Co-ordinates	Observation Well No.	Co-ordinates
1	(04, 16)	21	(16, 14)
2	(06, 16)	22	(17, 13)
3	(08, 17)	23	(18, 12)
4	(09, 16)	24	(16, 11)
5	(11, 15)	25	(17, 10)
6	(12, 17)	26	(14, 12)
7	(13, 16)	27	(12, 12)
8	(05, 14)	28	(10, 12)
9	(07, 14)	29	(11, 11)
10	(09, 14)	30	(13, 10)
11	(04, 13)	31	(12, 08)
12	(06, 13)	32	(11, 09)
13	(08, 13)	33	(10, 07)
14	(11, 13)	34	(09, 10)
15	(13, 14)	35	(08, 09)
16	(14, 16)	36	(07, 07)
17	(15, 18)	37	(05, 07)
18	(16, 16)	38	(05, 09)
19	(15, 15)	39	(06, 10)
20	(18, 15)	40	(04, 11)

4.1 The Management Time Period at 2.5 years

Performance evaluation of the proposed methodology is carried out by specifying the optimal number of monitoring wells as 30, 25, and 20 for 2.5 years and 7.5 years management periods, respectively. The optimal design of the monitoring wells network model is solved by simulated annealing for this time period, and the optimal monitoring well locations are identified. The 40 potential monitoring well locations for the 2.5 years management period are shown in Fig. 2a. The pollutant concentration at 20, 25, 30 wells is extrapolated in Surfer 7.0 computer software to determine

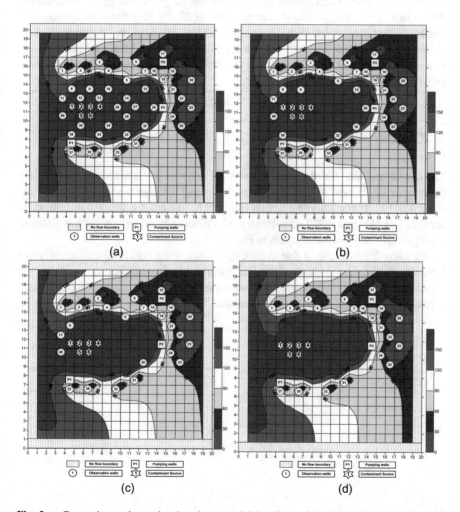

Fig. 2 **a** Contaminant plume showing the potential locations of the 40 wells for 2.5 years of management time periods. **b–d** Contaminant plume showing the optimal locations of the 30, 25, and 20 wells for 2.5 years management time periods, respectively

a total estimate of the contaminant mass in the study area after 2.5 years, which is obtained as 1540 kg, 3720 kg and 5900 kg respectively. The optimized location of 30, 25, and 20 wells shown in Fig. 2b–d for 2.5 years management time periods, respectively.

4.2   Management Time Period at 7.5 years

The 40 potential monitoring well locations for the 7.5 years management period are shown in Fig. 3a. The pollutant concentration at 20, 25,30 wells is extrapolated in Surfer 7.0 computer software to determine a total estimate of the contaminant mass in the study area after 7.5 years, which is 1910 kg, 5680 and 6000 kg respectively. The optimized location of 30, 25, and 20 wells shown in Fig. 3b–d for 7.5 years management time periods, respectively.

5   Conclusion

The model predicts that the mass estimate of error at 2.5 years management time periods for 20, 25, and 30 wells are 0.824, 0.575 and 0.326 respectively and the mass estimate of error for 7.5 years management time periods for 20, 25, and 30 wells are 0.782, 0.568, and 0.317 respectively. The mass estimate of error at the different management time period is shown in Fig. 4.

It is clear from the Fig. 4 that in both the management eras mass estimation error indicates diminishing pattern when the optimal number of observation wells are expanded. The mass estimation error likewise demonstrates a diminishing pattern when the management time period increases.

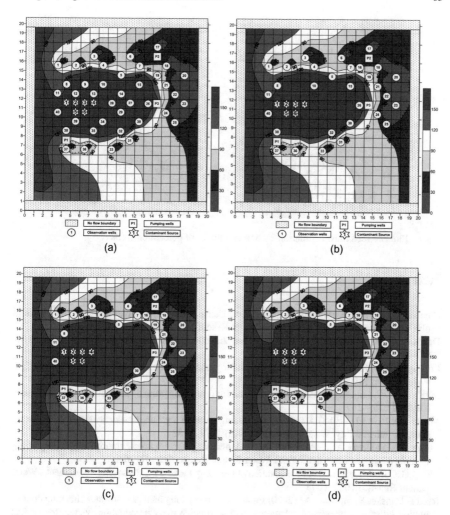

Fig. 3 **a** Contour showing optimal location of the 40 wells for 7.5 years management period. **b–d** Contaminant plume showing the optimal locations of the 30, 25 and 20 wells for 7.5 years management time periods respectively

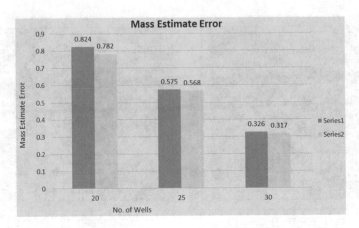

Fig. 4 Plot between mass estimation Error and number of perception wells

Reference

Cieniawski SE, Eheart JW, Ranjithan S (1995) Using hereditary calculations to explain a multi objective groundwater observing issue. Water Resour Res 31(2):399–409

Deb K (2002) Optimization for building outline, 2nd edn. Prentice Hall of India, New Delhi. ISBN 978-8120346789

Loaiciga HA (1989) An optimization approach for groundwater quality monitoring network design. Water Resour Res 25(8):1771–1782

McKinney DC, Loucks DP (1992) Network plan for foreseeing groundwater sullying. Water Resour Res 28(1):133–147

Meyer PD, Brill ED (1988) A strategy for finding wells in a groundwater checking system under states of vulnerability. Water Resour Res 24(8):1277–1282

Meyer PD, Valocchi AJ, Eheart JW (1994) Monitoring system configuration to give beginning location of groundwater defilement. Water Resour Res 30(9):2647–2659

Pinder GF, Bredehoeft JD (1968) Utilization of the computerized PC for aquifer assessments. Water Resour Res 4(5):1069–1093

Reed P, Minsker B, Valocchi AJ (2000) Financially savvy long haul groundwater checking configuration utilizing a hereditary calculation and worldwide mass introduction. Water Resour Res 36(12):3731–3741

Singh D (2008) Optimal monitoring network design for contamination detection and sequential characterization of contaminant olumes with feedback information using simulated annealing and linked kriging. Ph.D. thesis submitted at Indian Institute of Technology, Kanpur, India

Singh D (2015) Groundwater monitoring network design: an optimal approach. L. A. Distributing, Deutschland

Singh D, Datta B (2014) Optimal groundwater monitoring network design for pollution plume estimation with active sources. J Geomate 6(2):864–869

Singh D, Datta B (2016) Linked optimization model for groundwater monitoring network design. In: Urban hydrology, watershed management and socio-economic aspects. Springer International Publishing, pp 107–125. ISBN 978-3-319-40194-2

Zheng C, Wang PP (1999) MT3DMS: a modular three-dimensional multispecies transport model for simulation of advection, dispersion, and chemical reactions of contaminants in groundwater systems; documentation and user's guide total 169 pages, Contract Report SERDP-99-1. US Army Engineer Research and Development Center, Vicksburg, Mississippi

Study of Methods Available for Groundwater and Surfacewater Interaction: A Case Study on Varanasi, India

Padam Jee Omar, Nikita Shivhare, S. B. Dwivedi, Shishir Gaur, and P. K. S. Dikshit

Abstract In the present scenario, the conservation of water is the main concern of human beings. The depletion rate of water is rapid for surfacewater as well as for groundwater. In most of the landscapes, surfacewater and sub-surface water are inter-connected to each other, and the interaction between sub-surface water and surfacewater is a complex phenomenon. Remarkable advances in groundwater (sub-surface water) flow modelling have been influenced by the water demand which enables to prediction impacts of human activities on groundwater systems and associated environment. The main objective of this paper is to find out the different methods to predict the exchange of fluxes between groundwater and surfacewater. A case study of Varanasi district, India has been taken. In this, landfill sites were selected, and the impact of leachate parameters on the river water (surfacewater) was observed. Inter-relationship of groundwater with surfacewater was found and how much time it takes in reaching to the river. The steady state of the groundwater model is conceptualized by using groundwater flow modelling program. The conceptualization of the groundwater model needed so much input data, and compilation of this data is another tedious task. Combining with Geographic Information System (GIS) technology Groundwater Modeling System (GMS) provided good visualization interface for the user and played a significant role in groundwater evaluation and management. Various processes affect the groundwater and surface water interaction, such as flood recharge, evapo-transpiration from open/shallow water-bodies, interception of water by wetlands, para-fluvial flow. This study helps in choosing the appropriate modelling tools for groundwater and surface water interaction, done by striking the right balance between groundwater and surfacewater processes. The technique used in selecting the most suitable tool of the groundwater model is based upon the various parameters and processes associated with the groundwater/surfacewater. Depending upon the level of complexity of the model, various software packages are used.

P. J. Omar (✉) · N. Shivhare · S. B. Dwivedi · S. Gaur · P. K. S. Dikshit
Department of Civil Engineering, Indian Institute of Technology (Banaras Hindu University), Varanasi 221005, India
e-mail: sss.padam.omar@gmail.com

Keywords Groundwater modeling · Surface-water and ground-water interactions · Varanasi · Transport flow

1 Introduction

The paper is aimed to review the various process associated with groundwater and surfacewater interactions. Various issues related to the managing of groundwater and surfacewater exchange, are mainly addressed here. The procedures stirring groundwater–surfacewater (GW-SW) flux change are difficult and comprise of, groundwater interactions by wetlands hyporheic exchanges, flood inundation, evapo-transpiration from various water bodies, parafluvial flow, and bank storage.

This groundwater-surfacewater exchange phenomenon can be categorized into three important processes, and this is known as three stages of complexity.

Stage 1: Established on experimentally based relations, which derived from conclusions of various field experiments/observations.

Stage 2: Compared with the first-order model, working on finer Spatio-temporal scales has high logical resolution and process complexity.

Stage 3: Case-based, stochastic processes and based on that deterministic models.

After analyses of three stages of complexity, the development of the following model is proposed:

1. A stage 1 complexity, stream scale, 'Groundwater–Surfacewater connection' model, which works as a groundwater relation to the streams of low elevation. The predicted result of this approach is responsible for groundwater–surfacewater exchange at the stream level.
2. A stage 2 complexity, sub-stream scale, 'Floodplain course,' was done, and this model leads to the gradually model the flood plain inundation, bank storage, and rate of evapo-transpiration for the water bodies. The results of this complex model lead to GW-SW exchanges with the capability to relate to environmental response models.

This paper is aimed to summarize all possible prospects of the groundwater and surfacewater interactions. Along with this, the paper focuses on conjunctive use of groundwater/surfacewater and water reservoirs management, where groundwater–surfacewater involvement is a major ingredient. This study is also helpful for environmentalist because groundwater/surfacewater interaction leads to the mixing of their water qualities.

2 Background of the Study

Here the process is defined as being groundwater driven, surface water-driven, and groundwater–surface water driven. This categorization of the different processes is based on the interaction of groundwater and surfacewater. Figure 1 shows the

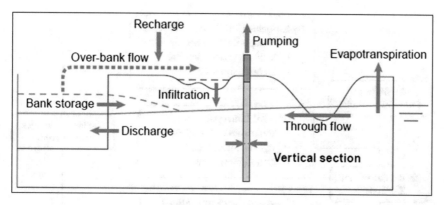

Fig. 1 Groundwater–surface water interaction

interaction of surfacewater (river) and groundwater (aquifer) and a gaining river stream.

2.1 Process Driven by Groundwater

River stream reduction caused due to pumping is explained as the decrement of water flow due to the induced incursion of river channel water into the under groundwater or the capture of groundwater expulsion to the river channel (Theis 1941; Sophocleous 1997). The situation of river-aquifer interface is connected to proper-use water management of reservoirs and hydrology of the riparian zone (Hantush 2005). This notion is applicable only to river water channels linked to the underground surface through a fully saturated substance. Groundwater pumping close to rising streams impacts water availability, quality, and quantity of water, and water supply management. The environmental effects of stream reduction are quite significant in low lying river channels (Hunt 2003). While pumping out, groundwater starts to decline, and the cone of depression is succeeding towards a close river channel. The start of stream depletion is marked by the infiltration of surface water (stream/river or runoff) into the aquifer because when the depression cone touches the stream, the amount of groundwater discharge to the river channel decreases. Due to long periods of continuous pumping, the cone of depression attains its ultimate figure. The quantity of pumping balanced by river channel reduction in the steady-state situation will rely on a number of aspects, such as the nearness of the borehole to the river channel as compared to the length between the bore and other points and diffused groundwater recharge sources. Stream reduction in the case of leaking aquifers partly supports groundwater removal from a pumping well (Zlotnik 2004). Some of the more important impacts that may affect flow reduction include the flow stoppage through the aquatic aquifer hydraulic conductivity contrast, flow partial osmosis, and aquifer heterogeneity quantification. The time taken to the reach steady state is essentially

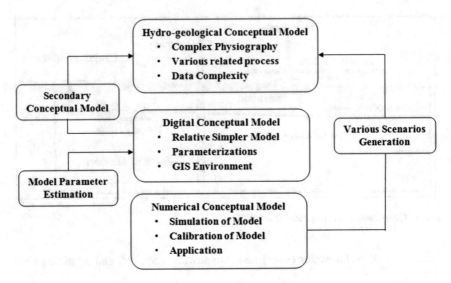

Fig. 2 Flow chart of the process of groundwater and surfacewater interactions

noted. Glover and Balmer (1954) provided an analytical solution which is utilized in order to point out the time frame for aquifer (groundwater) and river (surfacewater) decrease when pumping is carried out.

Sophocleous et al. (1995) used mathematical model (MODFLOW) (McDonald and Harbaugh 1988) to determine the precision of the surface and sub-surface inter-actions solutions; the three most significant aspects analyzed, which refer to the multidimensional behavior of the flow conditions of the aquifer are: (1) congestion of stream-bed as specified by river-aquifer hydraulic conductivity contrast; (2) degree of partial channel penetration; and (3) aquifer are diverse in character or content. It was finally agreed that some serious effort should be done to enumerate with appro-priate precision the surface and sub-surface hydraulic conductivity contrast. Figure 2 shows the process of groundwater and surface water interactions in the flow chart manner. It considers the whole process into three conceptual models. The first one is the hydro-geological conceptual model, second is a digital, conceptual model, and the last one is a numerical conceptual model. Under each conceptual model, there are various processes which affect the groundwater and surfacewater interactions. Parameter estimation and creation of various scenarios is also an essential part of the modelling. In various scenarios, results come out differently, and the model behaves differently when changing in the parameters.

2.2 Processes Driven by Surfacewater

2.2.1 Overland Flow and Through-Flow

The flow pathways of the basin that consist of straight rainfall, through-flow or shallow sub-surface flow, overland flow, and groundwater flow greatly influence the channel flow response to rainfall. The water that moves above the surface either in the form of quasi-laminar flow or as flow connecting the streams in the surface, while inter-flow can be related to the sub-surface flows that move tangentially to the (small) streams via sub-surface, is described as Overland flow (Ward and Robinson 2000). 'Quickflow' is known as the accumulation of channel precipitation, rapid through flow and overland flow. Spatial deviation in vegetation, surface topology, geology, and characteristics of soil, influence the overland flow (Jenkins et al. 1994). When the infiltration capacity of the soil is lower than the rainfall intensity, overland flow is said to occur (Horton 1933), or it occurs when soil becomes saturated and low lying storages are filled, leading to descending hill flow (Hewlitt 1961). Wardlow et al. (1994) done an integrated-basin management model study on the River Allen and separated the overland flow, surface flow, groundwater flow, and interflow (through flow).

Interflow and surface runoff was combined by Kunkel and Wendland (2002), in the study of river Elbe of the water balance in the long term. Studies by Muller et al. (2003), define that hydrological processes such as rainfall interception, depression storage, evapotranspiration, infiltration, etc., using hydrograph separation techniques, come together all sub-surface processes into a single 'base flow' module. Studies where quantification of interflow is important, models such as TOPMODEL (Moore and Thompson 1996; Walter et al. 2002) are utilized in order to forecast shallow interflow of perched groundwater.

2.2.2 River Flow Attenuation

Redistribution of river flow hydrographs at some interval points along a stream proves that the storage effects of water in the rivers control the passage of flow events. Declining in maximum flows and an increment in a time lag results in the hydrograph attenuation. According to Pilgrim (1987), it is dependent on the total volume of the flow event as compared to the storage volume, and also on some variable factors of the system such as hydraulic resistance, streamflow, and slope. Another important term related to this phenomenon is 'Flow routing.' It can be defined as the determination of the timing and flow magnitude at various points along the river channel from known upstream hydrographs.

2.2.3   In-Stream Storages

When explained in terms of channel flow events, enhancing the in-channel storage increases the streamflow attenuation, i.e., it results in increased time-lags and decreases peak flow. Pilgrim (1987) described various other methods along with the Level Pool method. These methods are used to estimate the model of the effects of in-stream storages on hydrographs. The process of determining the effect of inflow storages on the river aquifer exchanges, rather than the flood hydrology, of a stream, is complicated (Smakhtin 2001). Levitt et al. (2005) recently explored the effect of inflow expansion on the mixing or exchange of river (surfacewater) and aquifer (groundwater) in the Los Alamos Canyon. In order to enhance recharge to the surrounding aquifers, instream storages are constructed; the fluxes so obtained are known as 'artificial recharge' (Abu-Taleb 2003). Shaw (1988) and Eichert et al. (1982) gave a general overview of reservoir operation. In several cases, mixing between groundwater and reservoir is incorporated in the analyses of reservoir operation, and the recharge of groundwater as a result of leakage through reservoir can greatly impact the subsurface hydrology and further impact interactions between river and aquifer (Joeckel and Diffendal 2004; Dassi et al. 2005; Karamouz et al. 2003; Dawoud and Allam 2004).

2.2.4   Off-Stream Storages

Almost all reservoir and embankments undergo seepage, either through the base of the reservoir and abutments or through the bank itself in the case of rock-filled or earth reservoirs (US Army Corps of Engineers 1993). Certain studies analyzed the effect of seepage from off-flow storages water on both groundwater–surface water mixing and groundwater hydrology, although Ruiz and Rodriguez (2002) had already described dynamic modelling of reservoir-groundwater interactions. Surface water embankments are used to boost up groundwater storage with the help of artificial recharge which is a result of reservoir leakage.

2.2.5   Groundwater–Surfacewater Driven Processes

Water, in a different type of wetlands, is within either static or moving, fresh or algae water, brackish or saline. Bullock and Acreman (2003) find out about 6% of the world's land area is covered by Wetlands, and their impact on the water cycle can't be negligible. Winter et al. (1998) explained that the main propeller of wetland hydrology is: rainfall, evapo-transpiration from various sources, and exchange of surfacewater and groundwater. Mainly the inflow of water into the wetlands is from rainfall, continuously seepage into the river, overbank floods, and inflow of groundwater; whereas the outflows comprises of evapo-transpiration, effluent seepage from the river, groundwater outflow and surface runoff (Andersen 2004). Todd et al. (2006)

stated that inside the groundwater flow system, Wetland hydrology is greatly influenced by their position and order. The interchange of water between wetlands with groundwater and surface water are also impacted by the geologic features of their soil beds, and the behavior of the weather (Winter 1999).

However, despite these ongoing advances in the field of numerical models, modeling practices are still needed more research to integrate and incorporate the complexity of the interaction of groundwater and surfacewater. In the interaction of groundwater and surface water, river bed performs the major role, and it is greatly simplified and simulated as static and homogenous (Sophocleous 2002; Partington et al. 2017; Irvine et al. 2012).

3   A Case Study on Varanasi, City, India

For a better understanding of this concept, one case study has been taken. In India, Municipal Solid Waste (MSW) was dumped in open areas, especially in low lying areas. Open dumping of MSW is inclined to groundwater contamination on account of landfill leachate production. There are so many studies that focus on the impact of landfill leachate on the groundwater, but few studies focus on the impact of landfill leachate on the river. In the present study, the impact of the leachate generated from landfill sites in Varanasi on the Ganga River was investigated as the city is situated on the bank of Ganga River. The impact of three quality parameters of leachate is considered, namely Biochemical Oxygen Demand (BOD), Total Dissolved Solids (TDS), and Nitrate. The municipal landfill sites that are considered in the study area are located in different villages of Varanasi, namely Tatepur, Chaudripur, and Ramnagar. Selected sites were situated near to the Ganga River. Leachate and groundwater samples were collected from these landfill sites. The steady-state of the groundwater model is conceptualized by using groundwater flow modelling program. GMS includes a comprehensive graphical interface to the groundwater model and solute transport model. All input map layers, such as boundary conditions, river path, pumping and recharge well locations, map of recharge rate, map of evapotranspiration rate, location of observation wells, and map of borehole data, have been prepared for the analysis in the GIS environment. LULC map was also prepared using ERDAS Imagine, remote sensing software, based on supervised classification. For preparing, LULC map, LANDSAT 8 data was used. Data related to groundwater modelling was collected from different government organizations like Central Ground Water Board (CGWB), Jal Nigam, Varanasi, Central Water Commission (CWC) Varanasi and Indian Meteorological Department (IMD), Pune. GIS has the capability to manage a large volume of spatial data from different sources. Combining with GIS technology, GMS provided a good visualization interface for the user and played a significant role in groundwater evaluation and management (Omar et al. 2019). After the execution of MODFLOW and MT3DMS with the help of these input files, GMS can read back the output for post-processing. For two parameters, model calibration was also done. MT3DMS model is also executed to study the effect of variation of concentration of TDS, BOD and nitrate on the river Ganga with a simultaneous variation of longitudinal dispersivity in the study area. The concentration of quality parameters was increased to 100 and 200% of the actual value of concentration at the landfill site. This variation was studied for different cases of longitudinal dispersion, i.e., 10, 20, and 30 m/d. Mass transported to the Ganga River was also computed using a mass transport model for all the three parameters.

Contamination in the groundwater occurs when groundwater interacts with the pollutant, which is released to the ground. This contamination leads to groundwater pollution. A contaminant plume is created when the pollutant is released in the ground. Plume is the resulting body of polluted water in the aquifer, and its migrating ends called plume fronts.

In the ground, water is moving freely, and its movement combined with dispersion led to the spreading of pollutants in a much wider area. For the analysis of the movement of plume, a Groundwater model or hydrogeological model may be used. Groundwater pollution analysis may focus on hydrogeology, hydrology, soil characteristics, site geology, and contaminant nature. On the ground surface, there are various sources from where pollution can occur, such as leaking sewers, landfill sites, petrol stations, effluent from waste-water treatment plants, and fertilizers used exceedingly on agriculture fields. Studies on the river-aquifer interactions have focused on the physical, chemical, and biological processes and revealed various mechanisms (Boano et al. 2014; Constantz 2016; Brunner et al. 2017; Omar et al. 2020). In 2007, Srinivasa Rao et al. studied the temporal variation in groundwater quality to assess the intensity, extent of sulphide pollution, and its migration behavior in the groundwater system in the industrial area of Gajukawa, Visakhapatnam where zinc smelter, fertilizer, and oil refinery are the dominant industries. Thangarajan (1999) presented the groundwater transport model for the entire Palar river basin. In this regional level modelling, the influence of entire Palar river tributaries in the groundwater system is taken into account. This study emphasized the importance of transport models for the polluted river basins. Mohan and Muthukumaran (2004) applied a conceptual model for the prediction of solute transport in sub-surface flow in the lower Palar river basin, which is contaminated due to the disposal of untreated effluent by the tanneries situated in and around Ranipet area. Molykutty et al. (2005) developed a regional mass transport model for the river basin of the upper Palar River using Visual MODFLOW. In this study, the groundwater flow model was built-up to determine the hydraulic heads and flow velocity in the watershed. TDS concentration in the groundwater for different scenarios was presented using the mass transport model. Asadi et al. (2007) used Geomatics techniques to monitor the groundwater quality in relation to the land-use/land cover for a part of Hyderabad, India. In this study, Physio-chemical analysis of groundwater samples of the study area was carried out at a specified location, and spatial distribution maps of selected water quality parameters were prepared, using the curve fitting method in ArcView GIS software. Raju et al. (2009) estimated the groundwater quality in the lower Varuna river basin, which passes through Varanasi. They used the Piper diagram for graphical treatment of major ion chemistry, which helps in identifying hydro-geochemical facies of groundwater, and the dominant hydro-chemical facies is Ca-Mg-HCO$_3$ with an appreciable percentage of the water having mixed facies. Warhate and Chavan (2012) studied the effect of industrial effluents on water. The authors found that the pollution of Indian rivers if goes beyond a certain limit will increase the cost of water treatment for the maintenance of water quality. The study further reveals that the condition may be more worst if the entire aquatic system is thrown out of gear, which leads to biological imbalance causing ecological disaster in the biosphere.

3.1 Study Area

The Ganga River is the principal river of Varanasi flowing incised into its narrow valley from south to north direction. In the present study, the impact of the landfill leachate generated from the landfill sites in Varanasi district, near to River Ganga, on the groundwater was investigated. Geomorphologically, the study area is located in the central Ganga plain of the Indian sub-continent. The Gangetic plain was formed by the regular deposition of the silt carried out by the Ganga river. Soil is very fertile and consists of a loamy type of soil. The topography of the study area is irregular type, an elevation ranges from 33 to 101 m. Maximum elevation was found in the southern part of the study area. The average annual rainfall for the area is 1021 mm, which falls in the rainy seasons. In the study area, the water level in the Ganga River has been decreased significantly due to the construction of the dams in the upstream side of the river and the unregulated extraction of the water.

Varanasi is a religious place, along with a favorite destination for tourists. Due to rapid urbanization, there is a sudden population explosion in the study area. This increased and floating population will generate a sufficient amount of municipal solid waste (MSW). Dumping and incineration of MSW is the main issue for the study area. Dumped MSW can generate leachate when it comes to contact of the water for a long time. For the study, three landfill sites were identified at different locations. Site 1 is at Tatepur village at Varanasi; site 2 is at Chaudripur village near Varanasi, and site 3 is at Ramgarh village near Varanasi, as shown in Fig. 3. Three quality parameters were considered in the study, which was Bio-Chemical Oxygen Demand (BOD), Total Dissolved Solids (TDS), and Nitrate. The concentration of these parameters at the landfill sites depends on the physio-chemical concentration of the leachate generated from the sites and on the type of solid waste dumped. The type of solid waste dumped may include that municipal solid waste and industrial waste.

For the values of these parameters, a field survey was carried out, and samples of leachate and groundwater were collected. Different samples were collected from these sites, and value of concentration for each parameter is calculated in the lab. At site 1 Tatepur village, TDS concentration was found 5430 mg/l, BOD concentration was found 1425 mg/l, and Nitrate concentration was found 327 mg/l. At site 2 Chaudripur villages, TDS concentration was found 4800 mg/l, BOD concentration was found 1250 mg/l, and Nitrate concentration was found 240 mg/l. And at site

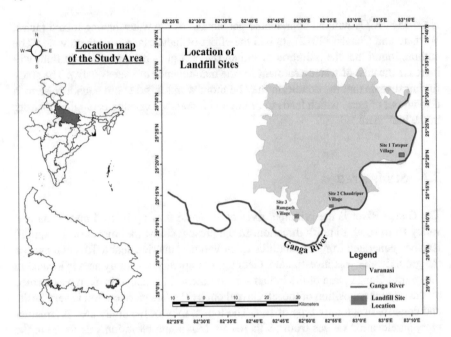

Fig. 3 Location map of the area of study

Table 1 Concentration of quality parameters at different landfill sites

Quality parameters (mg/l)	Site-1 Tatepur village	Site-2 Chaudripur village	Site-3 Ramgarh village
TDS	5430	4800	5120
BOD	1425	1250	1330
Nitrate	327	240	310

3 Ramgarh Village, TDS, BOD, and nitrate concentration was found 5120 mg/l, 1330 mg/l, and 310 mg/l, respectively. The characteristics value of these leachate parameters is presented in Table 1.

3.2   Data Used and Methodology

This model requires different input in the form of layers that have been created on a GIS platform, which are further used to develop a conceptual flow model. The different layers include data of boundary condition, top and bottom layers of the aquifer, recharge map, hydraulic conductivity map (horizontal K_x, K_y and vertical K_z), the value of well discharge, agriculture pattern, river parameters like the stage of the river, sedimentation load, canal layout, surface water bodies, etc. In addition to the above data, the model is also fed with the data of water consumption by crop, by livestock and by the population of the area. Water demand data was collected by carrying out the intensive field survey for the concerned area of study, and some data was procured from the concerned departments. For calculation of the total agricultural area and other land use areas, Image classification of the satellite imagery had also been done. In image classification, Landsat 8 satellite imagery was used, which was downloaded from the USGS website (https://earthexplorer.usgs.gov/). Landsat 8 satellite has two sensors Operational Land Imager (OLI) and the Thermal Infrared Sensor (TIRS). OLI sensor collects the information using nine spectral bands in different wavelength range, and TIRS provides two thermal bands of 100 m resolution. Out of eight bands of OLI sensor, three bands were stacked. After stacking the bands, the area of the study was subset using remote sensing software. For image classification, the maximum likelihood algorithm was applied. Using these classified images, the agricultural area was computed. Based on the fifteen bore wells available in the area concerned, aquifers of the Ganga basin is conceptualized as a single-layer model.

Water demand for livestock is calculated, and it comes out 0.255 Mm^3 per month. The population of livestock in the study area is gathered from the livestock website of Uttar Pradesh. Water demand for other factors such as by population and agriculture field was also calculated. The summation of all demands comprises the total water demand for the area of study.

The groundwater flow model was conceptualized by using information from the geological, climatic, and hydro-geological parameters of the concerned area. Simulation of groundwater flow for steady-state condition was done using MODFLOW program. The groundwater flow model is developed using MODFLOW 2005 based software GMS 10.2 as a graphical interface. The development of the groundwater model is done. The thickness of the aquifer varies from 150 to 280 m. The hypothetical unconfined aquifer is simulated using a single layer in the computational grid. The grid formed for the model development consists of 210 rows and 210 columns, each cell measuring 250 m by 250 m in plan view. For simplicity, the elevation of the top and bottom of each layer is considered flat. All input data sent to the grid. Calibration of the model was also done for two model parameters, namely, recharge rate and hydraulic conductivity. For calibration, fourteen observation wells were identified over the area, and the groundwater level was collected. Using this observed data, check whether the model represents the real field condition or not. If the residual error between observed and simulated head is within the limiting range, the model was accepted otherwise change the parameter value and again run the model until residual error minimizes.

For transport simulation, three constant contaminant sources were assumed. Species from all three landfill sites that were considered in contamination of river water are Bio-Chemical Oxygen Demand (BOD), Total Dissolved Solids (TDS), and Nitrate. The concentration of each of these species was taken from the secondary resources and mentioned in Table 2. Taken into all these parameters and boundary conditions, steady-state flow simulation model is developed using MT3DMS module, which is inbuilt in GMS software. MT3D is a modular groundwater mass transport

Table 2 Value of various parameters used in model preparation

S. No.	Parameter	Value
1	Hydraulic conductivity, K	18 m/d
2	Horizontal anisotropy	1
3	Vertical anisotropy (Kh/Kv)	3
4	Longitudinal dispersivity	20
5	Porosity (θ)	0.3
6	Ratio of vertical transverse dispersivity to longitudinal dispersivity	0.1
7	Ratio of horizontal transverse dispersivity to longitudinal dispersivity	0.2

model in three dimensional based on the finite difference method (FDM). It was used for the simulation of advection, dispersion, and chemical reactions of dissolved constituents in groundwater systems.

In developing the model, the storage term(s) of the groundwater flow equation is set to zero. This is the only part of the flow equation that depends on the length of time, i.e., the total number of the stress period. That's why, in a steady-state simulation, stress time does not affect the head calculation.

3.3 Results and Discussion

After the successful development of the solute mass transport model, results were executed for each site. Here, results for one site are presented, and for others, it was tabulated in Table 3. Results of TDS variation for Site 1 Tatepur were presented in detail. The concentration of TDS, BOD, and nitrate at Site 1 of Tatepur was found 5430 mg/l, 1425 mg/l, and 327 mg/l, respectively. With this value of all parameters, the solute transport model was executed for contaminant transport to predict

Table 3 Concentration of TDS, BOD and nitrate (mg/l) in river Ganga in year 2020 and 2030

Site name	Quality parameter	Year 2020	Year 2030
Tatepur	TDS concentration (mg/l)	650	2100
	BOD concentration (mg/l)	210	420
	Nitrate concentration (mg/l)	45	90
Chaudripur	TDS concentration (mg/l)	625	1425
	BOD concentration (mg/l)	170	850
	Nitrate concentration (mg/l)	75	125
Ramgarh	TDS concentration (mg/l)	2200	3200
	BOD concentration (mg/l)	750	850
	Nitrate concentration (mg/l)	180	210

the traveling time of plume and pollution load (due to TDS, BOD, and Nitrate) reaching to river Ganga. The model is executed for the period of 1825 days (2005–2010), 5475 days (2005–2020), and 9125 days (2005–2030). Figure 4a–c shows the movement of TDS plume in the year 2010, 2020, and 2030, respectively. With the movement of plume towards the River Ganga, the concentration at the points along the trajectory is increasing. With an actual value of longitudinal dispersivity (20 m/d), it was observed that the pollution load had already reached river Ganga in 4015 days (in 2016). From the study, it was also found out that in 2020, the concentration of TDS in the River Ganga will be 650 mg/l, and in 2030, the concentration will be 2100 mg/l. Similarly, BOD concentration will be 210 mg/l and in 2020 and 420 mg/l in 2030. Nitrate concentration will be 45 mg/l and in 2020 and 90 mg/l in 2030.

Site 1 Tatepur shows the greatest movement of TDS for ten-year time duration, from 2020 to 2030 as compared to the other two sites. For BOD concentration, the Chaudripur site shows maximum movement, for the same time duration. Site 3 Ramgarh shows the maximum nitrate concentration among the three sites.

4 Concluding Remarks

The proverb 'horses for courses' fits rightly in the modelling world; it stands for "different people are suited to different things"; here, it means choosing the critical model. If we have conceptualized and organized our problem situation correctly, then we can choose the suitable modelling tool, this essentially means that we are aware of all the processes involved in modelling and how they will interact with one another. Then the second step is choosing (or developing) a modelling tool that has the ability to model the chosen processes at the necessary spatial and temporal scales. The temporal scales mentioned here are very much related to the processes themselves; say, for example, water processes at the surface are fast whereas groundwater processes are comparatively slower. The spatial scales are governed by the problem statements that are being put, i.e., if we want the response of a whole catchment area, a river reach, or some area of riparian land in the closeness of the river stream. The spatial scale will govern the degree of model complexity where larger-scale models normally take a subjective approach, and the models built on a smaller scale adopt a more physical technique. This impacts the need for data requirements with the temporal scales normally needing much larger data as compared to the spatial scales. The setting of the relief features where the model is to be used also has a lasting and significant impact on our choice of the model. Take, for example, water flow modelling in case of defected medium require two-side porosity models; strata systems need models which can handle heterogeneity and relief features with very large and flat flood plains require an efficient modelling generator in order to handle evapo-transpiration and over flooding of banks.

The three dimensional groundwater flow and transport processes were approximated using numerical simulation models, which are available in a comprehensive graphical interface form in the groundwater modelling system (GMS). Input data

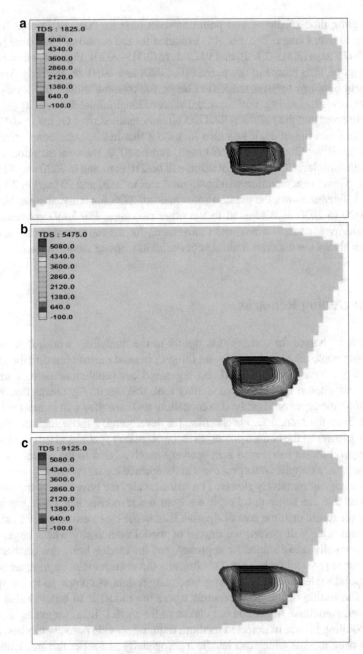

Fig. 4 a Variation of TDS at landfill site 1 Tatepur village in 2010. **b** Variation of TDS at landfill site 1 Tatepur village in 2020. **c** Variation of TDS at landfill site 1 Tatepur village in 2030

for groundwater model and solute transport model was generated by the groundwater modelling system (GMS), which is saved to a set of files. After input data, MODFLOW was executed; some files were generated as output to MODFLOW, which acts as input files to MT3DMS model. After the execution of MT3DMS model, contaminant transport with respect to time is predicted.

During the transport of TDS plume from the landfill site to the river Ganga, it contaminates the potential aquifer lying along its way and thus makes water unfit for various use. The water from these aquifers should be treated before using it if TDS level increases beyond 500 mg/l to make it unfit for daily purposes. After the successful execution of the simulation model, it can be concluded that TDS, BOD, and nitrate generated from the leachate site at Tatepur, Chaudripur, and Ramgarh Village are polluting River Ganga since 2016, 2014 and 2007 respectively. If this pollutant keeps on reaching the river at such rate, it is inevitable to have no aquatic life, no water will be available for irrigation use as well as for industries, and no water will be available for domestic use from the holy river as the cost of treatment would be very high. In order to avoid the above scenario to take place, landfill sites shall be constructed at a large distance from the river, and proper treatment of the dump should be done at the site.

Acknowledgements We would like to acknowledge the support of the Department of Civil Engineering, IIT (BHU) Varanasi. We are also thankful to our lab staff of the department who spent extra time helping us with the field survey.

References

Abu-Taleb MF (2003) Recharge of groundwater through multi-stage reservoirs in a desert basin. Environ Geol 44:379–390

Andersen HE (2004) Hydrology and nitrogen balance of a seasonally inundated Danish floodplain wetland. Hydrol Process 18:415–434

Asadi SS, Vuppala PA, Reddy M (2007) Remote sensing and GIS techniques for evaluation of groundwater quality in municipal corporation of Hyderabad (Zone-V), India. Int J Environ Res Public Health 4(1):45–52

Boano F, Harvey JW, Marion A, Packman AI, Revelli R, Ridolfi L, Wörman A (2014) Hyporheic flow and transport processes: mechanisms, models, and biogeochemical implications. Rev Geophys 52:603–679. https://doi.org/10.1002/2012RG000417

Brunner P, Therrien R, Renard P, Simmons CT, Hendricks Franssen HJ (2017) Advances in understanding river-groundwater interactions. Rev Geophys 55(3):818–854

Bullock A, Acreman M (2003) The role of wetlands in the hydrological cycle. Hydrol Earth Syst Sci 7:358–389

Constantz J (2016) Streambeds merit recognition as a scientific discipline. Wiley Interdisc Rev Water 3(1):13–18

Dassi L, Zouari K, Faye S (2005) Identifying sources of groundwater recharge in the Merguellil basin (Tunisia) using isotopic methods: implication of dam reservoir water accounting. Environ Geol 49:114–123

Dawoud MA, Allam AR (2004) Effect of new Nag Hammadi Barrage on groundwater and drainage conditions and suggestion of mitigation measures. Water Resour Manage 18:321–337

Eichert BS, Kindler J, Schultz GA, Sokolov AA (1982) Methods of hydrological computations for water projects. UNESCO, Paris, p 124

Glover RE, Balmer GG (1954) River depletion resulting from pumping a well near a river. Am Geophys Union Trans 35:368–470

Hantush M (2005) Modelling stream aquifer interactions with linear response functions. J Hydrol 311:59–79

Hewlitt JD (1961) Watershed management. In: Report for 1961 southeastern forest experiment station. US Forest Service, Ashville, pp 62–66

Horton RE (1933) The role of infiltration in the hydrologic cycle. Am Geophys Union Transcripts 14:446–460

Hunt B (2003) Unsteady stream depletion when pumping from semiconfined aquifer. J Hydrol Eng 8:12–19

Irvine DJ, Brunner P, Franssen H-JH, Simmons CT (2012) Heterogeneous or homogeneous? Implications of simplifyingheterogeneous streambeds in models of losing streams. J Hydrol 424–425:16–23. https://doi.org/10.1016/j.jhydrol.2011.11.051

Jenkins A, Peters NE, Rodhe A (1994) Hydrology. In: Moldan B, Cerny J (eds) Biochemistry of small catchments: a tool for environmental research. Wiley, Chichester, pp 31–54

Joeckel RM, Diffendal RF (2004) Geomorphic and environmental change around a large, aging reservoir: lake C. W. McConaughy, Western Nebraska, USA. Environ Eng Geosci 10:69–90

Karamouz M, Zakraie B, Khodatalab N (2003) Reservoir operation optimization: a non-structural solution for control of seepage from tar reservoir in Iran. Water Int 28:19–26

Kunkel R, Wendland F (2002) The GROWA98 model for water balance analysis in large river basins-the river Elbe case study. J Hydrol 259:152–162

Levitt DG, Newell DL, Stone WJ, Wykoff DS (2005) Surface water–groundwater connection at the Los Alamos Canyon weir site: part 1. Monitoring and tracer tests. Vadose Zone J 4:708–717

McDonald MG, Harbaugh AW (1988) A modular three-dimensional finite-difference ground-water flow model, Book 6, Chapter A1. US Government Printing Office, Washington

Mohan S, Muthukumaran M (2004) Modelling of pollutant transport in ground water. J Inst Eng (India) Part EN Environ Eng Div 85:22–32

Molykutty MV, Jothilakshmi M, Thayumanavan S (2005) Groundwater flow modelling of upper Palar basin. Indian J Environ Prot 25(10):865–872

Moore RD, Thompson JC (1996) Are water table variations in a shallow forest soil consistent with the TOPMODEL concept? Water Resour Res 32:663–669

Muller K, Deurer M, Hartmann H, Bach M, Spiteller M, Frede HG (2003) Hydrological characterization of pesticide loads using hydrograph separation at different scales in a German catchment. J Hydrol 273:1–17

Omar PJ, Gaur S, Dwivedi SB, Dikshit PKS (2019) Groundwater modelling using an analytic element method and finite difference method: an insight into lower ganga river basin. J Earth Syst Sci 128(7):195

Omar PJ, Gaur S, Dwivedi SB, Dikshit PKS (2020) A modular three-dimensional scenario-based numerical modelling of groundwater flow. Water Resour Manage 34:1913–1932. https://doi.org/10.1007/s11269-020-02538-z

Partington D, Therrien R, Simmons CT, Brunner P (2017) Blueprint for a coupled model of sedimentology, hydrology and hydrogeology in streambeds. Rev Geophys 55(2):287–309. https://doi.org/10.1002/2016RG000530

Pilgrim DH (1987) Flood routing. In: Pilgrim DH (ed) Australian rainfall and runoff: a guide to flood estimation. The Institution of Engineers, Australia, pp 129–149

Raju NJ, Ram P, Dey S (2009) Groundwater quality in the lower Varuna river basin, Varanasi district, Uttar Pradesh. J Geol Soc India 73(2):178–192

Ruiz HL, Rodriguez JBM (2002) Explicit modeling of a reservoir and an aquifer. Ing Hidraulicaen Mexico 17:89–97

Shaw EM (1988) Hydrology in practice, 2nd edn. Van Nostrand Reinhold, London, p 539

Smakhtin VU (2001) Low flow hydrology: a review. J Hydrol 240:147–186

Sophocleous M (1997) Managing water resources systems: why safe yield is not sustainable? Ground Water 35:561

Sophocleous M (2002) Interactions between groundwater and surface water: the state of the science. Hydrogeol J 10:52–67

Sophocleous M, Koussis A, Martin JL, Perkins SP (1995) Evaluation of simplified stream aquifer depletion models for water rights administration. Ground Water 33:579–588

Thangarajan M (1999) Modeling pollutant migration in the upper Palar river basin, Tamil Nadu, India. Environ Geol 38(3):209–222

Theis CV (1941) The effect of a well on the flow of a nearby stream. Am Geophys Union Trans 22:734–738

Todd AK, Buttle JM, Taylor CH (2006) Hydrologic dynamics and linkages in a wet land dominated basin. J Hydrol 319:15–35

US Army Corps of Engineers (1993) Engineering and design—seepage analysis and control for dams with CH 1. Manual no. 1110-2-1901, Washington, 214 p

Walter MT, Steenhuis TS, Mehta VK, Thongs D, Zion M, Schneiderman E (2002) Refined conceptualization of TOPMODEL for shallow subsurface flows. Hydrol Process 16:2041–2046

Ward RC, Robinson M (2000) Principles of hydrology, 4th edn. McGraw-Hill, London, p 450

Wardlow RB, Wyness A, Rippon P (1994) Integrated catchment modeling. Surv Geophys 15:311–330

Warhate SR, Chavan TP (2012) Evaluation of effect of industrial effluents on water. Sci Revs Chem Commun 2(3):220–224

Winter T (1999) Relation of streams. Lakes, wetlands to groundwater flow systems. Hydrogeol J 7:28–45

Winter T, Harvey J, Franke O, Alley W (1998) Groundwater and surface water: a single resource. USGS Circular 1139. US Geological Survey, Denver, Colorado

Zlotnik VA (2004) A concept of maximum stream depletion rate for leaky aquifers in alluvial valleys. Water Resour Res 40:W06507. https://doi.org/10.1029/2003WR002932

River Bank Filtration in Indo-Gangetic Basin

Aseem Kumar Thakur, Chandra Shekhar Prasad Ojha, Vijay P. Singh, Vidisha Kashyap, and B. B. Chaudhur

Abstract In Indo-Gangetic Basin (IGB), while high flow condition occurs, the rivers carry a high suspended load, and while low flow condition occurs, many rivers are contaminated with pathogenic bacteria. The groundwater is affected by a high concentration of arsenic affecting the districts lying on either side of River Ganga in India, which is also reported as a serious environmental and health problem. The study area that is dominated by the presence of arsenic and fluoride are not suitable for domestic consumption. With developing technologies, we can abstract the good quality of water by River Bank Filtration (RBF). When a production well is pumped in the vicinity of a River, there is drawdown in pumping well. Due to a change in the hydraulic gradient, effective stress is developed in the River bed. Alternatively, hydrometeorology can be useful for many cities and towns in India, which are situated on river banks and have favorable hydrogeological conditions. During the passage of river water to the production well, using mass balance relationships, the contribution

A. K. Thakur (✉)
Government Polytechnic Bhagalpur, Bhagalpur, Bihar 812003, India
e-mail: aseem_thakur@yahoo.co.in

C. S. P. Ojha
Department of Civil Engineering, I.I.T. Roorkee, Roorkee, Uttarakhand 247667, India
e-mail: cspojha@gmail.com

V. P. Singh
Department of Biological and Agricultural Engineering and Zachry Department of Civil Engineering, Texas A and M University, Scoates Hall, 2117 TAMU, College Station, TX 77843-2117, USA
e-mail: vsingh@tamu.edu

V. Kashyap
VIT Vellore, Vellore, Tamil Nadu 632014, India
e-mail: kashyapvidisha@gmail.com

University of Utah, Salt Lake City, UT, USA

B. B. Chaudhur
Government Polytechnic Khagaria, Khagaria, Bihar, India
e-mail: bbchaudhur@gmail.com

© The Author(s), under exclusive license to Springer Nature Switzerland AG 2021
M. S. Chauhan and C. S. P. Ojha (eds.), *The Ganga River Basin: A Hydrometeorological Approach*, Society of Earth Scientists Series,
https://doi.org/10.1007/978-3-030-60869-9_6

of the groundwater to abstracted water is assessed. Detailed analysis of the flow field, including the drawdown of the groundwater near the abstraction, well is avoided, and only a partial background is included. In Indo-Gangetic Basin, it is possible to estimate the relative contribution of groundwater and river water to the abstracted water using authentic data of the last fifteen years with such a flow-field analysis. Keeping hydrometeorology as an important parameter, the water quality data of different RBF sites are studied and managed.

Keywords Arsenic · Fluoride · Mass balance relationships · River bank filtration · River Ganga

1 Introduction

The Indo-Gangetic basin (IGB) is one of the world's most important freshwater resources and home to the largest surface-water irrigation system is getting extensively contaminated. Figure 1 shows the extant of River Ganges in Indo-Gangetic River Basin (IGB), an alluvial aquifer system. Contamination is, in fact, beginning to outstrip the rapid depletion of groundwater in the area. The IGB system is formed by eroded sediments from the Himalayas, and it contains a high population. There are many aquifers that are contaminated by various human activities. The depletion of groundwater due to its over-extraction in IGB causes arsenic problems, salinity problems, fluoride, iron, and human-made use of fertilizer and chemical problems. The latest research findings concerns expressed by United Nations World Water Report in 2015 that India, China, Nepal, Bangladesh, and Pakistan (four of these are in the

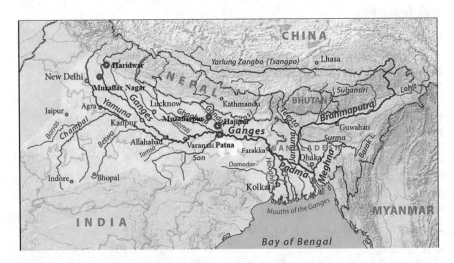

Fig. 1 The extant of River Ganges in Indo-Gangetic basin, an alluvial aquifer system

IGB) alone account for almost half of the world's total groundwater use. It is due to uncontrolled groundwater abstraction.

During riverbank filtration, groundwater and surface water are subjected to a combination of physical, chemical, and biological processes such as filtration, dilution, sorption, and biodegradation, which improves the raw water quality. The two immediate benefits of River Bank Filtration (RBF) are the minimized need for adding chemicals like disinfectants and coagulants to surface water to control pathogens and the decreased cost to the community without increased risk to human health (Ray et al. 2002a). RBF is a widely adopted method for water supply, e.g., groundwater derived from infiltrating river water provides 50% of potable water supplies in the Slovak Republic, 45% in Hungary, 16% in Germany, and 5% in the Netherlands (Hiscock and Grischek 2002). RBF is considered as an efficient treatment mechanism for the removal of various physical, chemical and biological contaminants present in the surface water (Ray et al. 2002b; Prommer et al. 2003; Berger 2002; Price et al. 1999). Transport of surface water through alluvial aquifers is associated with a number of water quality benefits, including removal of microbes, pesticides, total and dissolved organic carbon (TOC and DOC), nitrate, and other contaminants (Ray 2003; Kim et al. 2003; Weiss et al. 2002; Kuehn and Mueller 2000; Cosovic et al. 1996; Bourg and Bertin 1993). The concentrations of anions and cations in river water and riverbank filtrate from vertical as well as horizontal collector wells vary, i.e., depending upon site conditions, well type, location, and the contaminant itself (Weiss et al. 2002). Thakur and Ojha (2005) analyzed the water quality in pre-monsoon and monsoon seasons of the River Ganga and five closely located production wells within a 15 km stretch of the bank of River Ganga at Patna and observed that RBF was effective in removing E. coli and turbidity. The sediment load increase in the source water may affect the turbidity, and this may also be reflected in the abstracted water. The variation of turbidity in the source water and the abstracted water was studied by Ojha and Thakur (2010). Ojha and Thakur (2011a) described a model using the data of influent concentration and bank filtrate using travel time. Evaluation of probabilistic approach for simulating pathogen removal at a RBF site in India was developed by Ojha and Thakur (2011b). Ojha et al. (2013) observed that filtration kinetics, as well as the SCS-CN approach, is found equally effective in simulating filtrate quality at the RBF site in Haridwar. Thakur et al. (2013) developed a two-tier model, which includes the effect of the clogged layer to obtain an equivalent filtration coefficient. This coefficient is found to be linearly related to the natural logarithm of the concentration of pathogens in the source water. Thakur et al. (2017) focused on the use of a mass balance equation to simulate the contribution of groundwater or river water to the RBF water in the arsenic affected area.

In light of the above introduction, the following objectives of this study were considered: (i) To select five RBF sites and study water quality parameters in pre-monsoon, monsoon and post-monsoon seasons to assess the various water quality in river water, groundwater and the abstracted water from a pumping well which has the contribution from riverbank filtrate and groundwater; and (ii) to develop a model to assess the variation of water quality.

2 Study Area

In this study, the focus is to assess the quality based on hydrometeorology as an important parameter, the water quality data of five different RBF sites at Haridwar, Muzaffar Nagar, Patna, Hajipur and Muzaffarpur as shown in Fig. 1 are studied. However, prior to the development of any water quality model, it is relevant to present certain basics related to pumping well near a river. Figure 2 shows a well, near a recharge boundary. The pattern of streamlines and equipotential lines are also shown in the figure. As abstraction from the well begins, the cone of depression starts developing; and it may reach the recharge boundary (river in the present case) after some time. Irrespective of the progression of the cone, the drop in the level at the boundary will always be zero.

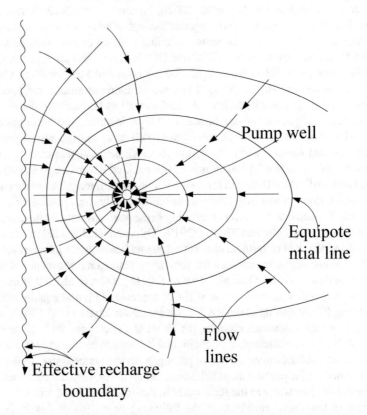

Fig. 2 Abstraction well near a river shows effective recharge boundary

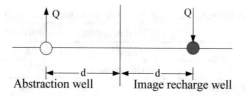

Fig. 3 Representation of abstraction well and image recharge well

Consideration of observation taken only from a single well will indicate that there is some drawdown at the recharge boundary. However, in reality, the drawdown is zero. Thus, to get zero drawdowns, a certain amount of replenishment of water level has to be brought in. This is intelligently done by bringing in another well which is recharging in nature and is located on the other side of the boundary. It is noted that the recharge well has to be on the other side of the boundary. The recharge well is also known as image well despite reverse nature.

Thus, Fig. 2 is replaced by Fig. 3 for well field analysis. The drawdown at any point in the flow domain is simply the algebraic sum of the drawdown due to abstraction as well as its image well, which is a recharge well. The knowledge about streamlines and equipotential lines can be utilized to know the relative contributions of river water and groundwater in the abstraction well. As the aquifer parameters are not known for the site under consideration, the relative contribution of groundwater and river water is made using a mass balance relationship.

2.1 Mass Balance Relationships

In view of the above, we postulate that the quality of the water which comes out at an abstraction well is influenced by the flow Q_{gw} towards abstraction well from groundwater as well as Q_r from river water. If the quality of the groundwater is C_{gw} and river water quality is C_r, the quality C_w in the well water discharging flow Q_w can be assumed to satisfy the following mass balance relationships:

$$C_w Q_w = C_{gw} Q_{gw} + C_r Q_r \tag{1}$$

$$Q_w = Q_{gw} + Q_r \tag{2}$$

Through experiments, it is known that river water carries no arsenic concentration. Thus, for the analysis of arsenic quality parameter, Eq. 1 simplifies to

$$C_w Q_w = C_{gw} Q_{gw} \tag{3}$$

Using Eqs. 2 and 3, one obtains

Table 1 Water quality measurement at RBF site Haridwar, India (30-11-2005) to (08-03-2007)

Instant parameters	Method of testing	Ganga river	Production well
pH	pH meter	7.9–8.5	7–7.56
Turbidity	HACH instrument	18–750	0.119–0.8
Electrical conductivity (μmhos/cm)	Conductivity meter	135–260	505–655
D.O. (mg/l)	D.O. meter	8.0–9.1	0.3–2.7
Total dissolved solid (mg/l)	Conductivity meter	90–205	322–450
Ca^{++} (mg/l as $CaCO_3$)	HACH instrument	30–90	135–210
Mg^{++} (mg/l)	HACH instrument	5.0–35	11–110
Chloride (mg/l)	HACH instrument	2.0–10	10–25
Sulphate (mg/l)	HACH instrument	6.0–35	19–35
Alkalinity (mg/l)	Titration	36.0–105	190–290
Total coliform	Test-tube technique	4.0–9300	2–57
Fecal coliform	Test-tube technique	9.0–7500	3–9

$$Q_{gw} = \frac{C_w}{C_{gw}} Q_w \tag{4}$$

$$Q_r = \frac{C_{gw} - C_w}{C_{gw}} Q_w \tag{5}$$

These relationships are expected to work well for a steady-state situation.

3 Analysis of Field Data

Five RBF sites such as Haridwar (30-11-2005 to 08-03-2007), Muzaffar Nagar (04-03-2006 to 30-10-2006), Patna (from 2006 to 2007), Hajipur (06-10-2010 to 06-12-2011) and Muzaffarpur (17-08-2010 to 12-10-2011) where water quality studies were carried. The data of the RBF water quality of Haridwar is shown as Table 1.

The data of RBF water quality of Muzaffar Nagar is shown as Table 2.
The data of RBF water quality of Patna is shown as Table 3.
The data of the RBF water quality of Hajipur is shown in Table 4.
The data of RBF water quality of Muzaffarpur is shown as Table 5.

4 Results and Discussions

Observations made from Tables 1, 2, 3, 4 and 5, it has been observed that during RBF the polluted river water travel with the bank filtrate towards the pumping well.

Table 2 Water quality measurement at RBF site Muzaffarnagar (U.P.) (04-03-2006) to (30-10-2006)

Instant parameters	Method of testing	Kali river	Pumping well
pH	pH meter	6.9–7.7	6.7–7.8
Turbidity	HACH instrument	4.1–9.9	0.16–2.2
Electrical conductivity (μmhos/cm)	Conductivity meter	453–599	477–606
D.O. (mg/l)	D.O. meter	0–4.1	0–1.2
Total dissolved solid (mg/l)	Conductivity meter	295–350	355–380
Ca^{++} (mg/l as $CaCO_3$)	HACH instrument	35–110	120–210
Mg^{++} (mg/l)	HACH instrument	15–35	25–110
Chloride (mg/l)	HACH instrument	5.0–15	10–29
Sulphate (mg/l)	HACH instrument	21–40	25–55
Alkalinity (mg/l)	Titration	205–279	238–250
Total coliform	Test-tube technique	93–2400	2–28
Fecal coliform	Test-tube technique	23–1500	2–11

It is possible that certain type of impurities such as turbidity, total coliform, etc. may get attenuated, whereas certain types of impurities such as Total Dissolved Solids, Ca^{2+}, Mg^{2+}, Chloride$^-$, Total Hardness, etc. may get increased.

During the present study, different indicators of impurities such as turbidity and total coliform and arsenic have been taken into account to study their variations in the river as well as the pumped water. In the monsoon season, the discharge in the river is high, and it carries a high load of silt with it. The peak values of turbidity of the river and the peak value of production well is taken. The data of the turbidity of Patna was not available in Table 3. However, data of turbidity at RBF site Patna has been taken of the month of July 2004 from Thakur and Ojha (2005). Figure 4 shows the variation of turbidity during RBF sites in India. It has been observed that due to RBF, turbidity has been removed up to 99–99.9%.

During low flow conditions, in the pre-monsoon season, the total coliform is high. The peak values of total coliform in RBF sites have been taken into account. Figure 5 shows the variation of total coliform during RBF sites in India. It has been observed that due to RBF, total coliform has been reduced up to 99–99.9%.

Table 3 Water quality measurement at RBF site Patna (2006–07)

Parameter	Ganga river (R1…R4)		Production well (T1…T8)		Dug well (W1…W3)	
	Pre-monsoon	Monsoon	Pre-monsoon	Monsoon	Pre-monsoon	Monsoon
pH	6.9–8.4	7.8–8.2	7.1–7.8	7.5–7.9	7.0–8.0	7.4–7.7
Water temp. (°C)	18.0–33.6	29.4–30.7	18.0–29.0	28.3–28.7	18.1–28.3	28.0–28.5
Electrical conductivity ($\mu S\ cm^{-1}$)	303–549	168–296	505–681	219–940	734–1547	210–1532
Dissolved oxygen (mg L^{-1})	4.8–9.0	2.9–5.6	0.7–7.4	0.0–6.6	1.8–6.9	3.5–5.5
Total alkalinity (mg L^{-1})	110–230	57–284	151–238	57–277	194–332	287–294
Total hardness (mg L^{-1} as $CaCO_3$)	105–506	90–123	84–364	142–218	148–503	356–446
Ca^{2+} (mg L^{-1})	43–46[3]	22[1]	67–84[9]	70–74[4]	97–104[3]	n.d.
Mg^{2+} (mg L^{-1})	13–18[3]	5[1]	16–20[9]	15–16[4]	26[1]	n.d.
Na^+ (mg L^{-1})	17–29[3]	10[1]	25–35[9]	27–29[4]	83[1]	n.d.
K^+ (mg L^{-1})	4–5[3]	3[1]	2–7[9]	3–5[4]	55[1]	n.d.
Cl^- (mg L^{-1})	3–21	4–9	3–69	3–41	39–83	52–114
SO_4^{2-} (mg L^{-1})	12–29	9–42	2–31	3–74	18–57	29–60
NO_3^- (mg L^{-1})	<1[2]	n.d.	0.7–7.7[5]	1–15[4]	55–83[2]	n.d.
DOC	1.9–2.1[2]	4.9[1]	0.2–2.8[4]	0.6–1.6[4]	0.6[1]	n.d.
Total coliform (MPN/100 mL)	24,000–160,000	90,000–160,000	8–170	8–300	170–2200	800–5000

MPN most probable number, n = 180, [1…9] n = 1…9, n.d. = not determined (Sandhu et al. 2011)

Table 4 Water quality measurement at RBF site Hajipur (06-10-2010–06-12-2011).

Instant parameters	River Gandak	Pumping well	Groundwater
Turbidity	18–595	0.20–0.5	0.15–0.21
Arsenic	0	5–10.5	47.8–66.6
Total coliform	45–500	2–4	1–2

Using in detail data of arsenic of RBF site at Hajipur, Thakur et al. (2017), short-form shown in Table 4 and using mass balance Eqs. (4) and (5) to compute the contribution from groundwater as well as river water. The analysis of observed and relevant data is Table 6.

Figure 6 shows the observed variation of arsenic concentration in river, ground-water and abstraction well.

Figure 7 shows the monthly variation of computed flow components. The abstraction is assumed as continuous as it is continuously used by the users. Thus, any intermediate disruption of abstraction is considered as insignificant.

It is apparent from Fig. 6 that there is a drastic reduction in the concentration of arsenic in the abstracted well water. This is only possible through dilution achieved through river water. The stage in the river was recorded to see if there was any influence of river water depth on the recharge from the river, however this concept of constant head discussed herein. Using Darcy's law, the flow taking place from the river is to be proportional to the hydraulic gradient. The hydraulic gradient will depend on the level of water in the river to the drawdown or the head available at the abstraction well. Considering that the abstraction is continuous, it is reasonable to assume that the head at the good face should be constant. Of course, the hydraulic conductivity can also be modified after a long run of operation if the turbid water filtrates through the porous bed. However, to begin with, it is assumed that the hydraulic conductivity does not alter.

5 Conclusions

The following conclusions can be drawn from this study:

1. Due to RBF, turbidity and total coliform present in river water reduce up to 99–99.9%.
2. A mass balance relationship is capable of simulating arsenic concentration in the abstracted water. The analysis is based on the assumption that no arsenic is picked up from the porous media as river water travels through the porous media to reach the abstraction well.

Table 5 Water quality measurement at RBF site, Muzaffarpur (Bihar) (17-08-2010–12-10-2011)

S. No.	Instant parameters	Desirable limit	Permissible limit	Method of testing	River Burhi Gandak	Pumping well
1	pH	6.5–8.5	No relaxation	pH meter	7.4–9.1	7.0–8.2
2	Turbidity (N.T.U.)	5.0	10.0	Nephalometric	20–100	0.18–0.45
3	Conductivity (μmhos/cm)	–	–	Conductivity meter	600–670	720–760
4	D.O. (mg/l)	1.0–9.0	–	D.O. meter	2.0–5.0	0.1–0.25
5	Total dissolved solid (mg/l)	500.0	2000.0	Conductivity meter	410.0–428.0	425.0–465.0
Chemical parameters						
6	Total hardness (mg/l as CaCO₃)	300.0	600.0	EDTA method	215.0–305.0	265.0–350.0
7	Ca⁺⁺ (as Ca, mg/l)	75.0	200.0	EDTA method	18.0–32.0	36.0–48.0
8	Mg⁺⁺ (as Mg, mg/l)	30.0	100.0	EDTA method	9.8–20.0	19.0–30.0
9	Chloride (mg/l)	250.0	1000.0	Titration	19.0–29.0	39.0–51.8
10	Alkalinity (as CaCO₃), mg/l	200.0	600.0	Titration	225.0–290	265.0–325.0
11	Iron (as Fe), mg/l	0.3	1.0	Phenanthroline method	0.01–0.025	0.025–0.032
12	Nitrate (as NO₃), mg/l	45.0	No relaxation	UV-method	0.31–0.49	0.03–0.039
13	Sulphate (as SO₄), mg/l	200.0	400.0	Turbidimetric method	4.0–4.85	4.0–4.9

(continued)

Table 5 (continued)

S. No.	Instant parameters	Desirable limit	Permissible limit	Method of testing	River Burhi Gandak	Pumping well
14	Fluoride (as F), mg/l	1.0	1.5	SPANDS	0.02–0.19	0.53–0.69
15	Arsenic (as As), mg/l	0.01	0.05	SDDC method	n.d.	n.d.
Bacteriological parameters						
16	Total coliform (MPN/100 ml)	<1.0	–	Test-tube technique	110–1600	2.0–4.0

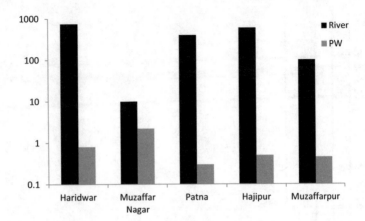

Fig. 4 Variation of turbidity during RBF sites in India

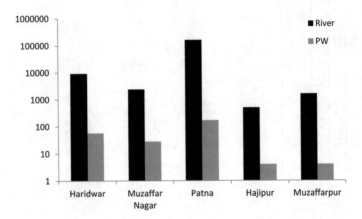

Fig. 5 Variation of total coliform during RBF sites in India

Table 6 The analysis of observed and relevant data

Date	Q_r (l/min)	C_r	Q_g (l/min)	C_{gw}	Q_w (l/min)	C_w
06.10.2010	15.2	0	2.1	52.2	17.4	6.5
06.11.2010	14.8	0	2.5	54.2	17.4	8
05.12.2010	14.9	0	2.5	56.8	17.4	8.2
06.01.2011	14.9	0	2.4	60.4	17.4	8.5
06.02.2011	15.4	0	2.0	61.5	17.4	7.2
05.03.2011	15.3	0	2.0	64.2	17.4	7.6
07.04.2011	15.0	0	2.4	66.6	17.4	9.2
06.05.2011	14.4	0	2.9	61.4	17.4	10.5
06.06.2011	15.9	0	1.4	58.2	17.4	5
06.07.2011	15.9	0	1.4	50.4	17.4	4.2
06.08.2011	15.8	0	1.6	47.8	17.4	4.4
06.09.2011	15.7	0	1.6	54.6	17.4	5.2
07.10.2011	15.4	0	2.0	55.8	17.4	6.5
06.11.2011	15.0	0	2.3	56.4	17.4	7.6
06.12.2011	14.7	0	2.6	55.2	17.4	8.5

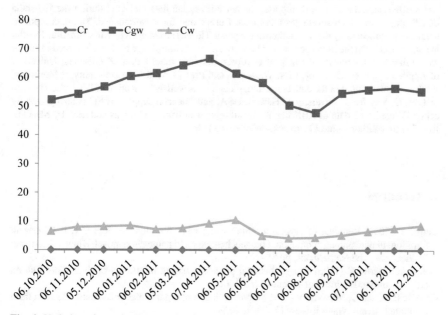

Fig. 6 Variation of arsenic concentration in river, groundwater and abstraction well

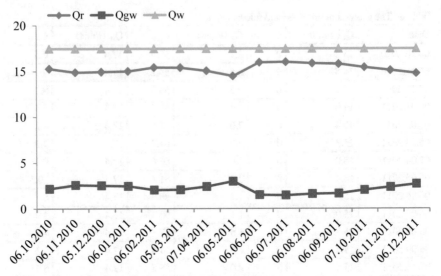

Fig. 7 Computation of river and groundwater flow contribution

Acknowledgements The work reported herein utilizes the data collected during the EU-India ECCP project in which several partners from Europe and India participated. The data reported herein were collected by the first author in the paper. He was also helped by Mr. Cornellius Sandhu in data collection in the initial stages, and he would like to acknowledge his help. The second author would like to acknowledge the help of European partners, mainly Prof. T. Grischek, University of Applied Sciences, Germany, Prof. W. Rauch, and Prof. B. Wett from University of Innsbruck, Austria, for their input in the data collection, modeling as well as a selection of RBF site. The help of Prof. C. Ray from University of Hawaii, USA, and financial support of EU is also gratefully acknowledged. The data of RBF site at Muzaffarpur is utilized herein was collected by Mr. C.B. Rai. The first author would like to acknowledge the help.

References

Berger P (2002) Removal of cryptosporidium using bank filtration. In: Ray C (ed) Riverbank filtration: understanding contaminant biogeochemistry and pathogen removal. Kluwer Academic Publications, The Netherlands, pp 85–121

Bourg AC, Bertin C (1993) Biogeochemical processes during infiltration of river water into an alluvial aquifer. Environ Sci Technol 27:661–666

Cosovic B, Hrsak D, Vojvodic V (1996) Transformation of organic matter and bank filtration from a polluted stream. Water Res 30(12):2921–2928

Hiscock KM, Grischek T (2002) Attenuation of groundwater pollution by bank filtration. J Hydrol 266(3–4):139–144

Kim SB, Corapcioglu MY, Kim DJ (2003) Effect of dissolved organic matter and bacteria on contaminant transport in riverbank filtration. Contam Hydrol 66(1–2):1–23

Kuehn W, Mueller U (2000) Riverbank filtration—an overview. J AWWA 92(12):60–69

Ojha CSP, Thakur AK (2010) River bank filtration in North India. In: ASCE conference proceedings. Published June 15, 2010. World Environmental and Water Resources Congress: Challenge of

Change, Providence, Rhode Island, USA, 16–20 May 2010, pp 782–791. https://doi.org/10.1061/41114(371)87

Ojha CSP, Thakur AK (2011a) Turbidity removal during a subsurface movement of source water: a case study from Haridwar, India. ASCE J Hydrol Eng 16(1):64–70

Ojha CSP, Thakur AK (2011b) Evaluating a probabilistic approach for simulating pathogen removal at a RBF site in India. ASCE J Pract Periodical Hazard Toxic Radioactive Waste Manage 15(2):64–69 (2011)

Ojha CSP, Thakur AK, Singh VP (2013) Modelling of river bank filtration: experience from RBF site in India. J Groundwater Res (JGWR) 2(01):46–55. ISSN 2321-4783

Price ML, Flugum J, Jeane P, Tribbet-Peelen L (1999) Sonoma county finds groundwater under direct influence of surface water depends on river conditions. In: Proceedings, international riverbank filtration conference. National Water Research Institute, Fountain Valley, California

Prommer H, Barry DA, Zheng C (2003) MODFLOW/MT3DMS based reactive multi-component transport model. Groundwater 42(2)

Ray C (2003) Modeling RBF efficacy for migrating chemical shock loads. J AWWA 96(5):114–128

Ray C, Melin G, Linsky R (eds) (2002a) Riverbank filtration—improving source-water quality. Kluwer Academic Publishers, Dordrecht, pp 1–18

Ray C, Grischek T, Schubert J, Wang JZ, Speth TF (2002b) A perspective of riverbank filtration. J AWWA 94(4):149–160

Sandhu C, Grischek T, Schoenheinz D, Prasad T, Thakur AK (2011) Evaluation of bank filtration for drinking water supply in Patna by the Ganga river India, Chapter 12, In: Ray C, Shamrukh M (eds) Riverbank filtration for water security in desert countries. https://doi.org/10.1007/978-94-007-0026-0_12. © Springer Science+Business Media B.V.

Thakur AK, Ojha CSP (2005) A preliminary assessment of water quality at selected riverbank filtration sites at Patna, India. In: Ray C, Ojha CSP (eds) Riverbank filtration—theory, practice and potential for India. In: Proceedings of two-day international workshop on riverbank filtration, 1–2 Mar 2004, IIT Roorkee, India. Water Resources Research Center, University of Hawaii, Manoa, USA (Cooperative Report CR-2005-01), pp 113–130

Thakur AK, Ojha CSP, Singh VP, Gurjar RB, Sandhu C (2013) Removal of pathogens by river bank filtration at Haridwar, India. J Hydrol Process 27(11):1535–1542 (2013)

Thakur AK, Ojha CSP, Singh VP, Chaudhar BB (2017) Potential of river bank filtration in arsenic—affected region in India: case study. J Hazard Toxic Radioactive Waste. https://doi.org/10.1061/(ASCE)HZ.2153-5515.0000363. ©2017ASCE. ISSN 2153-5493

Weiss WJ, Bouwer EJ, Ball WP, O'Melia CR, Arora H, Speth TF (2002) Reduction in disinfection byproduct precursors and pathogens during riverbank filtration at three midwestern United States drinking-water utilities. In: Riverbank filtration: improving source water quality, pp 147–173

River Discharge Study in River Ganga, Varanasi Using Conventional and Modern Techniques

Shiwanshu Shekhar, Manvendra Singh Chauhan, Padam Jee Omar, and Medha Jha

Abstract The present study attempts a new approach, Acoustic Doppler Current Profiler (ADCP), to measure the discharge and other parameters like velocity depth, etc. It also compares the same with the conventional method. ADCP discharge measurements were made at the eight sections of the river Ganga at Varanasi. The result obtained was analyzed by Win-River II Software. This study also compares the sediment distribution along with the upstream and downstream of Samne Ghat Bridge, Varanasi. It was observed that ADCP measures six out of eight sections more precisely as compared to conventional methods. Changes due to aggradations and degradations of the sediment particles along shores may damage the bridge, and river banks may be unstable. After the construction of the bridge, the river profile will be affected due to disturbance in the flow of the river water. This disturbance leads to the change of the present channel of the river.

Keywords River Ganga · ADCP · Discharge · River cross-section · Varanasi

1 Introduction

Automatic and continuous monitoring of river discharge is often performed by an indirect approach that calculates discharge at all stream stages on the basis of the stage-discharge relationship or rating curve. With the development of sonar related devices, a direct method for discharge measurement can be possible. Acoustic Doppler Current Profiler (ADCP) is a very important technique for direct river

S. Shekhar · M. Jha
Department of Civil Engineering, Indian Institute of Technology, (BHU), Varanasi 221005, India

P. J. Omar (✉)
Department of Civil Engineering, Motihari College of Engineering, Motihari 845401, India
e-mail: sss.padam.omar@gmail.com

M. S. Chauhan
Department of Civil Engineering, Holy Mary Institute of Technology and Science, Hyderabad, Telangana 501301, India

© The Author(s), under exclusive license to Springer Nature Switzerland AG 2021
M. S. Chauhan and C. S. P. Ojha (eds.), *The Ganga River Basin: A Hydrometeorological Approach*, Society of Earth Scientists Series,
https://doi.org/10.1007/978-3-030-60869-9_7

101

discharge measurement, which is based on the doppler principle (Morlock 1996; Simpson and Oltmann 1990; Simpson 2001). Direct measurements of discharge used to develop the rating curve made by using a large variety of current sensors, which include propellers, electromagnetic sensors, floats, image technique, radio current meters, and an ADCP. The direct discharge measurements using current sensors are not reliable and unsafe under some flow conditions like large floods (Rantz 1982). ADCP uses acoustic pulses to measure water velocities and depths. The specifications of this device indicate that it has sufficient resolution and precision to calculate river discharge measurements in water as shallow as 4 ft. With the help of a moving boat, ADCP is moved across the river to sketch the river cross-section. It is also equipped with a Geographical Positioning System to facilitate the coordinates of the cross-sections, and the sonar signal is used to acquire velocity readings in the water column.

Data is further analyzed by software WinRiverII. Morlock (1996) discussed the developments in Acoustic Doppler Current Profiler (ADCP) technologies, which have made instruments potentially useful for making measurements of discharge in river sand large streams. The software also calculates concurrent backscatter movement, then calculates and displays velocities in the water column. By the time the ADCP has finished, velocity measurements at intervals as small as 0.2 mm, from the surface to substrate, have been obtained. Wang and Huang (2005) discussed "Horizontal Acoustic Doppler Current Profiler (H-ADCP) for real-time open channel flow measurement: flow calculation model and field validation H-ADCP is an effective tool for river or open channel discharge monitoring (either real-time or self-contained deployment) (Huang 2006). Nihei and Sakai (2007) had given an idea about a new monitoring system using horizontal acoustic Doppler current profiler (H-ADCP) measurements and river flow simulation to attain accurate and continuous monitoring for river discharge at a low cost. The H-ADCP can measure the velocity profile along a horizontal line. In the numerical simulation, the measured velocities were interpolated and extrapolated for a river cross-section. As part of the simulation, the river flow model was developed with a new approach for data assimilation to reflect rationally measured velocities in numerical simulations.

The ADCP works by transmitting "pings" of sound at a constant frequency into the water. (The pings are so highly pitched that humans and even dolphins can't hear them.) As the sound waves travel, they ricochet off particles suspended in the moving water and reflect back to the instrument. Due to the Doppler Effect, sound waves bounce back from a particle moving away from the profiler having a slightly lower frequency when they return. Particles between the waves which, the profiler sends out, and the waves it receives is called the Doppler shift. The instrument uses this shift to calculate how fast the particle and the water around it are moving. Sound waves that hit the particle far from the profiler take a longer time to come back than waves that strike close to it. By measuring the time, it takes for the waves to bounce back and the Doppler shift; the profiler can measure current speed. Current speed is measured at different depths with each series of pings. This procedure assumes that the same density in the water column nearby is mainly determined by temperature, i.e., that the salinity has a preconfigured constant value.

In discharge measurement, there are always some uncertainties associated, and quantification of this uncertainty is a challenge for the researchers (Moore et al. 2016). A technique for estimating the uncertainty in discharge measurement is the field inter-laboratory experiment (Le Coz et al. 2016; Gore and Banning 2017). A relationship was established between computed uncertainties for different cross-sections based on some criteria such as flow shallowness and measured discharge ratio. The uncertainty of the discharge measurements varies among the cross-sections (Nihei and Kimizu 2008; Despax et al. 2017).

2 Study Area

The Ganga River was selected as a study area for research analysis of the Varanasi district. Varanasi is situated at the bank of Ganga River. The holy river Ganga is the largest river in India. The flood discharge value obtained from the Central Government organization, Central Water Commission (CWC), Varanasi, which collects the river data. The soil type of this area is very fertile as the Ganga river deposit the silt carried out from the different places. Ganga River is one of the highest sediment load-carrying rivers. Flood discharge value varies from 134.5 cumecs to the maximum value of the 4318.3 cumecs. In this present study, 6.5 km length was surveyed along the river reach, covering almost all ghats and two bridges one is Vishwasundari Bridge, and the other is Samne Ghat Bridge (Fig. 1).

Total reach was divided into eight parts at an equal distance of around 800 m. All eight parts were named as Profile P1 to P8. After that, for further analysis, two sections were taken near Vishwasunderi Ghat and Samne Ghat Bridge, i.e., upstream and downstream each section at a distance of 500 m. Central Water Commission (CWC), Varanasi, provides one cross-section data for Rajghat. There is an observation station of CWC at which observation was made for the river cross-section data, discharge and river gauge. Starting from the SamaneGhat, river cross-section was taken from the sand side of the River to Ghat Side with the help of ADCP. This procedure was done for all the eight stations from P1 to P8. Based on the data recorded by ADCP, eight cross-sections were prepared.

3 Methodology

ADCP is a very important tool for river discharge measurement. This equipment is further analyzed by software WinRiverII. The "heart" of WinRiverIIis the measurement file (*.mmt). A measurement file is created by running the measurement wizard. The measurement control window helps keep track of the files used in the measurement and provides a quick way to access program controls by right-clicking on items in the list. For more information about the measurement control window, the following step was done.

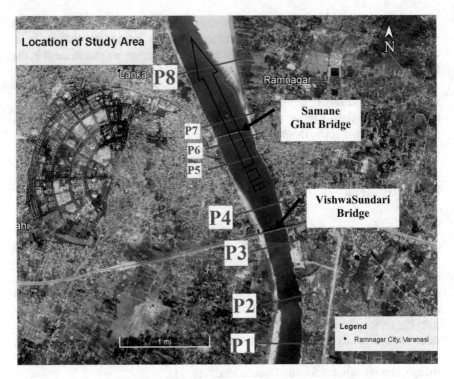

Fig. 1 Location of eight profiles in the River Ganga

3.1 Configuration Information

The Field Configuration for each transect contains the settings used to collect the data. No changes can be made to this information once a transect has started. A Playback Configuration is created when a transect is reprocessed Data.

- QA/QC information (ADCP tests, Moving Bed Test, etc.)
- The Measurement Control window shows a list of all the transects and support files.

ASCII out files—These files contain ASCII text where post-processing takes place. During playback, we can subsection, average, scale, and process data. We can also write this data to an ASCII file and can use these files in other programs (spreadsheets, databases, and word processors).

- Navigation ASCII—These files contain ASCII data collected from an external navigation device during data acquisition. WinRiver II reads the navigation data from a user-specified serial port.

- Depth Sounder ASCII—These files contain ASCII data collected from an external depth sounder device during data acquisition. WinRiver II reads the depth data from a user-specified serial port.
- External Heading ASCII—These files contain ASCII data collected from an external heading device during data acquisition. WinRiver II reads the heading data from a user-specified serial port.

Creating an ASCII file:

- Start off with one item.
- Create a subsection of one ensemble to test the ASCII file.
- Add one item at a time. Check the output against the appropriate tabular display or mouse over the velocity contour screen as appropriate.
- Reprocess each time you make a change.

After this, all notepad data are extracted in excel format, and plotting of graph for various cross-sections takes place. Figure 2 shows the graph between water depth and width of the river water in Sect. 1. The graph depicts that as ADCP moves from one side of the river to the other bank, the depth of water increases up to 13 m. After that, depth remains constant and further reduces up to the end of the river bank.

In order to use WinRiver II, Workhorse ADCP must meet the following criteria. Workhorse ADCP must have the bottom track upgrade and installed. The bottom track and high resolution of WinRiver II can also be used with a River Ray and Stream Pro ADCP if the computer has a Bluetooth connection. In StreamPro Firmware version 31.07, the Long Range mode is extended up to 6 m. For taking the reading, ADCP was tied up with the moving boat and start the software along with ADCP. Starting from the one bank, ADCP starts to record the profile section measurements and simultaneously display it on the WinRiver II. In moving the boat from one bank to another, precaution must be taken and move the boat in a single line. Reaching to

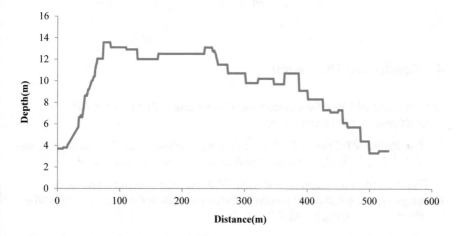

Fig. 2 Result of profiling from ADCP

the other bank, stop the ADCP, and final cross-section has been ready. Similarly, this process was repeated for the other seven sections. Discharge data by the conventional method was taken from the CWC Varanasi. The percentage difference in discharge by both methods was computed. This difference can explain which method is more accurate and reliable.

Conventional methods generally involve the use of previous stage/discharge relations. Historically, a rating is constructed by making measurements of river discharge and plotting the discharge value against the stage of the stream at the time of the measurement. The conventional methods involve measuring depth, width, and velocity at a number of vertical sections across a stream. Depths are measured by sounding with heavyweights, and velocity is measured with rotating-cup current meters. As water flows past the meter, the meter cups rotate at speeds proportional to the current velocity. The product of depth, width, and velocity is the discharge.

By the conventional method, the discharge was computed using the basic relationship between area, the velocity of flow, and discharge.

$$Q = A * V$$

where, A is the area of cross-section (m^2), V is the flow velocity (m/s), and Q is the river discharge (m^3/s).

For area and flow velocity, data was obtained from the CWC, Varanasi, at the same place where the discharge was taken by ADCP.

For P1 profile, A = 5267.35 m^2 and V = 0.245 m/s.

Hence, Q = 1290.5 m^3/s.

The difference in the discharge value measured from both methods, ADCP and from conventional method was computed as:

$$\Delta Q = \frac{1224.93 - 1290.5}{1290.5} * 100 = -5.08\%$$

4   Results and Discussion

Measurement of River Discharge from Cross-section P1 to P8 by ADCP and by Conventional Method respectively:-

1. **For Profile P1**: From ADCP, the Discharge measured for the first section was 1224.93 m^3/s. Table 1 shows more details obtained from the ADCP.

Figure 3 shows the graph plotted using WinRiver II. For comparative analysis, the discharge was also computed by conventional methods at the same locations where the discharge was taken by ADCP.

For Profile P2: Discharge measured from ADCP was 1243.67 m^3/s, and detailed information about P 2 profile is shown in Table 2.

Table 1 Discharge at Profile P1

Discharge (Ref.BT) left to right		
No. of ens.	212	[s]
Start time	05:25:15	
Duration	235.59	
Total Q	1226.83	[m³/s]
Top Q	28.45	[m³/s]
Measured Q	976.15	[m³/s]
Bottom Q	220.33	[m³/s]
(T + M+B) Q	1224.93	[m³/s]
Left dist.	4.00	[m]
Left vel.	0.158	[m/s]
Left depth	2.97	[m/s]
Left area	5.95	[m/s]
Left Q	0.67	[m³/s]
Right dist.	8.00	[m]
Right vel.	0.150	[m/s]
Right depth	2.92	[m/s]
Right area	11.68	[m/s]
Right Q	1.24	[m³/s]
Width	533.24	[m]
Total area	5267.35	[m²]
Q/area	0.233	[m/s]
Flow speed	0.245	[m/s]
Flow dir.	69.09	[°]
Course MG	164.31	[°]
Avg. boat speed	2.285	[m/s]

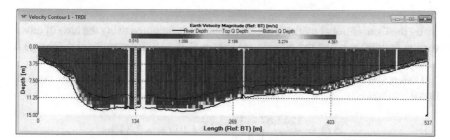

Fig. 3 Velocity Contour for P1

Table 2 Discharge at Profile P2

Discharge (Ref.BT) left to right		
No. of ens.	369	[s]
Start time	05:10:15	
Duration	214.65	
Total Q	1249.64	[m^3/s]
Top Q	40.96	[m^3/s]
Measured Q	960.89	[m^3/s]
Bottom Q	241.82	[m^3/s]
(T + M + B) Q	1243.67	[m^3/s]
Left dist.	7.00	[m]
Left vel.	0.223	[m/s]
Left depth	4.74	[m/s]
Left area	16.58	[m/s]
Left Q	2.62	[m^3/s]
Right dist.	24.00	[m]
Right vel.	0.262	[m/s]
Right depth	1.51	[m/s]
Right area	18.16	[m/s]
Right Q	3.36	[m^3/s]
Width	621.47	[m]
Total area	4344.43	[m^2]
Q/area	0.288	[m/s]
Flow speed	0.305	[m/s]
Flow dir.	66.07	[°]
Course MG	152.69	[°]
Avg. boat speed	2.786	[m/s]

By the Conventional method, the discharge was computed using the area of cross-section is 4400 m^2, and the current velocity is 0.29 m/s. Conventional discharge is 1285.29 m^3/s (CWC, Varanasi).

Difference between discharge measured from both ADCP and from the conventional method is

$$\Delta Q = \frac{1243.67 - 1285.29}{1289.29} * 100 = -3.34\%$$

For Profile P3: Discharge measured by ADCP was 1211.54 m^3/s, and discharge measured by the conventional method was 1260.045 m^3/s with an area of 4200.15 m^2. This profile is having a flow velocity of 0.3 m/s.

Difference between discharge measured from both ADCP and from the conventional method is

$$\Delta Q = \frac{1211.54 - 1260.045}{1211.54} * 100 = -4.00\%$$

Similarly, differences in discharge by the ADCP method and the conventional method were calculated for all the remaining five stations. The results of that calculation are shown in Table 3. Figure 4 shows the velocity contour for P2 profile, and Fig. 5 shows the variation of depth with distance for the same profile. Table 3 comprises the discharge measurement by ADCP and by conventional methods along with the difference in discharge by both methods for all eight profile stations. Out of eight stations, six stations have differences of more than 4%. This shows that the ADCP method is more accurate as compared to the available conventional method.

After analyzing the discharge summary, this paper also studies the sediment load due to the construction of the bridge and the variation in flow on both sides of the river. A comparison of the cross-sections of the upstream and downstream of SamaneGhat Bridge, at a distance of 500 m from the bridge, was also done. Figures 6 and 7 shows

Table 3 Comparison between discharge measured by ADCP and conventional method

Site name	Discharge measured by ADCP (m³/s)	Discharge measured by a conventional method (m³/s)	The difference in discharge by ADCP and by conventional method (%)
P1	1224.93	1290.5	−5.08
P2	1243.67	1285.29	−3.34
P3	1211.54	1260.54	−4.00
P4	1238.7	1289.6	−4.11
P5	1216.58	1282.56	−5.42
P6	1222.75	1288.58	−5.38
P7	1268.46	1285.5	−1.34
P8	1289.33	1298.8	−0.73

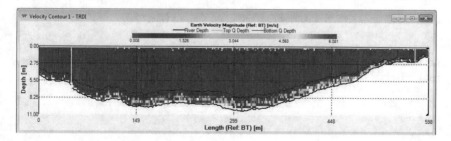

Fig. 4 Velocity contour for Profile P2

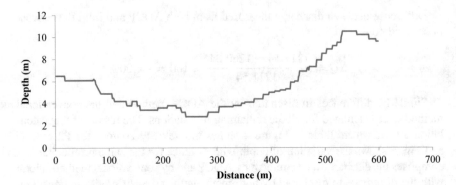

Fig. 5 Depth versus distance for Profile P2

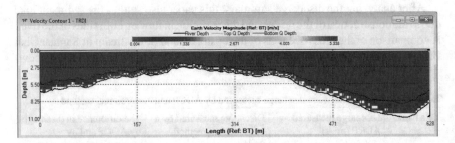

Fig. 6 River cross-section at the upstream side of the Samane Ghat Bridge

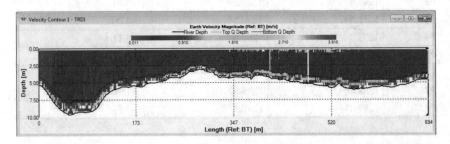

Fig. 7 River cross section at downstream side of the Samane Ghat Bridge

the cross-section of the river at the upstream side and downstream side of the bridge, respectively.

4.1 Cross-Section at the Upstream Side

The total distance measured from the Sand side to the Ghat side was 628 m dividing the cross-section into three parts at a length of 157, 314, and 471 m. From Fig. 6, it can be said that river depth is lowest in the middle around 200 m, and the depth was found 2.78 m. This is the minimum depth in the upstream cross-section of the river.

Total discharge measured was 1275.327 m³/s. The bottom and top discharges were 237.00 m³/s and 49.522 m³/s, respectively. The boat angle was 52.02° in taking the measurements. The total time taken from the sand side to the boat side was 374.39 s.

4.2 Cross-Section at the Downstream Side

The total distance measured from the Sand side to Ghat side was 694 m dividing the cross-section into three parts at a length of 173, 347, and 520 m. From Fig. 7, it is observed that the variation of depth is lowest in the middle, i.e., around 255 m, and this depth is 3.28 m.

The total discharge measured was 1271.366 m³/s. The bottom and top discharges were 244.416 m³/s and 59.409 m³/s, respectively, where the boat angle made by taking these discharge was 53.76°. The total time taken from the sand side to the boat side was 367.80 s.

These results show the effect of the bridge on both sides of the river. Combining and comparing both the results of the upstream side and downstream side between distance versus depths. This result was shown in Fig. 8.

From Fig. 8, it is clear that there is a reduction of sediment load from the starting point of the upstream sand side, whereas it's clearly seen that in upstream sediment load is increasing. When distance traveled in upstream was 1 m river depth was 6.5 while in downstream it was 5.1, it was an initial point where difference level was quite similar to upstream. When distance traveled was 68.1 m, it was 6.2 m for the upstream side and 9.3 m for the downstream side. At a distance of 120 m, 173 m, 313.11 m, and 380.1 m river depth was 6.5, 3.8, 3.1, and 6.4 m for upstream side and 4.4, 5.7, 3.9 and 3.8 m for the downstream side. This is the region where both profiles are changing, and degradation/aggradation is taking place.

Further, after going at a distance of 435.7, 483.3 m, it was 4.6, 8.1 m for upstream side and 6.5, 5.3 m for the downstream side. Now, the difference is coming depth around the cross-section of downstream is decreasing while preceding towards Ghat side, whereas depth is increasing while going towards Ghat in case of upstream.

The major difference has come at a distance of 565.3 m. Here, depth was 10.7 m for the upstream side and 5.4 for the downstream side. These are the changes due to the construction of the bridge, which is the cause for degradation and aggradation of sediment load.

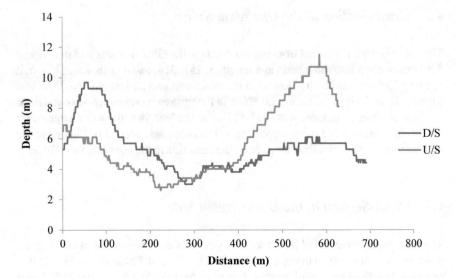

Fig. 8 Combined distance versus depth of upstream side and downstream side

5 Conclusion

Discharge measurements were compared by conventional and ADCP method at river Ganga in Varanasi. In which, three ADCP discharge measurements were within 5% of the conventional discharges computed from the river Ganga, Varanasi. Five ADCP discharge measurements differed by more than 5% from the respective rating discharges; the maximum departure was 5.51% at P 1 station.

The evaluation of ADCP discharge measurements documented in this study indicates that ADCP can be successfully used for data collection under a variety of field conditions. ADCP is more efficient in saving time for discharge calculation, and it is proved to be economical in the long run.

Changes due to aggradation and degradation of sediment particles along shores may damage the bridge, and also river banks will be unstable. After the construction of the bridge, the river profile is changing, and the river may change its present channel. The Ganga River is a meandering river, and the bridge should be constructed exactly at the S-point where less scouring will take place. For, model studies it is required to take the cross-section of a river at least up to 0.5 km upstream and downstream of the site, and this software will be very useful for finding sections for different points.

Thus it can be clearly seen that in order to accurately calculate the parameters on which the river profile depends, conventional methods (current meter) simply fails thus there is a requirement of such advanced method like ADCP.

Erosion pattern of river Ganga due to this Samne Ghat Bridge may affect the stability of the Ramnagar Fort and Ramnagar area in a short course of time. From, safety point of view, proper evaluation is needed, and scouring should be checked at regular intervals of the time. Maximum depth should be near the Centreline, so less effect comes near the banks. Near the bends velocity is less so, it is required to take a proper site for the construction of the bridge.

References

Despax A, Le Coz J, Hauet A, Engel FL, Oberg KA, Dramais G, Blanquart B, Besson D, Belleville A (2017) Quantifying the site selection effect in the uncertainty of moving-boat ADCP discharge measurements. HMEM, Durham, New Hampshire, United States, p 8

Gore JA, Banning J (2017) Discharge measurements and streamflow analysis. In: Richard Hauer F, Lamberti GA (eds) Methods in stream ecology, vol 1, 3rd edn. Academic Press, San Diego, California, pp 49–70

Huang H (2006) River discharge monitoring using horizontal acoustic Doppler current profiler (H-ADCP) Teledyne RD Instruments, USA

Le Coz J, Blanquart B, Pobanz K, Dramais G, Pierrefeu G, Hauet A, Despax A (2016) Estimating the uncertainty of stream gauging techniques using in situ collaborative inter laboratory experiments. J Hydraul Eng 7(142):04016011

Moore SA, Jamieson EC, Rainville F, Rennie CD, Mueller DS (2016) Monte Carlo approach for uncertainty analysis of acoustic Doppler current profiler discharge measurement by moving boat. J Hydraul Eng 143(3):04016088

Morlock SE (1996) Evaluation of acoustic Doppler current profiler measurements of river discharge. U. S. Geological Survey Water-Resources Investigations Report 95-4218, Washington, DC

Nihei Y, Kimizu A (2008) A new monitoring system for river discharge with horizontal acoustic Doppler current profiler measurements and river flow simulation. Water Resour Res 44:W00D20. https://doi.org/10.1029/2008wr006970

Nihei Y, Sakai T (2007) ADCP measurements of vertical flow structure and coefficients of float in flood flows. Paper presented at the 32nd congress of the international association of hydraulic engineering and research, Venice, Italy, 1–6 July

Rantz SE (1982) Measurement and comparison of streamflow: volume 2. Computation of discharge. United States Geological Survey, (USGS) Water-Supply paper 2175

Simpson MR (2001) Discharge measurements using a broad-band acoustic Doppler current profiler. Open-File, Report 01-1. United States Geological Survey, Washington, DC

Simpson MR, Oltmann RN (1990) An acoustic Doppler discharge measurement system. In: Hydraulic engineering-proceedings of the 1990 national conference, pp 903–908

Wang F, Huang H (2005) Horizontal acoustic Doppler current profiler (H-ADCP) for real-time open channel flow measurement: flow calculation model and field validation. In: XXXI IAHR CONGRESS, vol 1, pp 319–328

Review on the Field Applications of River Training Structures for River Bank Protection

Alok Kumar and C. S. P. Ojha

Abstract Implementations of conventional river training structures such as spurs, marginal embankments or levees, revetments, and longitudinal dikes are exclusively costlier and comparatively less efficient than the recently developed river training structures. In addition to the effectiveness of the structures adopted, river experts are also interested in several aspects like the cost involved, navigation development, including enhancement of ecological and morphological diversity. Different head shapes of dike behave differently as compared to the conventional straight spurs and produce more satisfactory results to ascertain the objectives of river training in many critical situations. Permeable structures like reinforced cement concrete (RCC) jack-jetties, porcupines, and timber pile dikes can be used either as a separate system or in conjunction with impermeable structures to overcome the cost-related issues and also to create a better sustainable living environment for aquatic habitats. This chapter focuses on review for the field performances of the application of different permeable and impermeable-type structures in several river basins to emphasize the effectiveness and non-effectiveness of these structures. The outcomes may be helpful for potential river engineers in utilizing the information to select a better alternative as per their field requirement.

Keywords River training structure · Spur dike · L-head dike · Vane dike · Bendway weir · Jack-jetty · Porcupine · Pile dike · Field-applications

A. Kumar (✉) · C. S. P. Ojha
Department of Civil Engineering, Indian Institute of Technology Roorkee, Roorkee, Uttarakhand 247667, India
e-mail: alokiit777@gmail.com

C. S. P. Ojha
e-mail: cspojha@gmail.com

M. S. Chauhan and C. S. P. Ojha (eds.), *The Ganga River Basin: A Hydrometeorological Approach*, Society of Earth Scientists Series,
https://doi.org/10.1007/978-3-030-60869-9_8

1 Introduction

Different rivers and different reaches of the same river have different alignments, channel cross-section shape, bed and bank materials, and slope, including valley characteristics. River alignments can generally be categorized into straight, meandering, and braided channels (Petersen 1986). Straight channels are small reaches and transitory, as even minor disturbance in alignment or channel shape or presence of a temporary blockade can generate a transverse flow causing meandering. However, meandering channels comprise of a sequence of bends of alternate curvature linked by straight crossing reaches. Slopes are usually relatively flat. These channels are unstable, with banks caving in the downstream reaches of concave bends. However, braided channels have the tendency to separate and reconnect in braided reaches. The channel is wide, including poorly defined and unstable banks. Such rivers consist of comparatively steep slopes and transport a large sediment load. A typical sketch for the alignment of the straight channel, meandering channel, and the braided channel is shown in Fig. 1.

Riverbank stabilization structures are termed as structural obstructions, constructed for the protection of riverbank from collapsing. Every structure possesses a particular hydraulic and geomorphic process that is required to be studied very precisely to use them in a more effective manner. Hence, specific bank protection work suitable for the location of the river may not be suitable for other locations in

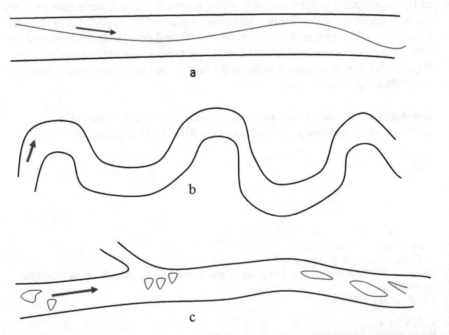

Fig. 1 Typical sketch for the alignment of **a** Straight channel, **b** meandering channel, and **c** braided channel

the same river or different rivers. The efficiency of a selection method for bank stabilization purposes is dependent on the properties of the method and characteristics of
the selected site. In fact, no such distinct method is applicable to every field condition. However, suitable measures may be adopted by assessing characteristics of the
method, ecological welfares, variations in river circumstances, geomorphic response
including early as well as ensuring costs (Bureau of Reclamation, U.S. Department
Technical Service Center 2015).

Bank stabilization and channel rectifications are divided as follows (Petersen
1986):

1. Revetments are constructed parallel to the flow direction and are implemented
   primarily for strengthening concave banks of bends.
2. Dikes are structures, constructed at some inclination to the current, and applied

   • to direct flow smoothly from a bend towards consequent bend located at the
     downstream,
   • to provide appropriate channel alignment by smoothening of sharp bends into
     a greater radius of curvature, and
   • to maintain flow in the required limited width in case of a broader channel.

3. Cutoffs are constructed channels, used primarily to cut off sharp bends for
   refining channel alignments and navigability, including reduction of potential
   bank erosion.

On the basis of the method adopted and materials of construction used, dikes are
further classified as permeable and impermeable-types (Zhang and Nakagawa 2008;
Mansoori et al. 2012). Impermeable dikes generally work by blocking and deflecting
the river flow. In addition to the objectives mentioned above,impermeable-type dikes
are incessantly providing a living environment to the aquatic habitats by generating
glow velocity zone around dike fields (dike field pools) in rivers such as Lower
Mississippi River (Shields 1984; ASCE Task Committee on Sediment Transport and
Aquatic Habitats 1992; Shields 1995). However, the permeable-type of dikes allows
the water to pass through it by reducing the velocity. Flow behavior in the case of
impermeable and permeable-type dikes is shown in Fig. 2.

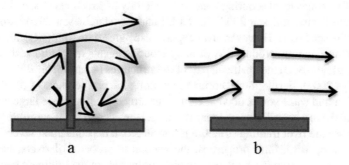

Fig. 2 Flow behavior in **a** Impermeable, and **b** permeable dikes

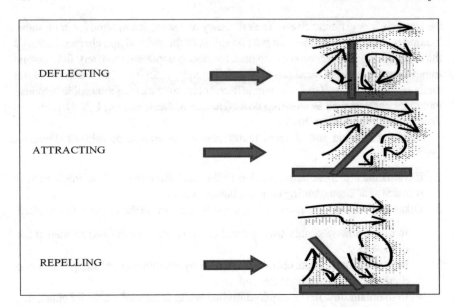

Fig. 3 Orientation and eddies generation around deflecting, attracting, and repelling type of dikes

Based on the orientation, dikes are classified as deflecting, attracting, and repelling types. A deflecting type of dike is constructed normal to the flow and performs by diverting the flow at its head, resulting in the creation of a wake zone downstream of it. An attracting type of dike orients downstream to the flow and works by attracting most of the flow towards its head and bank. However, repelling dike is inclined in the upstream direction and repels the flow away from it. Orientations of these dikes and generation of eddies are sketched in Fig. 3.

The formation and successive growth of local scour must be properly understood to the river engineers for the aspects regarding stability to the foundation of these river training structures. In fact, strong dynamic eddies produced as a consequence of placing structures against the flow direction cause enhancement of pressure and bed shear stress on the bed. Therefore, entrainment of sediment particles gets started, and finally creates local scouring zone in the vicinity of structures (Lane et al. 2004; Koken and Constantinescu 2008a, b; Kirkil and Constantinescu 2010; Koken and Constantinescu 2011). However, river experts also agree with the generation of local scour by the dynamics produced by primary scouring agents such as horseshoe vortex and downflow around the upstream part of the structures (Ansari et al. 2002; Dey and Barbhuiya 2005). Furthermore, vortex tubes along the profile plane of the detached shear layer and wake vortex downstream to the structure also induce large bed shear stress and make a contribution to local scouring. However, as far as equilibrium depth of scour around river training structure is concerned, it depends upon several factors like flow properties, fluid properties, the geometric property of rivers, geometric properties of river training structure adopted, properties of the sediment bed as well as properties and amount of suspended sediments carrying with river flow.

2 Problem Statement

In recent decades, conventional river training structures implemented for the purpose of stream bank erosion inhibition, control of flood, including navigability development, have become exorbitant because of several factors involved. Therefore, conventional river training structures are practically unreasonable, when day by day continuously spreading regions of channel banks affected by scouring and erosion problems need immediate counteractive action. The increasing requirement for providing protection works to the affected banks in different reaches has engrossed attention for the implementation of developing economical as well as environmental-responsive river training structures. This chapter emphasizes the field applications of some of the conventional and recently used river training structures to highlight the applicability of these structures according to the practical field requirements.

3 River Training Structures and Field Applications

3.1 Impermeable Structures

Impermeable structures are made of local soil, stones, gravels, rocks, or gabions and perform by blocking and deflecting the river flow. On the basis of flow condition, impermeable dikes are further classified as submerged and unsubmerged type. The submergence ratio is basically treated as the ratio of dike height to the total depth of approaching flow. In general, the submerged condition is not desirable for this type of dikes, as the overtopping flow may cause severe erosion and damage to the downstream bank.

3.1.1 Spur Dikes

Spur dikes are the impermeable-type of conventional river training structures projected at a reasonable distance from the bank into a stream to deflect flowing water away from the bank. It prevents erosion of the bank and provides a more desirable width and channel alignment. After deflecting the devastating flow away from the affected bank and producing sediment deposits, spur dikes have the ability to protect the stream bank with more efficiency and economically than riprap or revetment works. They are constructed of sand, gravel, or soil as per the availability in the bed of the river. Also, they are protected on the top and sides by durable stone pitching. However, they are also made up of timbers packed with stone or rubble masonry.

Mazumder (2014) reported on the implementation of spur dikes in River Kosi that emerges from the Himalayas and traveled approximately 468 km distance through Tibet, Nepal and India, prior to its meeting with Ganga River at Kursela, Bihar. River

Kosi is having a history of shifting its course sometimes eastward or westward by 150 km distance during a period of 202 years from 1731 to 1933. An 1150 m long barrage having a crest height of 1.53 m was constructed at Hanuman Nagar place close to the border of India and Nepal. Flood embankments with length 144 and 125 km in east and west, respectively, were constructed on both sides of the river in the year 1961. The variation in the embankment spacing was kept in the range of 4–6 km along the reaches. A study by Mazumder (2014) may be referred for detailing of spurs constructed, and the braches history occurred after the construction of flood embankments.

Both the embankments, left as well as right, were tried to be protected by the construction of 378 impermeable spurs in Kosi River. However, embankments got breached on different locations because of spurs failure from time to time. The most recent breaching of 2.2 km reach at Kusaha in the left flood embankment, situated at 12 km upstream of the barrage in the year 2008, was caused by the failure of two long spurs. These two spurs were situated at 12.1 and 12.9 km, and their length was 200 m and 269 m, respectively. The failure of spurs occurred at a discharge of about 4081 cumecs, which was too less than 1/6th of the design flood discharge of 26,922 cumecs. The cause of failure of spurs was attributed to the fact that the length of spurs at the locations was too long and exceeded the maximum allowable constriction of one-fifth of Lacey's waterway. In fact, an increase in constriction length of spur dike provides enhanced protection in the downstream; however, it also causes the formation of an amplified zone of velocity in the vicinity of the dike, resulting in its failure.

Wu et al. (2005) reported for the implementation and effects of certain river training structures like spur dikes, short spurs, and revetment works that played an important role in the flood control and river management in Lower Yellow River, China. The Yellow River is China's second-largest river and consists of a drainage area of 7,95,000 km^2. Yellow River has a history of carrying 1.6 billion tons of sediment load per year measured from 1919 to 1960 pertaining to the highest load of sediment in the world. Almost 25, 50, and 25% silt load are deposited in the lower reaches, estuary, and sea, respectively. Therefore, the rising of bed level in the present scenario is approximately 10 m above the neighboring land in some of the reaches. With an increase in precipitation during flood conditions, the water level in the river rises and causes the breaking of levees. Different river training structures were attempted by the authorities to control the situation from time to time. Nine thousand ninety-six numbers of spur dikes, including short spurs, were constructed in Lower Yellow River by the year 1997. A sum total of 323 numbers of river training works that consist of 134 numbers for protection of levees and 189 numbers for the guidance of flow were constructed after the year 1950. In the transitional as well as meandering reaches, river training structures were also constructed. The implementation of these river training structures resulted in the proper regulation of the main channel.

Shields et al. (1998) reported that spur dikes are more preferable than longitudinal stone revetments for creating a suitable living environment for aquatic habitats. In Goodwin Creek, Mississippi River, eleven numbers of stone spur dikes were constructed perpendicular to the existing longitudinal stone revetment protection. The existing revetment protection was about 170 m of length. On the sand bars existing on the other bank, willow posts were implanted. The development of an appropriate environment for the suitability of aquatic habitats was monitored for four years in reaches with and without spur dikes. It was observed that base flow stoned bank line, the width of water including the availability of pool habitat increased modestly in the reaches of stone revetment associated with spur dikes, however, only local effects were noticed on the depth. The implementation of spur dikes provided shelter and protection against high velocities to the aquatic habitats. With the widening of the base flow channel, it migrated into the willow posts nearer to the stream, thus created a small zone of aquatic habitats of width varying between 0.1 and 1.0 m.

3.1.2   L-Head Dikes

L-Head dikes are impermeable-type of dikes and made up of local soils, but preferably stone gabions using factory-made double twisted hexagonal wire crates. L-Head dikes are also called as trail dikes. It consists of an added structural feature extending towards downstream from the main dike parallel to flow line and works by dispersing the energy of flow over a greater area around the dike. Further, it checks the movement of sediment downstream of the dike by tumbling the production of intensified eddies there. Damage to the main body can be evaded by this structural feature, as this feature absorbs the strong eddies towards it that are primarily responsible for causing scour (Pokrefke 2013). L-Headdikes can be used for the enhancement of spacing between dikes, deterioration of the intensity of scour-prone eddies on the stream side of the dike, and for increasing the effects of dike field towards the downstream.

Solani River is a momentous tributary of River Ganga that flows towards the south east direction. This river possesses a remarkable increase in discharge during the monsoon period. Solani River is fundamentally having the propensity of producing substantial losses to the neighboring lowland areas by eroding the bank as well as changing of the river course during the rainy season. The catchment of river including Dhanori escape, generates runoff that bears a capacity of causing disaster of about 9860 ha of agricultural land. Flooding of river disturbs around 60 villages of Hardwar and Muzaffarnagar districts in Uttarakhand, India.

Nayak et al. (2017) reported that at Bhagwanpur site nearby Roorkee in Uttarakhand, India, 2 mega watt capacity solar power plant is located very close to the left bank of Solani River. The height of channel banks at this site ranges between 3 and 4 m along with 1H: 1 V slope of the banks. Prior to the rainy season of the year 2012, Solani River was having a bulging tendency toward the left bank and threatening to the Power Plant. Therefore, to avoid damages to the plant, six numbers of L-Head dikes were implemented along the reach before the monsoon season of the year 2012. However, in spite of the implementation of these L-Head dikes, some portions of the nearby bank area of the plant got damaged. Thus, to strengthen the bank and protection of transmission lines of power plant further, two additional L-Head dikes, with configuration ratios L1:L2 equal to 15 m:7.5 m and 20 m:10 m, were constructed at

Fig. 4 High intensity flood during the initial phase of monsoon season, 2014 in Solani River

the bank in front of the plant prior to the rainy season of the year 2014. L1 and L2 represent the dike length perpendicular to the flow and parallel to the flow, respectively. In spite of the implementation of these dikes, an initial monsoon flood of the year 2014 (Fig. 4) succeeded in shifting the bank by about 30 m towards the plant (Fig. 5).

Since this flood occurred at the early phase of the monsoon period, it was not possible to extend further the length of L-Head dikes or provide additional dikes, as the construction of these dikes in muddy flowing water was not possible. Therefore, the existing system of L-Head dikes was supplemented with RCC jetty-field consisting of dimension 1.5 m × 0.1 m × 0.1 m. Jetty-field was installed with a single row of a retard as well as a diversion line in one tier each to inhibit further bulging of the river towards the plant during successive floods. To emphasize the effectiveness of joint system of L-Head dikes and RCC jack-jetties, survey work was also accompanied prior to and after the monsoon season of the year 2014 by Sharma and Kumar (2012, 2014).

From the survey results, it was observed that the sedimentation of river land ensued with a considerable average accretion of 1 foot. The conjunctive field application of L-Head dikes and jetty-field in Solani River made the devastating flow currents away from the affected bankand restricted the further bank erosion in front of the plant. The application of these structures also succeeded in the transmission lines protection of power plant. The safeguard to the transmission lines was a massive achievement, since it was supposed to break the electricity supply to power grid

Fig. 5 Scouring of the left bank during monsoon season, 2014 in Solani River. Left bank in front of the plant got shifted by about 30 m by initial monsoon flood of the year 2014 (Fig. 4)

prior to the application of installed training structures. The site condition in Feb, 2017 is shown in Fig. 6 to justify the successful implementation of these structures.

Blancett and Jarchow (2009) reported that L-Head dikes were chosen as the combination of most cost-effective and eco-friendly structures to stabilize the problem of channel distending towards the left side bank of Kansas River, situated at 70 km upstream from the convergence of Kansas and Missouri River. The variation of channel width in this area ranges from 180 to 360 m and height from 6 to 9 m in the channel banks. Channel slope consisted mainly of sandy soil and stood at a slope of 1H: 1 V. Bank materials were having a tendency of falling down into the channel flow with the subsidence of water level. Satisfying the expectation of technical panel, the adopted L-Head dike system is working well at present, as the structures diverted the channel flow away from the bank, initiated the vegetation along the slopes of the channel, and produced significant sedimentation between L-Head dikes and the bank.

3.1.3 Vane Dikes

Vane dikes consist of dike segments placed riverward from the surviving bank, with an appropriate gap in between the subsequent dikes. The gap between the dikes is generally about 50–60% of the individual vane length. Basically, all the vanes in the

Fig. 6 Huge Siltation and formation of new bank at location-A (in front of power plant) in the year, 2017. Site condition of location-A in the year, 2017 can be compared with the condition in year, 2014 (Fig. 5)

system retain equal length. The optimum angle of attack to the vane that generates minimum scour hole should not be greater than 20° (Odgaard and Kennedy 1983). Vane dikes are usually installed in series, protruding transversely from the outer bank into the stream, therefore causing the decrease in velocity at the outer bank, but increase along the inner bank and center of the channel (Scurlock et al. 2012).

In case of Lower Mississippi River near Providence, Los Angeles, vane dikes were installed as bank protection and river training measures, but these were not so effective for the accumulation of sediments and were converted to the L-Head dikes with a provision of tie bank to the river bank (Pokrefke 2013). This modification was accepted as significant shoaling of materials was observed between the vanes as well as in the zone landward of the dikes. After the implementation of such L-Head dikes, the joint structures are performing well.

Odgaard and Wang (1991) reported for the successful implementation of vanes at the bridge site of River West Fork Cedar situated in Butler County, Iowa. A six span bridge consisted of 150 m length and 9 m width with the level of road surface 5 m above the river bed running at low flow condition. However, in the upstream of this site, width at its top was about 30–40 m, whereas the depth of the river was varying between 1.9 and 2.1 m. The problem was initiated at the bridge construction itself in the year 1970 after the straightening and widening of the river. However, a significant excavated region of the portion upstream from the bridge site was filled and found vegetated up to the year 1984. Due to the encroachment produced by the sand bars in four spans along left side abutment, the flow was thrown towards the right side abutment, causing undermining and erosion of the bank. For the purpose of river training and resolving the issues related, 12 vane systems were installed in the year 1984. All of the vanes were installed with sheet piles that were vertically driven into the river bed. Vanes were oriented at a 30° angle to the direction of current in the primary channel. The length of sheet piling was kept at 3.7 m, whereas the elevation of the top was at 0.6 m above the river bed. The purpose of installing these vanes was to reduce the approaching depth of flow and velocity towards the right side bank with the intensification of these flow parameters along the centerline of the river. The vane systems resulted in the formation of a stabilized protective berm along with the erosion-prone bank and also started maintaining the cross-section of bed shape analogous to the designed during the construction of the bridge. The vane systems resulted in the bed aggregation by 0.6–1.0 m around the vane-affected region. Quantity wise, vane systems induced aggradation of approximately 276 m^3 per vane in the affected reach. The reason behind the improved aggradation in the downstream portion from the installed vanes was attributed to the fact that vanes diverted the primary flow towards the middle of the river so that a considerable amount of velocity reduction occurred along the bank.

3.1.4　Bendway Weirs

Bendway weirs are the recently developed impermeable-type of river training structures, that are constructed as a submerged weir and aligned along upstream to serve the purpose of developing and maintaining navigability of the channel. Bendway weirs produce enhanced benefits to the aquatic habitats than the conventional structures like rip-rap, revetment works, gabions (Knight and Cooper 1991; Hildrew 1996; Kuhnle et al. 1999; Shields et al. 2000; Kuhnleet al. 2002; Kinzli and Myrick 2010). The principle behind the bendway weir is that it forces the flow to move towards the point bar situated on the inside portion of bend, and therefore utilizes the energy of highly dynamic flow for widening of the navigation channel. Bendway weirs should be constructed by considering several aspects like the height of bendway weir to avoid interference to the tows, bend radius, geometry of point bar as well as the distribution of velocity (Pokrefke 2013).

Jia et al. (2009) reported that six numbers of submerged weirs were implemented in Mississippi River in year 1995 for developing navigation. These weirs were constructed on the outer bank portion of Victoria Bend. The submerged weirs diverted the flow towards the centre of channel and improved navigation as well. However, some of the bendway improvement results were not found satisfactory. Jia et al. (2009) conducted a numerical study to assess the realignment of surface flow produced by these bendway weirs in the Victoria bend. They found that the flow pattern around bendway weirs was improved but efficiency of individual weir depends upon its alignment, morphology of the local channel, along the condition of flow. In the case of Victoria bend, because of the occurrence of channel deposition around the first weir and returning of flow from the point bar around last weir increased the helical current, and thus these two weirs were found not as effective as other weirs.

3.2 Permeable Structures

Permeable structures are built of local steel angles, wood, timber piles, reinforced cement concrete (RCC). They work on the principle of permitting the water to pass through them by reducing the velocity. On the basis of flow condition, permeable dikes are further classified as submerged and unsubmerged type. The crest of permeable dikes should be intended as level unless the height of the bank or other exceptional circumstances direct to the usage of a crest with sloping design (Julien 2006).

3.2.1   Jack-Jetties

Jack-jetties are the permeable-type of river training structures and consist of 3 steel angles/reinforced cement concrete poles tied together and laced them with wires. It works by reducing the velocity and accumulating the sediment and debris during floods, thus resulting to the formation of a new bank. The jack-jetty system functions well if placed appropriately in concave bend of banks, and the incoming flow carries full of suspended sediments. Restriction created by the jetty-field to sediment, debris, and small trees results to the accretion of land in the jetty-field.

Basically, the concept about the jack-jetty is that a particular or erected unit of the model is said to be a jack, and while jacks are tied together using a cable, it is called as jetty-field. Jetty-lines are termed as diversion lines when they are placed in a direction parallel to the channel bank. However, when the lines are laid at some angle to the bank, they are termed as retards. Therefore, the jetty field is nothing but a combination of retard and diversion lines.

A reinforced cement concrete jack of 1.5 m length, 0.1 m width with 0.1 m thickness meshed with 4 mm galvanized iron wire is shown in Fig. 7. Four numbers of longitudinal bar consisting of 6 mm diameter tor steel tied with 6 mm diameter stirrups @150 mm c/c are generally provided in each of the RCC members of the jack. After placing in the position of RCC jack-jetties at the site, they should be properly connected by a steel or nylon rope for working the whole system as a unit. Trapping of small trees and debris in Solani River, India by jack- the jetty field is shown in Fig. 8.

Grassel (2002) reported briefly about the jack-jetty field that jetty-field is a permeable-type of dike system, which performs more economically than the impermeable-type of dikes. According to his report, Kellner initiated some experiments in a stream nearby Topeka, Kansas in the year 1920. The jacks were constructed

Fig. 7 RCC Jack-Jetty of 1.5 m length, 0.1 m width, and 0.1 m thickness meshed with 4 mm galvanized iron wire

Fig. 8 Trapping of small trees and debris by jetty-field in Solani River, India

of 3 numbers of willow poles. Willow poles were connected with each other at their mid-point. Further, Kellner laced the willow poles with wire to keep them in the extended position. Later on, willow poles were replaced by angles. Grassel (2002) also reported that in the early 1950s, installation of jetty-system at five places in Arkansas River and two places in Rio Grande River was completed. Up to 1953s, seven numbers of additional installations were completed in the Rio Grande. High flows were faced by the Purgatoire River, Higbee, and Arkansas River, Manzanola, just after the installation of jetty-system. However, the prime purpose of the installation for ascertaining bank protection was resolved, as no damages to the system were encountered, thus indicating the efficacy of the system.

The Middle Rio Grande Conservancy District (MRGCD) was created in 1925. Diversion dams, levees, and drainage canals were constructed by the high involvement of the U.S. Army Corps of Engineers and the Bureau of Reclamation within ten years of its creation. In order to rehabilitate the MRGCD, levees, jetty-system and some other training structures were constructed by "The Bureau of Irrigation and Drainage Systems" that was working to the part of Rio Grande Channelization Project. The Espanola floodway witnessed widespread rehabilitation of levee sand channel straightening with jetty-systems. Application of these structures converted Rio Grande to an improved storage and water conveyance system (Lagasse 1980; Crawford et al. 1993; Najmi 2001; U.S. Army Corps of Engineers 2003).

Jack-jetties were also used in Middle Arkansas and Middle Rio Grande, consisting of steep slopes and high velocity reach, for the successful transportation of coarse sediments into the dike fields. Petersen (1986) reported that jetty-field constructed on Russian River, California in the year1956, collected a huge amount of debris and sediments, resulting to the establishment of native vegetation.

Nayak (2012) described in her Ph.D. thesis that jetty-field was implemented in the Ganga River at Nakhwa site, located at 11 km downstream of Varanasi, to inhibit stream bank erosion during year 2007–2009. At the project site, a streamflow portion was diverting along the left side subsidiary channel, and the main channel was flowing along the right side of erosion affected bank, thus causing the channel insufficient for navigation. After the implementation of RCC jack-jetty system in the main as well as subsidiary channel, erosion of the main channel on the right bank inhibited, and the subsidiary channel got choked almost in five successive years.

3.2.2 Porcupines

Porcupines are basically the prismatic-type of permeable structure, which comprises six RCC members, joined using iron-made nuts and bolts. Porcupines may be laid along and across to the direction of flow for inducing gradual and steady obstruction to the current. This results to the reduction of velocity locally followed by induction of the sedimentation process.

Porcupine System has been implemented in River Brahmaputra and Ganga, India, with the history of obtaining worthy outcomes. The Brahmaputra is the largest river in the Indian subcontinent, which possesses an active and energetic tendency regarding its channel pattern (Akhtar et al. 2011). At Bonkual in the upstream of Kaziranga, India, the porcupine system, with approximately 1 km of length, was implemented during the period of the year 2005–2006 in River Brahmaputra. The system was built-up of 3 m length porcupine members. They were arranged in 3 staggering row sand protruded from the affected bank to adjacent sand bar. Prior to the implementation of the Porcupine system, the embankment was continuously eroding downstream to the location of porcupines, and the severity of erosion was drifting towards upstream. The adopted system was chosen as a preliminary one and found much success in restricting the prolonged and critical erosion state of the affected reach. Presently, the porcupine system is behaving well to maintain the bed level after facing high floods and draw-downs of successive years. However, in the downstream to the location of the installed porcupine system, some additional porcupine system was also implemented at Agaratali and Arimora during the year 2009. The porcupine systems installed at these locations are also performing well (Sarma and Acharjee 2012; Aamir and Sharma 2015).

Kharya and Kumar (2012) reported for the dual problem of flood inundation and the bank erosion in Majuli Island, Assam, India. Overflow water of Subansiri and Brahmaputra rivers had been causing bank erosion on both sides of the island since 1954. It was noticed that erosion of banks occurred mainly during the receding stage of the floods, when surplus sediments got deposited in form of sand bars, and channel was instigated to change in the direction of flow. Suitable river protection measures, including RCC porcupines were adopted during year, 2005–2012 to solve the issues related. Significant reduction of erosion with increased sedimentation was observed in most of the vulnerable reaches by the implementation of RCC porcupines system.

3.2.3   Pile Dikes

Pile dikes are the type of permeable dikes, normally constructed of double rows of vertical timber piles situated on either side of a horizontal timber spreader. The pile dikes prevent the creation of a solid, dam-like structure through its permeability. In fact, solid structure could generate turbulent outer flows that can destabilize the foundation of structure. Timber pile dikes are used in rivers containing low sediment burden.

The basic features of timber pile dikes are described as below (Pokrefke 2013):

1. Constructed of two or three rows of vertical timber piles with spacing 0.75 m c/c located on either side of a horizontal timber spreader.
2. The foundation of dikes and contiguous river bank regions are protected using stone blanket against the scouring action of water and debris.
3. Stones must be clean and angular rocked. The density of stone should be greater than 2400 kg per cum, whereas size should vary between 9 kg–180 kg and 45 kg–272 kg for low and high currents, respectively.

Pokrefke (2013) reported that the timber pile dikes were used in Lower Columbia River since 1885 to regulate effective width of the river, to reduce shoaling and increased scour capacity, to decrease annual maintenance as well as for strengthening of river banks. Timber piles of length 10 m were used for the construction of pile dikes, whereas king-piles and outer dolphin piles were made up of 13 m length each. The outer marker dolphins contained 10 or more piles and tangled together using wire rope, including one or more king-piles. These were constructed by driving piles as near as possible to each other to provide ideal strength. Piles were driven using a barge mounted crane that was equipped with a vibratory or impact hammer. An inspection carried out in the year 2010, found noteworthy system-wide damages to the individual pile dike. However, many of the pile dikes were still accomplishing their originally proposed purpose, and were tumbling maintenance dredging along with providing protection to the dredged material disposal sites. In addition, these pile dikes were observed to be generating or protecting juvenile salmon habitats at several locations. Approximately 72% of the pile dikes were suggested for retention and repair, 23% were to be observed or alternatively removed, whereas 5% were carefully chosen for the further study.

4   Conclusion

This chapter focusses on the review for the field applications of different permeable and impermeable-type training structures implemented for the protection of riverbank and resolving other related issues. Applications of conventional structures like spurs, cut-offs, levees, pitching of banks, guide banks, pitched islands, revetments, and longitudinal dikes are becoming too expensive and comparatively less effective due to several factors involved. In the present situation, river experts are interested not only in the effectiveness of the structure but also in different aspects like cost factor, navigation or bank nourishment, augmentation of ecological diversity, including coastal engineering for bank and beach shielding are also considered.

Compared to the conventional straight type of dikes, different head shapes such as L-Head, T-Head, etc., are required to be implemented to ascertain the purpose of utilizing these structures in a more cost-effective and eco-friendly manner. Different head shape dikes constructed in submerged condition, are proved to establish lower velocity zone around dike fields by distributing highly dynamic eddies over a larger area. As a consequence of this, field pools created around dike provide shelter to aquatic habitats.

Permeable structures like RCC jack-jetties and porcupines can also be economically used in place of conventional impermeable structures to overcome the increase in the prices of steel and other construction materials. In several situations, these permeable structures are more successful in creating a living environment for aquatic habitats rather than conventional structures. However, in some situations, these jetties and porcupines can be used in combination with impermeable dikes. In fact, it is a better practice to construct impermeable dikes at important reaches, whereas other portions in between these impermeable dikes can be linked with permeable jetties and porcupines to make the system more effective and economical.

However, there is an imperative need to conduct experimental, numerical and field studies as well, on different combinations of permeable and impermeable structures, so that the outcomes of the study can be successfully implemented in practical field conditions. Model studies based on the evaluation of turbulence characteristics, coherence mechanism, appropriate layout of dike field, including scouring phenomenon for different head shapes of impermeable dikes, can reveal enhanced information to replace conventional structures.

Acknowledgements The authors would like to thankfully acknowledge the central library, IIT Roorkee, for providing uninterrupted access to related research papers.

References

Aamir M, Sharma N (2015) Riverbank protection with Porcupine systems: development of rational design methodology. ISH J Hydraul Eng 21(3):317–332

Akhtar MP, Sharma N, Ojha CSP (2011) Braiding process and bank erosion in the Brahmaputra River. Int J Sedim Res 26(4):431–444

Ansari SA, Kothyari UC, Ranga Raju KG (2002) Influence of cohesion on scour around bridge piers. J Hydraul Res 40(6):717–729

ASCE Task Committee on Sediment Transport and Aquatic Habitats, Sedimentation Committee (1992) Sediment and aquatic habitat in river systems. J Hydraul Eng 118(5):669–687

Blancett J, Jarchow P (2009) Kansas River bank stabilization and post-project conditions. In: World environmental and water resources congress 2009: Great Rivers, pp 1–11.

Bureau of Reclamation, U.S. Department Technical Service Center, Drew CB, Lisa F, Cassie CK, Sculock SM (2015) Bank stabilization and design guidelines. Report No. SRH-2015-25

Crawford CS, Cully AC, Leutheuser R, Sifuentes MS, White LH, Wilber JP (1993). Middle Rio Grande ecosystem: Bosque biological management plan. Middle Rio Grande.

Dey S, Barbhuiya AK (2005) Time variation of scour at abutments. J Hydraul Eng, 131(1):11–23.

Grassel K (2002) Taking out the jacks: issues of jetty jack removal in bosque and river restoration planning. Water Resource Programme, University of NewMexico, Publication No. WRP-6.

Hildrew AG (1996) Whole river ecology: spatial scale and heterogeneity in the ecology of running waters. Large Rivers 10(1–4):25–43

Jia Y, Scott S, Xu Y, Wang SSY (2009) Numerical study of flow affected by bendway weirs in Victoria Bendway, The Mississippi River. J Hydraul Eng 135(11):902–916

Julien PY (2006) River mechanics. Cambridge University Press, USA

Kharya A, Kumar P (2012) RCC porcupines: an effective river bank protection measure–a case study of protection of Majuli Island. In: India water week: water, energy and food security: call for solutions, 10–14 April 2012, New Delhi

Kinzli KD, Myrick CA (2010) Bendway weirs: could they create habitat for the endangered Rio Grande silvery minnow. River Res Appl 26(7):806–822

Kirkil G, Constantinescu G (2010) Flow and turbulence structure around an in-stream rectangular cylinder with scour hole. Water Resour Res 46(11):W11549

Knight SS, Cooper CM (1991) Effects of bank protection on stream fishes. In: Proceedings of the 5th Federal interagency sedimentation conference, vol 13, pp 34–39.

Koken M, Constantinescu G (2008a) An investigation of the flow and scour mechanisms around isolated spur dikes in a shallow open channel: 1. Conditions corresponding to the initiation of the erosion and deposition process. Water Resour Res 44(8):W08406

Koken M, Constantinescu G (2008b) An investigation of the flow and scour mechanisms around isolated spur dikes in a shallow open channel: 2. Conditions corresponding to the final stages of the erosion and deposition process. Water Resour Res 44(8):W08407

Koken M, Constantinescu G (2011) Flow and turbulence structure around a spur dike in a channel with a large scour hole. Water Resour Res 47(12):W12511

Kuhnle RA, Alonso CV, Shields FD (1999) Geometry of scour holes associated with 90 spur dikes. J Hydraul Eng 125(9):972–978

Kuhnle RA, Alonso CV, Shields Jr FD (2002) Local scour associated with angled spur dikes. J Hydraul Eng 128(12):1087–1093.

Lagasse PF (1980) An Assessment of the Response of the Rio Grande to Dam Construction: Cochiti to Isleta Reach. A Technical Report for the US Army Corps of Engineers, Albuquerque District, Albuquerque, New Mexico

Lane SN, Hardy RJ, Elliott L, Ingham DB (2004) Numerical modeling of flow processes over gravelly surfaces using structured grids and a numerical porosity treatment. Water Resour Res 40(1):W01302

Mansoori AR, Nakagawa H, Kawaike K, Zhang H, Safarzadeh A (2012) Study of the characteristics of the flow around a sequence of non-typically shaped spur dikes installed in a fluvial channel. Ann. Disas Prev Res Inst, Kyoto Univ. 55(B):453–458.

Mazumder SK (2014) Protection of flood Embankments by Spurs with reference to Kosi River. Paper presented and published in the proceedings of the HYDRO-2011 held at SVNIT, Surat, Dec 29–30.

Najmi YV (2001) The middle Rio Grande conservancy district and bosque management: a planning framework and guidelines for restoration projects. Professional Project, Community and Regional Planning, University of New Mexico, Albuquerque, New Mexico, USA Google Scholar

Nayak A (2012) Experimental Study of RCC Jack Jetty systems for River training. Ph.D. Thesis, Indian Institute of Technology Roorkee, Roorkee, India

Nayak A, Sharma N, Mazurek KA, Kumar A (2017) Design development and field application of RCC Jack Jetty and Trail Dykes for River training. River System Analysis and Management, Springer, Singapore, pp 279–308

Odgaard AJ, Kennedy JF (1983) River-bend bank protection by submerged vanes. J Hydraul Eng 109(8):1161–1173

Odgaard AJ Wang Y (1991) Sediment management with submerged vanes. II: applications. J Hydraul Eng 117(3):284–302.

Petersen MS (1986) River engineering. Prentice-Hall, USA

Pokrefke TJ (2013) Inland Navigation Channel Training Works. Task Committee on Inland Navigation of the Waterways Committee, Manual of Practice 124

Sarma JN, Acharjee S (2012) A GIS based study on bank erosion by the River Brahmaputra around Kaziranga National Park, Assam, India. Earth System Dyn Discuss 3(2):1085–1106

Scurlock SM, Cox AL, Thornton CI, Baird DC (2012) Maximum velocity effects from vane-dike installations in channel bends. In: World Environmental and Water Resources Congress 2012: Crossing Boundaries, pp 2614–2626.

Sharma N, Kumar A (2012–2014) Consulting Project no.-WRC-1005/12-13 on Solani River Training for Solar Power Plant near Roorkee. IIT Roorkee, India.

Shields FD (1984) Environmental guidelines for dike fields. In Elliot CM (ed). In: Proceedings of the conference rivers' 83, River meandering, ASCE, NY, pp 430–441.

Shields Jr FD (1995) Fate of Lower Mississippi River habitats associated with river training dikes Aquat Conserv: Marine Freshw Ecosyst 5(2):97–108

Shields Jr FD, Knight SS, Cooper CM (1998) Addition of spurs to stone toe protection for warmwater fish habitat rehabilitation. J Am Water Resour Assoc 34(6):1427–1436

Shields Jr FD, Knight SS, Cooper CM (2000) Warmwater stream bank protection and fish habitat: a comparative study. Environ Manage 26(3):317–328

U.S. Army Corps of Engineers (2003) Method and cost evaluation report for the Middle Rio Grande Bosque Jetty Jack Removal Evaluation Study. U.S. Army Corps of Engineers Report, Albuquerque District, Albuquerque, New Mexico, USA.

Wu B, Wang G, Ma J, Zhang R (2005) Case study: river training and its effects on fluvial processes in the Lower Yellow River, China. J Hydraul Eng 131(2):85–96

Zhang H, Nakagawa H (2008) Scour around spur dyke: recent advances and future researches. Ann Disas Prev Res Inst, Kyoto Univ 51(B):633–652.

Comparison of Conceptual and Distributed Hydrological Models for Runoff Estimation in a River Basin

S. Sreedevi, A. Kunnath-Poovakka, and T. I. Eldho

Abstract Hydrologic models are important tools for accurate water resource assessment. Distributed models are commonly used for this purpose as its output provides detailed information about catchment hydrology. High data requirements and complex structure limit the application of the distributed models at the majority of locations. In comparison, conceptual models are simple and require minimum inputs for runoff generation. In this study, the performance of a physically-based spatially distributed model SHETRAN is compared with two simple lumped conceptual rainfall-runoff models, namely Australian Water Balance Model (AWBM) and GR4J, at Vamanapuram river basin in India. The results showed comparable performance of SHETRAN and AWBM. The study concludes that conceptual models are best suited for data-scarce regions, and the choice of distributed and lumped models for hydrologic studies are dependent on data availability and output requirements

Keywords River basin · Runoff · SHETRAN · AWBM · GR4J · Distributed models · Conceptual models

1 Introduction

Hydrologic models are mathematic representations of the hydrologic cycle or part of the hydrologic cycle. They play an important role in studies related to water resource management, flood and drought management, the impact of climate change, and land-use changes. Hydrologic models were classified in many ways based on their structure, application, time-scale, etc. Distributed and lumped model classification is one among them. Distributed models are capable of generating spatially varying outputs in response to distributed information about catchment physical and meteorological properties. The lumped model considers catchment as a homogenous unit and computes the output at the outlet. Distributed models are common in practice due

S. Sreedevi · A. Kunnath-Poovakka · T. I. Eldho (✉)
Department of Civil Engineering, Indian Institute of Technology (IIT) Bombay, Powai, Mumbai 400076, India
e-mail: eldho@civil.iitb.ac.in

© The Author(s), under exclusive license to Springer Nature Switzerland AG 2021
M. S. Chauhan and C. S. P. Ojha (eds.), *The Ganga River Basin: A Hydrometeorological Approach*, Society of Earth Scientists Series,
https://doi.org/10.1007/978-3-030-60869-9_9

135

to its detailed output about the hydrological properties of the catchment (Mourato et al. 2015; Wagner et al. 2011). However, the lack of information about catchment meteorological and non-meteorological data in the vast majority of the world limits distributed model applications. Conceptual rain-runoff models are gaining much attention these days due to its fewer input requirements and computational efficiency (Boughton et al. 2007; Traore et al. 2014). Many studies are reported across the globe in recent times to understand the efficacy of conceptual models in comparison with distributed models (Kumar et al. 2015; Tegegne et al. 2017). Most of the studies have reported positively for conceptual models in runoff modeling studies. In the present study, the performance of a physically-based distributed model SHETRAN was compared with two conceptual rainfall-runoff models, Australian Water Balance Model (AWBM) and GR4J. The study was conducted in the Vamanapuram river basin in Kerala, India. The models have been calibrated for the period 1981–1988 and validated for 1989–1994. The results are inter compared, and the efficiency of each model are discussed.

2 Literature Review

This review mainly refers to the development and application of SHETRAN, AWBM, and GR4J in different fields of hydrology, including water resource assessment, sediment yield, and impact of climate and land-use change. Physically-based spatially distributed (PBSD) hydrologic model, SHETRAN has a multitude of input parameters representing the hydrologic processes within a river basin. A major constraint in the wide application of such PBSD model is the requirement to calibrate a large number of model input parameters. Manual calibration is mostly performed on SHETRAN model by altering the model parameter values based on the physical reasoning (Bathurst et al. 2011; Birkinshaw et al. 2011). Zhang et al. (2013) has applied Shuffled Complex Evolution (SCE-UA) algorithm to automatically calibrate SHETRAN in Cobres basin. To manually calibrate model parameters in river basins with highly heterogeneous soil and landuse types, an appropriate step before model calibration would be to carry out a sensitivity analysis to identify the most significant parameters to be altered. Anderton et al. (2002) performed the detailed sensitivity analysis of SHETRAN subsurface component and demonstrated the highly interactive parameter space within SHETRAN. Đukić and Radić (2016) quantified the effect of change in SHETRAN model parameters on runoff and sediment yield of Lukovska River catchment in Serbia by varying them one at a time and simplifying the calibration procedure. SHETRAN model has been applied over different locations to study the impact of climate change on runoff (Mourato et al. 2015), estimating sediment yield (de Figueiredo and Bathurst 2007; Op de Hipt et al. 2017) and estimating land subsidence (Shrestha et al. 2017).

AWBM and GR4J are two simple conceptual lumped rainfall-runoff models, which relate the catchment daily rainfall and evapotranspiration (ET) to stream discharge. AWBM was developed by Boughton (2004), for water balance studies in Australia. The model was mainly applied for runoff estimation in ungauged catchments of Australia (Boughton and Chiew 2007). Zhang et al. (2016) used this model to study the impacts of human activities and climate change at Poyang Lake Basin, China. Kumar et al. (2015) used this model for a multi-model simulation of the Mahanadi river in India and found that AWBM was one of the best. GR4J model by Perrin et al. (2003) was the modified version of GR3J model developed by Edijatno et al. (1999). GR4J was used mainly for water resource assessment in different regions across the world (Nepal et al. 2015; Traore et al. 2014). Oudin et al. (2008) employed this model for runoff generation in ungauged catchments in France using the regionalization technique.

3 Study Area and Data

The Vamanapuram river basin with a catchment area of 787 km^2 is located mainly in Thiruvananthapuram district, with a small part falling in the Kollam district of Kerala, India. The river basin is bounded by latitudes of 8° 35′ 24″ N and 8° 49′ 13″ N and longitudes of 76° 44′ 24″ E and 77° 12′ 45″ E. Vamanapuram river basin has an elliptical shape with a network of tributaries following a dendritic drainage pattern. Upper Chittar and Manjaprayar are two tributaries of the river. The river has its origin from Ponmudi hills (belonging to Western Ghat ranges) and traversing a length of 88 km finally drains to Arabian sea at Mudalapallipozhi. The major part of the Vamanapuram river flows through midland terrain and the remaining through highlands and lowland areas. Figure 1 shows the location of the basin. The basin receives an annual average rainfall of 3200 mm. The area receives rainfall mainly during Southwest (June–September) and Northeast monsoons (October–November). The river basin is characterized by a hot climate with coastal belt in midland being a humid and cool hilly region for most of the time in the year. The average daily temperature varies from 26.2 to 28.8 °C within the basin.

The different datasets used in the study include elevation, land use, soil, meteorological data such as rainfall, potential evapotranspiration (PET), and temperature of the air, streamflow. The Advanced Space-Bone Thermal Emission and Reflection Radiometer (ASTER) DEM at 30 m resolution from Earth explorer (https://earthexplorer.usgs.gov/) gave the elevation of the study area. The daily rainfall data for the time period 1980 to 2010 were derived from gauge observations and Asian Precipitation Highly Resolved Observational Data Integration Towards Evaluation (APHRODITE) at 0.25°, further interpolated to 0.05° resolution using Inverse Distance Weighting (IDW) method, for application to SHETRAN model. The rain gauges were maintained by Hydrology Project-II, Trivandrum. The temperature dataset at Thiruvananthapuram station maintained by India Meteorological Department is utilized to calculate daily potential evapotranspiration for the time period

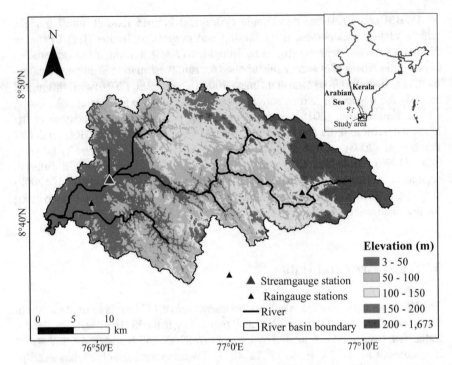

Fig. 1 Location of Vamanapuram basin with elevation, rain gauge stations and stream gauge station

1980–2010 using Blaney-Criddle method (Blaney and Criddle 1950). Daily stream-flow data at Ayilam station (76° 51′ 15″ E and 8° 42′ 55″ N) is procured from Central Water Commission for temporal coverage of 1980–2010. The location of rain gauge stations and stream gauge station are shown in Fig. 1.

Landuse maps at 100 m resolution for the time periods 1985, 1995, and 2005 prepared by National Remote Sensing center are available for entire India. These decadal land use maps are capable of representing changes at the regional scale with an accuracy of 90% and were validated using ground truth data and other high-resolution satellite images (Rosenbrock 1960). For the present study, land use map of 1985 is applied. Eight landuse types were represented by the landuse maps with rubber plantation (68%) and evergreen forest (17%) being the predominant types. The soil profile for each SHETRAN grid square comes from the HWSD database (FAO 2012). Topsoil texture (0–30 cm) and subsoil texture (30–100 cm) for each SHETRAN grid square are obtained from this dataset. Three soil layers are considered in this study comprising of the top, middle, and bottom layers. Soil properties (saturated conductivity, moisture content, van Genuchten parameters, specific storage) were calibrated for the study area. Information about the type and depth of the bottom layer is derived from lithologs in the Vamanapuram river basin. The basin belongs to Trivandrum district of Kerala, and the hydrogeology of this region is classified into crystalline formation and sedimentary formation. The crystallines comprise of khondalites, charnockites, migmatites, and intrusives occur at shallow or deep with or without fractures. Sedimentary formations include recent alluvium (composed of sand and clay), Tertiary formation (Warkali, Quilon, and Vaikom beds), and laterites (Rani 2013). Depth of top and middle soil layers are taken directly from HWSD database. The thickness of the bottom layer was decided based on soil depth map prepared by the interpolation of lithology data. Twenty-two soil profiles with three soil layers were thus generated. Thus, each SHETRAN grid has a unique soil profile. Detailed information about the aquifer properties in the study region was lacking and hence, was calibrated.

4 Methodology

In this study, two conceptual models (AWBM and GR4J) and a distributed model SHETRAN are used for rainfall-runoff modelling. A brief description of the three hydrologic models used in the study is given in this section.

4.1 SHETRAN Model

The physically based distributed model, SHETRAN (https://research.ncl.ac.uk/she tran/) is an improved version of SHE (Système Hydrologique Europeen) model (Abbott et al. 1986) with additional components of sediment (Wicks and Bathurst 1996) and a fully 3D subsurface water flow component (Parkin 1996). The model structure is shown in Fig. 2. The model solves the partial differential equations for flow and transport by finite difference methods. In order to implement this, the river basin is discretized in the form of rectangular computational grids with each surface grid having columns of cells extending downwards representing the soil/rock layer beneath. River network is depicted as a network of river links run along edges of grid squares. Surface/Subsurface flow is coupled in the model, allowing overland flow generated by excess rainfall over infiltration and by upward saturation of the soil column (Birkinshaw et al. 2011). The parameters used in SHETRAN are shown in Table 1. The effects of changes in climate and landuse could be effectively predicted using the model (Ewen et al. 2000). The hydrologic processes are represented as rainfall interception by modified Rutter model, the actual evapotranspiration calculated from PET (potential evapotranspiration) by the function of soil water potential, overland and channel flow processes by diffusive wave approximation of Saint–Venant equation, one-dimensional flow in the unsaturated zone and two-dimensional flow in the saturated zone by the Richard's equation and Boussinesq equation and river-aquifer interaction calculated from Darcy's equation. In the present study, the water flow component of SHETRAN model (v4.4.5) is utilized.

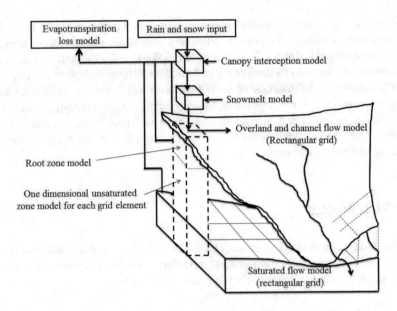

Fig. 2 SHETRAN hydrological component. Modified from Abbott et al. (1986)

Table 1 Parameters of the SHETRAN model for calibration and the optimal parameter values

No	Parameter	Low	High	Optimum	Description
1	kxt	0.001	10	5	Horizontal saturated hydraulic (m/day)
2	kxm	0.01	25	25	
3	kxb	0.0001	6	3	
4	kzt	0.001	10	3	Vertical saturated hydraulic conductivity (m/day)
5	kzm	0.001	5	0.3	
6	kzb	0.0001	2	0.1	
7	Sst	0.001	0.5	0.45	Specific storage (m^{-1})
8	Ssm	0.0005	0.1	0.02	
9	Ssb	0.0005	0.1	0.05	
10	Str1 (builtup)	0.01	0.5	0.5	Strickler overland coefficient ($m^{1/3} s^{-1}$)
11	Str2 (forest)	0.5	1.5	1.9	
12	Str3 (rubber)	0.5	1.5	1.9	
13	Str4 (grassland)	0.01	0.2	0.2	
14	Str5 (mixed trees)	0.5	1.8	1.8	
15	Str6 (barren land)	0.01	0.2	0.2	

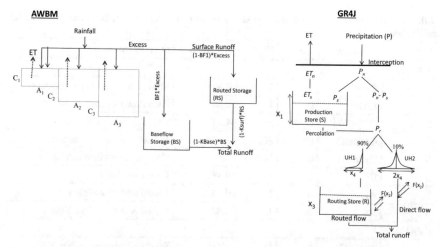

Fig. 3 Schematic Diagram of rainfall-runoff models, AWBM. (Modified from Boughton (2004)) and GR4J (Modified from Perrin et al. (2003))

4.2 Australian Water Balance Model (AWBM)

AWBM is a simple catchment water balance model, which computes runoff as a response daily rainfall and evapotranspiration (Boughton 2004). A schematic sketch of the water movement in AWBM is shown in Fig. 3. Rainfall and PET are the two major inputs to this model. It is an eight parameter (see Table 2) five-store model. AWBM has three surface water storages A1, A2, and A3, with respective capacities C1, C2, and C3. Rainfall first fills these storages, and excess water is distributed between baseflow and surface water routing store. The parameter baseflow index (BFI) determines the amount of water to be distributed between baseflow and routed storage. The parameters baseflow recession constant (Kbase) and surface flow recession constant (Ksurf) decides the baseflow and routed flow contribution to total runoff.

4.3 GR4J Model

GR4J is the simplest among the three models used in this study, developed by Edijatno et al. (1999). It is a two-store (production and routing store) four-parameter model (Fig. 3 and Table 2). Rainfall and PET are the two inputs to this model, as in the case of AWBM. The net rainfall (Pn) after initial losses enters the production store, and water is lost as evaporation and percolation from this store. The water exceeding the maximum water holding capacity of the production store (x_1) joins the percolation water, and 90% of it is drained as fast flow through the routing store, and rest 10% is directly contributed to total runoff as slow flow. The time lag for the hydrograph

Table 2 Parameters of the AWBM and GR4J models and the optimal parameter values after calibration

No	Parameter	Low	High	Optimum	Description
AWBM					
1	Ksurf	0	1	0.99	Surface flow recession constant
2	Kbase	0	1	0.68	Base flow recession constant
3	C_1 (mm)	0	50	4.84	Capacity of surface store 1
4	C_2 (mm)	0	200	88.74	Capacity of surface store 2
5	C_3 (mm)	0	500	500	Capacity of surface store 3
6	BFI	0	1	0.48	Base flow index
7	A_1	0	1	0.44	Partial area of surface store 1
8	A_2	0	1	0.42	Partial area of surface store 2
GR4J					
1	x_1 (mm)	1	1500	1500	Production store capacity
2	x_2 (mm)	−10	5	−2.06	Water exchange coefficient
3	x_3 (mm)	1	500	120.22	Routing store capacity
4	x_4 (days)	0.5	4	2.06	Time-base of the unit hydrograph

of these two flows is determined by the parameter x_4. The parameter x_3 represents the capacity of the routing store. The loss or gain of water during routed and direct flow is represented by parameter x_2.

4.4 Sensitivity Analysis, Model Calibration and Validation

A 1 km grid resolution was selected to model Vamanapuram river basin using SHETRAN. The spatial heterogeneity in different soil and landuse/landcover classes over the study area could be represented well by the grid and column discretization of SHETRAN. Soil types as given by HWSD database has been taken. Top soil layer consists of Sandy clay Loam, Clay loam and Loam whereas the middle soil layer comprises clay loam. The thickness of the top layer is fixed at 0.3 m and middle layer as 3 m whereas bottom layer thickness was deduced from soil depth map prepared by interpolation of lithology data. The parameters which were subjected to sensitivity analysis include saturated hydraulic conductivity (k_z and k_x), Saturated and residual moisture content (Θ_{sat} and Θ_{res}), van Genuchten parameters (α and n) and Specific storage (S_s) of the three soil layers and strickler overland coefficient of landuse type. The parameter ranges were obtained from the SHETRAN manual (SHETRAN 2013). The parameters were varied one at a time to understand the effect of change in model parameters on total discharge. The manual calibration of most sensitive model parameters, namely the Strickler overland coefficient, saturated hydraulic conductivity, and the specific storage, was carried out to arrive at the optimal model parameter sets for simulating streamflow.

AWBM and GR4J were calibrated in Source modeling platform using Shuffle Complex Evoluion (SCE) (Duan et al.1992) and Rosen brock optimization (Roy et al. 2015) algorithm. Source is a catchment and river modeling platform developed by eWater Australia. The source is Australia's national modeling platform, which can be used for building both simple and complex catchment, river, and urban models. GR4J and AWBM are two among many conceptual rainfall-runoff models included in Source. In this study, daily calibration was performed using the flow calibration wizard of Source using Nash Sutcliffe efficiency (NSE) as an objective function. The catchment has been considered as a single unit, and runoff was generated at the outlet. Model calibration was performed from 1981 to 1988 and validated for the next six years. The year 1980 is considered a warm-up period for the models.

The model performances were assessed both statistically and visually. The efficiency criteria such as NSE, correlation coefficient (R), percentage bias (PBIAS), and root mean square error (RMSE) was used to evaluate the different models. The NSE represents the match between observations and simulations. Its value ranges between $-\infty$ and 1. NSE of 1 represents a perfect match. The Pearson correlation coefficient was estimated in this study to understand the linear relationship between observations and predictions. PBIAS was calculated to understand the over or underestimations of models compared to observations. It was determined using Eq. (1).

$$PBIAS = \frac{\sum_{i=1}^{n} Q_{oi} - Q_{pi}}{\sum_{n=1}^{i} Q_{oi}} * 100 \tag{1}$$

where, Q_{oi} and Q_{pi} is observed and predicted discharge. The positive value of PBIAS represents an underestimation of simulations or vice versa. RMSE depicts bias between observations and simulations. A value of zero represents an exact likeness between observations and simulations. Visual inspection of model predictions was carried out with the help of hydrograph, cumulated flow chart, and flow duration curve.

5 Results and Discussion

The performance of the hydrologic models SHETRAN, AWBM, and GR4J were compared in this study. The models have been calibrated for the years 1981–1988 and validated from 1989 to 1994. Fifteen parameters of SHETRAN, eight parameters of AWBM, and four parameters of GR4J were calibrated, and the optimized parameter values obtained are given in Tables 1 and 2. The calibrated model streamflow predictions for both calibration and validation period were generated, and the

Table 3 Model performance statistics during calibration and validation for daily streamflow

Models	NSE		R		PBIAS		RMSE (cumecs)	
	Cal	Val	Cal	Val	Cal	Val	Cal	Val
SHETRAN	0.673	0.726	0.821	0.861	3.142	1.198	14.911	19.325
AWBM	0.709	0.750	0.843	0.870	−2.714	−11.315	14.055	18.432
GR4J	0.507	0.645	0.721	0.823	12.844	0.193	18.310	21.9913

Cal—calibration; Val—validation

performance indices, including NSE, R, PBIAS, and RMSE were estimated for each model, as presented in Table 3.

In terms of NSE, R and RMSE AWBM performs the best during both calibration and validation periods. SHETRAN also provides close results with better PBIAS during the validation period. It is observed from the flow duration curve (Fig. 4) that

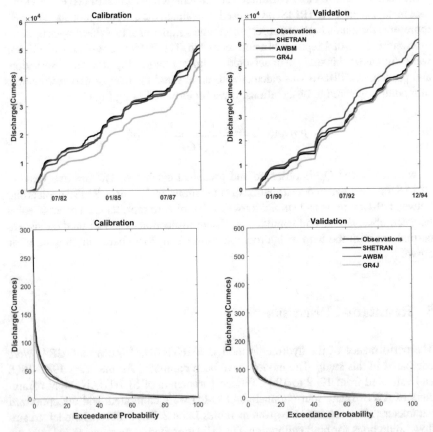

Fig. 4 Cumulative streamflow graph and flow duration curves for all the models during calibration and validation period

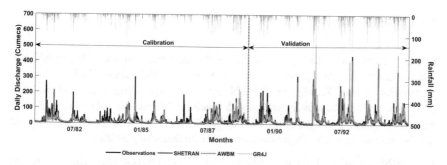

Fig. 5 Daily streamflow hydrograph

medium flows were over-estimated by AWBM, and that resulted in higher PBIAS during the validation period. Compared to SHETRAN and AWBM, GR4J performance was poor with NSE 0.5 during calibration and 0.6 during validation. GR4J was highly underestimating streamflow during the calibration period, as seen in the cumulative flow diagram in Fig. 4 and hydrograph in Fig. 5. It was also noted that GR4J provided the lowest PBIAS during the validation period, even though the RMSE was very high in contrast to the other two models. This PBIAS value was a result of the cancellation of positive and negative bias due to over and underestimation of GR4J predictions during different validation years (Fig. 5).

The present study found that AWBM and SHETRAN provide comparable streamflow estimates for Vamanapuram river basin. The models were able to catch the waviness in streamflow observations pretty well (Fig. 5). The peak flows were slightly underestimated during the calibration period. SHETRAN is a distributed model requiring much meteorological and non-meteorological information to simulate the results, whereas AWBM requires only rainfall and PET inputs to generate streamflow output. Distributed models are complex with many parameters to optimize during calibration, while conceptual lumped models are simple with a few parameters to calibrate. Due to these, conceptual models are computationally efficient. However, conceptual lumped models are not capable of producing streamflow as a response to spatially varying rainfall or other physical or meteorological inputs. AWBM considers catchment as a single unit and computes streamflow at the outlet. In addition, distributed models are able to generate different components of the hydrologic cycle along with streamflow. Lack of input data limits the use of distributed models in many parts of the world. This study recommends AWBM in data-scarce regions or for studies in which streamflow is the only output of concern.

6 Conclusions

In the present study, the performance of the distributed model SHETRAN was compared with the conceptual lumped models such as AWBM and GR4J at Vamanapuram river basin, Kerala. SHETRAN was calibrated using a manual trial and error process, and the conceptual models were calibrated using the eWater Source modeling platform using SCE and Rosenbrock algorithm. The model may be used in other basins also. The major conclusions from this study include:

- AWBM and SHETRAN performed well in the calibration and validation period compared to GR4J. AWBM was slightly better in terms of NSE.
- The applications of distributed models are limited by the data deficiency in many regions. The study revealed that conceptual models produce similar results with limited inputs. Therefore, this study recommends conceptual models in data-scarce regions.
- Conceptual models will be beneficial for water resource management, flood and drought risk mapping, and reservoir management studies, in which streamflow is the output of interest, due to its high computation efficiency and low data requirement.
- Conceptual models are incapable of generating results in response to spatially varying inputs and other components of hydrologic cycles, including soil moisture, baseflow, etc. However, a distributed model like SHETRAN will be useful for the same.

Acknowledgements The authors sincerely acknowledge India Meteorological department for providing meteorological data, the Central Water Commission for streamflow data, Hydrology Project II, Trivandrum for the rainfall data.

References

Abbott MB, Bathurst JC, Cunge JA, Connell PE, Rasmussen J (1986) An introduction to the European hydrological system—Systeme Hydrologique Europeen, "SHE", 1: history and philosophy of a physically based distributed modelling system. J Hydrol 87:45–59

Anderton S, Latron J, Gallart F (2002) Sensitivity analysis and multi-response, multi criteria evaluation of a physically based distributed model. Hydrol Process 16(2):333–353. https://doi.org/10.1002/hyp.336

Bathurst JC, Birkinshaw SJ, Cisneros F, Fallas J, Iroumé A, Iturraspe R, Novillo MG, Urciuolo A, Alvarado A, Coello C, Huber A, Miranda M, Ramirez M, Sarandón R (2011) Forest impact on floods due to extreme rainfall and snowmelt in four Latin American environments 2: model analysis. J Hydrol 400(3–4):292–304

Birkinshaw SJ, Bathurst JC, Iroumé A, Palacios H (2011) The effect of forest cover on peak flow and sediment discharge—an integrated field and modelling study in central-southern Chile. Hydrol Process 25(8):1284–1297

Blaney HF, Criddle WD, (1950) Determining water requirements in irrigated areas from climatological and irrigation data USDA-SCS-TP-96 Report 50, Washington, DC

Boughton W, Chiew F (2007) Estimating runoff in ungauged catchments from rainfall, PET and the AWBM model. Environ Model Softw 22(4):476–487. https://doi.org/10.1016/j.envsoft.2006.01.009

Boughton W (2004) The Australian water balance model. Environ Model Softw 19:943–956

de Figueiredo E, Bathurst JC (2007) Runoff and sediment yield predictions in a semiarid region of Brazil using SHETRAN. In: Proceedings of the PUB Kick-Off Meet. Bras. Brazil 2002. International Association of Hydrological Sciences (IAHS) Publications: Wallingford, UK

Duan Q, Sorooshian S, Gupta V (1992) Effective and efficient global optimization for conceptual rainfall runoffmodels. Water Resour Res 28(4):1015–1031

Đukić V, Radić Z (2016) Sensitivity analysis of a physically based distributed model. Water Resour Manag 30(5):1669–1684. https://doi.org/10.1007/s11269-016-1243-8

Edijatno NNO, Yang X, Makhlouf Z, Michel C (1999) GR3J: a daily watershed model with three free parameters. Hydrolog Sci J 44(2):263–277

Ewen J, Parkin G, Connell PE (2000) SHETRAN: distributed river basin flow modeling system. J Hydrol Eng 5(3):250–258

FAO/IIASA/ISRIC/ISSCAS/JRC (2012) Harmonized world soil database (version 1.2). FAO. Retrieved from http://www.fao.org/soils-portal/soil-survey/soil-maps-anddatabases/harmonized-world-soil-database-v12/en/

Kumar A, Singh R, Jena PP, Chatterjee C, Mishra A (2015) Identification of the best multi-model combination for simulating river discharge. J Hydrol 525:313–325. https://doi.org/10.1016/j.jhydrol.2015.03.060

Mourato S, Moreira M, Corte-Real J (2015) Water resources impact assessment under climate change scenarios in Mediterranean watersheds. Water Resour Manag 29(7):2377–2391 Springer Netherlands. https://doi.org/10.1007/s11269-015-0947-5

Nepal S, Zheng H, Penton DJ, Neumann LE (2015) Comparative performance of GR4JSG and J2000 hydrological models in the Dudh Koshi catchment of the Himalayan region. In Proceedings of the 21st international congress on modelling and simulation (MODSIM2015), Broadbeach, Queensland, Australia, pp 2395–2401

Op de Hipt F, Diekkrüger B, Steup G, Yira Y, Hoffmann T, Rode M (2017) Applying SHETRAN in a Tropical West African Catchment (Dano, Burkina Faso)- calibration, validation, uncertainty assessment. Water 9(2):101. https://doi.org/10.3390/w9020101

Oudin L, Andréassian V, Perrin C, Michel C, Le Moine N (2008) Spatial proximity, physical similarity, regression and ungaged catchments: a comparison of regionalization approaches based on French catchments. Water Resour Res 44(3):1–15. https://doi.org/10.1029/2007WR006240

Parkin G (1996) A three-dimensional variably-saturated subsurface modelling system for river basins. Ph.D. Thesis. University of Newcastle upon Tyne, UK

Perrin C, Michel C, Andréassian V (2003) Improvement of a parsimonious model for streamflow simulation. J Hydrol 279(1–4):275–289

Rani VR (2013) Ground water information booklet of Trivandrum District, Kerala State. Technical Reports: Series D, Central Groundwater Board, Kerala Region, Thiruvananthapuram, India

Roy PS, Roy A, Joshi PK, Kale MP, Srivastava VK, Srivastava SK, Dwevidi RS, Joshi C, Behera MD, Meiyappan P, Sharma Y, Jain AK, Singh JS (2015) Development of decadal (1985–1995–2005) land use and land cover database for India. Remote Sens 7(3):2401–2430

Rosenbrock HH (1960) An automatic method for finding the greatest or least value of a function. Comput J 3:175

SHETRAN (2013) Data requirements, data processing and parameter values. Available from: https://research.ncl.ac.uk/shetran/SHETRAN%20V4%20Data%20Requirements.pdf

Shrestha PK, Shakya NM, Pandey VP, Birkinshaw SJ, Shrestha S (2017) Model-based estimation of land subsidence in Kathmandu Valley Nepal. Geomatics, Nat Hazards Risk 8(2):974–996. https://doi.org/10.1080/19475705.2017.1289985

Tegegne G, Park DK, Kim YO (2017) Comparison of hydrological models for the assessment of water resources in a data-scarce region, the Upper Blue Nile River Basin. J Hydrol: Reg Stud 14:49--66 https://doi.org/10.1016/j.ejrh.2017.10.002

Traore VB, Sambou S, Tamba S, Diaw AT, Cisse MT, Fall S (2014) Calibrating the rainfall-runoff model GR4J and GR2M on the Koulountou river basin, a tributary of the Gambia river. Am J Environ Prot 3(1):36–44

Wagner PD, Kumar S, Fiene P, Schneider K (2011) Hydrological modeling with SWAT in a monsoon-driven environment: experience from the Western Ghats, India. Trans ASABE 54(5):1783–1790

Wicks JM, Bathurst JC (1996) SHESED: a physically based, distributed erosion and sediment yield component for the SHE hydrological modelling system. J Hydrol 175:213–238

Zhang Q, Liu J, Singh VP, Gu X, Chen X (2016) Evaluation of impacts of climate change and human activities on streamflow in the Poyang Lake basin, China. Hydrol Process 30(14):2562–2576. https://doi.org/10.1002/hyp.10814

Zhang R, Santos CAG, Moreira M, Freire PKMM (2013) Automatic calibration of the SHETRAN hydrological modelling system using MSCE. Water Resour Manag 27:4053–4068. https://doi.org/10.1007/s11269-013-0395-z

Monthly Variation in Near Surface Air Temperature Lapse Rate Across Ganga Basin, India

Richa Ojha

Abstract In hydrological and terrestrial models, the temperature at any ungauged location is commonly determined by interpolating surface air temperature observations obtained from a sparse network of temperature sensors using the standard environmental lapse rate (-0.65 °C/100 m). Though simplistic, this interpolation technique can often lead to incorrect results, as the standard environmental lapse rate may not account for temporal and regional variations. In this study, linear regression relationships are developed to estimate average minimum, average, and average maximum near-surface air temperature lapse rates (at monthly time scales) for the Ganga basin, India. Normal daily air temperature data from 178 stations and the latitude, longitude, and elevation of the stations were used for developing the regression equations. The computed temperature gradients exhibit seasonal variation. The lapse rates obtained for summer months are steeper when compared to lapse rates observed in winter months. The highest lapse rate is observed for the average maximum temperature in April month (-0.76 °C/100 m), whereas the lowest lapse rate is observed for the average minimum temperature in December month (-0.21 °C/100 m). The standard environmental lapse rate (-0.65 °C/100 m) is only observed for the monthly average maximum and average temperature for March and May, respectively. A very good agreement between the interpolated and the observed temperature is obtained for the monthly average temperature. The developed regression equations can be reliably used to predict the temperature in the Ganga basin.

Keywords Lapse rate · Surface temperature · Ganga basin · Monthly variation

R. Ojha (✉)
Department of Civil Engineering, Indian Institute of Technology, Kanpur,
Uttar Pradesh 208016, India
e-mail: richao@iitk.ac.in

© The Author(s), under exclusive license to Springer Nature Switzerland AG 2021   149
M. S. Chauhan and C. S. P. Ojha (eds.), *The Ganga River Basin: A Hydrometeorological Approach*, Society of Earth Scientists Series,
https://doi.org/10.1007/978-3-030-60869-9_10

1 Introduction

Precise characterization of near-surface air temperature (NSAT) variation in a region requires a dense network of temperature sensors with fine temporal resolution (Lundquist and Cayan 2007). As such, observations are limited, and temperature data at fine spatial and temporal resolution are mostly obtained by interpolating sparse temperature observations using the global standard environmental lapse rate (Rolland 2003).

The usage of a uniform lapse rate -0.65 °C/100 m is based on the premise that the environmental lapse rate varies between -0.6 and -0.65 °C/100 m. However, as the variation in near-surface temperature depends on location indicators, terrain characteristics, synoptic weather conditions and landuse/landcover (Blandford et al. 2008; Bharath et al. 2016; Kattel et al. 2013; Rolland 2003). Using a standard environmental lapse rate, implicitly ignores the role of local conditions in determining surface temperature variation. Moreover, a standard environmental lapse rate may also not represent the variations in temperature of a particular region across different seasons (Blandford et al. 2008).

Many studies based on temperature measurements have revealed that in complex terrains, surface lapse rates exhibit large spatial and temporal variability. For different regions of northern Italy, Rolland (2003) computed NSAT lapse rates for minimum, maximum, and average monthly temperatures using linear regression analysis and observed that the annual lapse rates varied between -0.54 and -0.58 °C/100 m. Blandford et al. (2008) using linear regression estimated daily and seasonal NSAT lapse rates for daily average, maximum and minimum temperatures for an area in Idaho. They observed that only the lapse rate for daily maximum temperature was close to the standard environmental lapse rate. Minder et al. (2010) computed surface lapse rates for Cascade mountains. The analysis revealed that annual average lapse rates vary in the range -0.39 to -0.52 °C/100 m. Bandyopadhyay et al. (2014) studied surface lapse rate variation in Arunachal Himalaya using temperature measurements from sixteen stations. The surface lapse rate values varied in the range -0.32 to -0.56, -0.44 to -0.54, and -0.36 to -0.52 °C/100 m for maximum, minimum, and average monthly temperatures, respectively, and exhibited seasonal variation. These studies indicate that surface lapse rate substantially differs from standard environmental lapse rate, and is dependent on the local topography and seasonal conditions.

The standard environmental lapse rate is commonly used in hydrologic and ecosystem models to generate gridded temperature data. As most of these models are used for a specific study area, using a standard environmental lapse rate can significantly affect their prediction capability (Gardner et al. 2009; Otto-Bliesner et al. 2006). White et al. (2002) for four sub-watersheds of California, USA, performed sensitivity analysis of runoff simulations to changes in the altitude of the 0 °C constant temperature surface (freezing level). A three times increase in runoff prediction was noticed for a 600 m rise in freezing level. Lundquist et al. (2008), for the North fork of the American river basin in California, USA, observed a 5% error in runoff predictions for a 100 m error in elevation estimates at which snow changes to rainfall. Therefore, it is necessary to estimate region-specific lapse rates.

In India, only a few studies have estimated local NSAT lapse rates, and most of the hydrologic studies utilize a standard environmental lapse rate. The Ganga basin, with its fertile soils, plays an important role in agricultural economy of India. As Ganga and its tributaries provide irrigation water to an extensive area, reliable prediction of hydrologic variables in this basin is crucial. Therefore, the prime objectives of this study are: (i) to develop linear regression equations for determining monthly surface air temperature at any ungauged location in the Ganga basin (ii) to quantify monthly values of lapse rates, and iii) to evaluate the prediction ability of the developed regression equations.

2 Study Area

The methodology developed in this study has been applied across the Ganga basin, India (latitude from 21° 6′ N to 31° 21′ N and longitude from 73° 2′ E to 89° 5′ E (Fig. 1). Ganga basin is bounded by the Himalayas in the north, Indus basin, and Aravalli ridge on the west, Vindhyas, and Chota Nagpur Plateau on the south. In the

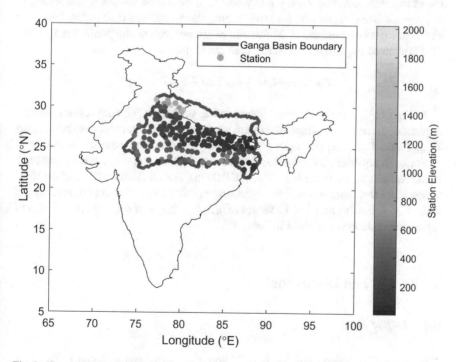

Fig. 1 The spatial distribution of the weather stations in the study area. The elevation of a station location with respect to average sea level is shown by different colors

east, river Ganga merges with the Brahmaputra and finally reaches the Bay of Bengal. The total catchment area of the basin is 861,452 km^2 (26% of the total geographical area of India), and it is the most populated river basin in the world. The annual water potential of the basin is 525 km^3, and it comprises 29.5% of the cultivable area of India. The climate of the basin is usually classified into four seasons (Subramanya 2005) (i) monsoon (June–September), (ii) post-monsoon (October–November), (iii) winter (December-February) and (iv) pre-monsoon (March–May).

3   Materials and Methods

Data of normal daily average temperatures (maximum, minimum, and average) for 178 districts within the Ganga basin for the period 1971–2000 were obtained from Indian Meteorological Department (IMD). As IMD mostly maintains weather stations at district headquarters, this data was considered adequate for the analysis. For the weather stations at the district headquarters, the latitude (*Lat*), longitude (*Lon*), and elevation above the average sea level (*Z*) were extracted from the site www.imdaws.com/550_aws.pdf.

Earlier studies (e.g., Rolland 2003; Gouvas et al. 2011; Kattel et al. 2013) have found that *Lat*, *Lon* and *Z* play a key role in temperature variation. Therefore, to develop the linear regression for temperature, three variables (*Lat* in °N, *Lon* in °E and *Z* in m) were considered. Multicollinearity analysis of the predictors by Ojha (2017) showed negligible colinearity between them.

$$T_{iN} = a_{iN}Lat + b_{iN}Lon + c_{iN}Z + d_{iN} \qquad (1)$$

In Eq. (1), T_{iN} is the air temperature at the surface in °C; i represents average minimum, average, or average maximum temperature (three temperature parameters $\overline{T}_{min}$, $\overline{T}$ and $\overline{T}_{max}$, respectively); N represents the 12 months from Jan to Dec.

For each equation, the four coefficients a_{iN}, b_{iN}, c_{iN}, and d_{iN} were computed. In the equations, c_{iN} is the lapse rate [°C/100 m]. For all combinations of the three temperature and months, multiple regression equations were developed, leading to a total of 36 equations and 4 × 12 values of a_{iN}, b_{iN}, c_{iN}, and d_{iN}. The parameters of regression equations are given in Table 1.

4   Results and Discussion

4.1   Lapse Rate

The near-surface lapse rates for $\overline{T}_{min}$, $\overline{T}$ and $\overline{T}_{max}$, are given in Table 1. The $\overline{T}_{min}$ lapse rates vary from −0.22 °C/100 m (in January) to −0.61 °C/100 m (in June),

Table 1 Results of multiple linear regression analysis between temperature, longitude, latitude and elevation

S. No	Month	$\overline{T}_{min}$					$\overline{T}$					$\overline{T}_{max}$				
		a	b	c	d	r^2	A	b	C	d	r^2	a	b	c	d	r^2
1	January	−0.65	0.16	−0.22	12.75	0.64	−0.87	0.03	−0.32	36.27	0.81	−1.04	−0.10	−0.42	59.12	0.83
2	February	−0.73	0.14	−0.29	18.99	0.69	−0.94	0.00	−0.41	43.57	0.86	−1.09	−0.14	−0.54	66.75	0.86
3	March	−0.84	0.03	−0.37	35.71	0.72	−1.00	−0.11	−0.50	59.68	0.89	−1.07	−0.24	−0.65	80.64	0.86
4	April	−0.79	−0.12	−0.48	52.73	0.76	−0.87	−0.30	−0.62	77.29	0.86	−0.86	−0.46	−0.76	98.75	0.77
5	May	−0.81	−0.36	−0.56	76.10	0.77	−0.91	−0.57	−0.65	103.22	0.83	−0.93	−0.75	−0.75	125.64	0.77
6	June	−0.45	−0.33	−0.61	66.16	0.74	−0.50	−0.52	−0.68	88.32	0.79	−0.48	−0.68	−0.74	106.77	0.76
7	July	−0.14	−0.17	−0.60	43.32	0.75	−0.06	−0.23	−0.64	50.43	0.78	0.01	−0.29	−0.68	57.83	0.76
8	August	−0.07	−0.08	−0.58	33.97	0.74	0.06	−0.09	−0.63	34.88	0.78	0.17	−0.11	−0.67	37.59	0.76
9	September	−0.23	−0.04	−0.59	33.97	0.78	−0.17	−0.15	−0.62	46.13	0.80	−0.11	−0.28	−0.66	59.48	0.77
10	October	−0.40	0.12	−0.50	21.61	0.77	−0.44	−0.12	−0.59	48.76	0.83	−0.44	−0.39	−0.68	76.89	0.80
11	November	−0.51	0.19	−0.34	12.62	0.68	−0.54	0.00	−0.48	36.38	0.82	−0.52	−0.21	−0.63	60.61	0.81
12	December	−0.53	0.18	−0.21	9.29	0.59	−0.67	0.05	−0.33	31.14	0.77	−0.76	−0.10	−0.46	53.10	0.80

In the table, a (in °C/°N) and b (in °C/°E) are the variations in temperature with latitude and longitude, respectively, c (in °C/100 m) is the temperature lapse rate, d (in °C) is the constant temperature at zero elevation, and r^2 is the coefficient of determination.

whereas $\overline{T}_{max}$ lapse rates vary from -0.42 °C/100 m (in January) to -0.76 °C/100 m (in April). Seasonal variability can be observed in monthly values of lapse rates for $\overline{T}_{min}$, $\overline{T}$ and $\overline{T}_{max}$. Steeper lapse rates are observed in summer months, whereas shallow lapse rates are observed in winter and monsoon months. This is in accordance to that observed in studies conducted for Mountainous regions (Blandford et al. 2008; Minder et al. 2010; Rolland 2003), but in contrast to that reported by Kattel et al. (2013) and Ojha (2017) for tropical regions i.e. southern Nepal, and India, respectively. Himalayan mountain ranges run parallel to Ganga basin in the north it is likely that they significantly affect its climatology. Earlier studies have suggested that in Mountainous regions, higher lapse rates in summer months can be observed due to dry convection whereas temperature inversions can lead to lower lapse rates in winter. The lapse rates for $\overline{T}_{min}$ are less than $\overline{T}_{max}$ for all the months. Moreover, it is worth noting that only $\overline{T}_{max}$ lapse rates for March month and $\overline{T}_{mean}$ lapse rates for May month are equal to the standard environmental temperature lapse rate of -0.65 °C/100 m. The monsoon and post-monsoon seasons experience heavy rainfall and cloud cover because of the southwest monsoon. Therefore, the lapse rates in these seasons are close to the moist adiabatic lapse rates.

4.2 Temperature Variation with Latitude and Longitude

The values of the coefficient b represent the monthly variations in $\overline{T}_{min}$, $\overline{T}$ and $\overline{T}_{max}$ with longitude, and varies in the range $(-0.36$ to $0.19)$ °C/°E, $(-0.57$ to $0.05)$ °C/°E and $(-0.75$ to $-0.1)$ °C/°E, respectively (see Fig. 2b). Longitude has more effect on temperature variation during pre-monsoon months and has the highest value in May (peak summer) month. In summer months, the main land experiences high temperatures compared to the coastal areas, the increased range of temperature variation (i.e., the difference in continentality) likely causes this pattern.

The coefficient a represents monthly variations in $\overline{T}_{min}$, $\overline{T}$ and $\overline{T}_{max}$ with latitude and varies in the range $(-0.84$ to $-0.07)$ °C/°N, $(-1$ to $0.06)$ °C/°N and $(-1.09$ to $0.17)$ °C/°N, respectively (see Fig. 2c). The values of the coefficient exhibit seasonal variability. The coefficient increases in magnitude during the winter season and attains a maximum in February and March. The lowest values are observed in June, July, and August months. The Bay of Bengal branch of southwest monsoon arrives in the Ganga basin early June, and by the end of June, the entire basin experiences heavy rainfall. Also, during monsoon months, the entire basin experiences high temperature due to shifting of the inter-tropical convergence zone (ITCZ). These factors together probably reduce the effect of latitude on temperature variations. The results indicate that compared to longitude, latitude plays a much more pronounced role in temperature variation. It is also worth noting, that the units of latitude and longitude coefficients are different as 1° latitude is not equivalent to 1° longitude.

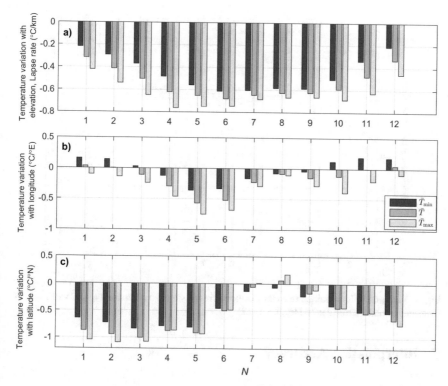

Fig. 2 Temperature variation with **a** elevation (in °C/100 m) **b** longitude (in °C/°E) and **c** latitude (in °C/°N) for monthly $\overline{T}_{min}$, $\overline{T}$ and $\overline{T}_{max}$. N represents months (from Jan to Dec, i.e. 1–12)

4.3 Coefficient of Determination

The coefficient of determination (R^2) was estimated to assess the efficacy of the developed linear equations (Fig. 3). The R^2 values show seasonal variation and are higher for monthly $\overline{T}$ (0.77 to 0.89), followed by $\overline{T}_{max}$ (0.76 to 0.86) and $\overline{T}_{min}$ (0.64 to 0.78), respectively. The R^2 reduces in magnitude during pre-monsoon and monsoon months for monthly $\overline{T}$, and $\overline{T}_{max}$. Sporadic day convection and influence of sea breezes in coastal areas can reduce daytime temperature predictability in pre-monsoon months, causing the low values of R^2.

Low values of R^2 for $\overline{T}_{min}$ are observed during winter months. The topography of the Ganga basin probably affects the temperature in winter. Due to cold air drainage, downslope and flat terrain areas are much colder than the upslope areas. This likely contributes towards low R^2. The lower values of R^2 suggest that local conditions such as synoptic weather conditions, topography, land use/land cover likely affect the temperature variations (Zheng et al., 1996; Gouvas et al. 2011).

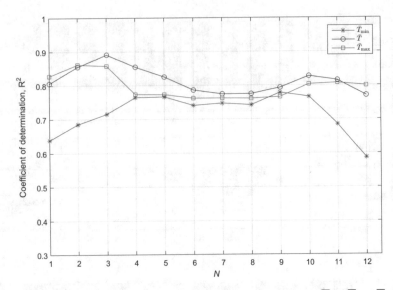

Fig. 3 Variation of R^2 obtained from the developed regression equations for $\overline{T}_{min}$, $\overline{T}$ and $\overline{T}_{max}$. N represents months (from Jan to Dec, i.e. 1–12)

5   Interpolation Accuracy

5.1   Variation of Residual Errors

The prediction ability of the developed regression equations was tested by comparing the observed monthly $\overline{T}_{min}$, $\overline{T}$ and $\overline{T}_{max}$ at each station in the study area with the interpolated temperature (obtained using the equations). For brevity, only the residual errors for January, April, July and October months are presented here. The residual errors for monthly $\overline{T}_{min}$ are shown in Fig. 4. The monthly residual errors for 95% of the stations vary between −2.1 to 2.9 °C, −2.7 to 2.6 °C, −3.3 to 1.75 °C, −2.9 to 2.6 °C, for the four months, respectively, and large residual errors are observed at only a few stations. In all the months, large positive residual errors are present (equation under-predicts) along with the eastern (close to the sea) and the western part of the Ganga basin (hilly topography due to Aravalli ridges), whereas in the intermediate areas low negative residual errors can be noticed. The intermediate region has a relatively flatter terrain compared to its surrounding; therefore, it is possible that strong nighttime radiative cooling and cold air drainage reduce the surface air temperature in this region. Large residual errors at few stations can be due to limitations of the linear regression model and the factors which were not accounted for in the regression analysis.

The residual errors observed for monthly $\overline{T}_{max}$ are larger than the errors obtained for monthly $\overline{T}_{min}$, especially in April month (Fig. 5). The values for the four months vary in between −4.0 and 3.6 °C, −5.9 and 4.2 °C, −5.4 and 2.9 °C, and −5.5 and

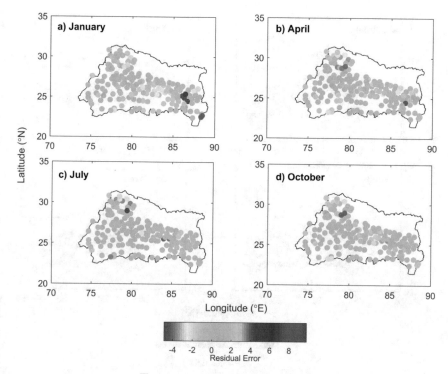

Fig. 4 Variation of monthly $\overline{T}_{min}$ residual errors in °C over Ganga basin in January, April, July and October months

3.6 °C, respectively. For 95% of the stations, the residual errors vary from −1.8 to 2.5 °C, −4.9 to 2.4 °C, −2.5 to 2.0 °C, −2.7 to 1.9 °C. In January month, the $\overline{T}_{max}$ is mostly over predicted (negative residual error) in the intermediate areas, probably due to cloud and fog formation during day time. Large positive residual errors are observed in April month due to strong radiative heating. Negative residual errors are also observed along with the coastal areas and hilly terrains. Coastal cooling (sea-breezes) and daytime cloud formation likely cause over-prediction of temperature in these areas.

5.2 Root Mean Square Error

The root mean square error (RMSE) between observed and interpolated temperature for all the stations in the Ganga basin was also examined for monthly $\overline{T}_{min}$, $\overline{T}$ and $\overline{T}_{max}$ (Fig. 6). The RMSE is higher for monthly $\overline{T}_{max}$ in pre-monsoon months, and is relatively smaller during monsoon months. The highest interpolation accuracy (lower RMSE) is observed for $\overline{T}$ in all the months. In winter months, $\overline{T}_{min}$ has the lowest interpolation accuracy.

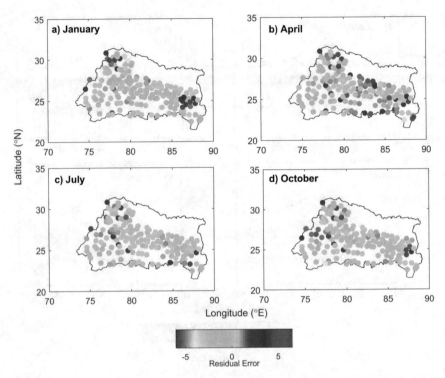

Fig. 5 Variation of monthly $\overline{T}_{max}$ residual errors in °C over Ganga basin in January, April, July and October months

6 Conclusion

In this study, 30 years of normal daily temperature data for 178 stations within Ganga Basin were used to develop linear equations between monthly $\overline{T}_{min}$ $\overline{T}$, and $\overline{T}_{max}$, and latitude, longitude, and elevation above the average sea level. The results suggest that the linear regression equations can be reliably used to interpolate monthly average temperatures at any ungauged location within the Ganga Basin. The regression equations for monthly $\overline{T}_{min}$ and $\overline{T}_{max}$ exhibit poor performance during winter and pre-monsoon months, respectively. This highlights the need for the inclusion of other local factors, such as surface characteristics and synoptic weather conditions in the analysis. Further, the zones within the basin where large residual errors are observed, the developed regression equations should be used with caution.

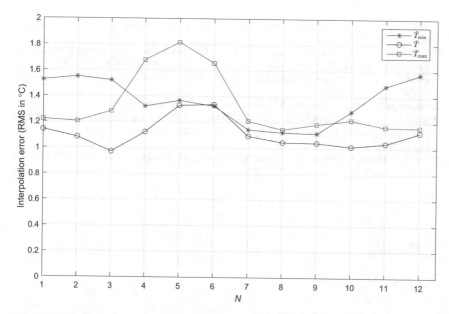

Fig. 6 Root mean square error between observed and interpolated temperatures. N represents months (from Jan to Dec, i.e. 1–12)

References

Bandyopadhyay A, Bhadra A, Maza M (2014) Monthly variations of air temperature lapse rates in Arunachal himalaya. J Indian Water Resour Soc 34(3):16–25

Blandford TR, Humes KS, Harshburger BJ, Moore BC, Walden VP, Ye H (2008) Seasonal and synoptic variations in near-surface air temperature lapse rates in a mountainous basin. J Appl Meteorol Climatol 47(1):249–261

Bharath R, Srinivas VV, Basu B (2016) Delineation of homogeneous temperature regions: a two-stage clustering approach. Int J Climatol 36(1):165–187

Gardner AS, Sharp MJ, Koerner RM, Labine C, Boon S, Marshall SJ, Burgess DO, Lewis D (2009) Near-surface temperature lapse rates over Arctic glaciers and their implications for temperature downscaling. J Clim 22(16):4281–4298

Gouvas MA, Sakellariou NK, Kambezidis HD (2011) Estimation of the monthly and annual average maximum and average minimum air temperature values in Greece. Meteorol Atmos Phys 110(3–4):143–149

Kattel DB, Yao T, Yang K, Tian L, Yang G, Joswiak D (2013) Temperature lapse rate in complex mountain terrain on the southern slope of the central Himalayas. Theoret Appl Climatol 113(3–4):671–682

Lundquist JD, Cayan DR (2007) Surface temperature patterns in complex terrain: Daily variations and long-term change in the central Sierra Nevada, California. J Geophys Res: Atmos 112(D11):1–15

Lundquist JD, Neiman PJ, Martner B, White AB, Gottas DJ, Ralph FM (2008) Rain versus snow in the Sierra Nevada, California: comparing Doppler profiling radar and surface observations of melting level. J Hydrometeorol 9(2):194–211

Minder JR, Mote PW, Lundquist JD (2010) Surface temperature lapse rates over complex terrain: lessons from the Cascade Mountains. J Geophys Res: Atmos 115(D14):1–13

Ojha R (2017) Assessing seasonal variation of near surface air temperature lapse rate across India. Int J Climatol 37:3413–3426

Otto-Bliesner BL, Marshall SJ, Overpeck JT, Miller GH, Hu A (2006) Simulating Arctic climate warmth and icefield retreat in the last interglaciation. Science 311(5768):1751–1753

Rolland C (2003) Spatial and seasonal variations of air temperature lapse rates in Alpine regions. J Clim 16(7):1032–1046

Subramanya K (2005) Engineering hydrology, 4e. Tata McGraw-Hill Education.

White AB, Gottas DJ, Strem ET, Ralph FM, Neiman PJ (2002) An automated brightband height detection algorithm for use with Doppler radar spectral moments. J Atmos Ocean Technol 19(5):687–697

Zheng X, Basher RE (1996) Spatial modelling of New Zealand temperature normals. Int J Climatol 16(3):307–319

Identification of Relationship Between Precipitation and Atmospheric Oscillations in Upper Ganga Basin

Lalit Pal and C. S. P. Ojha

Abstract The hydrology of a region is directly or indirectly dependent on atmospheric variables. Identification of large-scale climate circulations dominating temporal pattern in regional atmospheric variables becomes crucial for improved precipitation forecasting for the region. The livelihood of a large population living in the downstream reaches of Ganga basin is dependent on snow melt and precipitation received in the Upper Ganga basin. Therefore, the aim of this study is to identify trends along various precipitation time series (monthly, seasonally-based, seasonal and annual) over Upper Ganga basin using discrete wavelet transform (DWT) approach for the period of 116 years (1901–2015). Relationship between large-scale atmospheric oscillations and regional scale trends in precipitation series are analysed. In the results, insignificant increasing trend is observed in monthly, annual, monsoon and pre-monsoon time series. On the other hand, winter time series is found to be following significant decreasing trend. The temporal trend in monthly and seasonally-based precipitation series are found to be dominated by seasonal variations in Inter-Tropical Convergence Zone (ITCZ). Overall, El Niño-Southern Oscillations (ENSO) is having a dominant effect on most of the precipitation time series over UGB (monsoon, annual, and pre-monsoon). The study outcomes are particularly beneficial for hydro-meteorological analyses and climate impact assessment based studies in the region.

Keywords Precipitation · Atmospheric oscillations · Wavelet transform · MK test · Upper Ganga basin

L. Pal (✉) · C. S. P. Ojha
Department of Civil Engineering, Indian Institute of Technology Roorkee, Roorkee,
Uttarakhand 247667, India
e-mail: lalitpl4@gmail.com

C. S. P. Ojha
e-mail: cspojha@gmail.com

© The Author(s), under exclusive license to Springer Nature Switzerland AG 2021
M. S. Chauhan and C. S. P. Ojha (eds.), *The Ganga River Basin: A Hydrometeorological Approach*, Society of Earth Scientists Series,
https://doi.org/10.1007/978-3-030-60869-9_11

1 Introduction

Over the last 100 years, the global average surface temperature has experienced an increase of 0.85 °C which has altered the precipitation patterns and other hydrological systems, globally (Pachauri et al. 2015). As climatic variables directly or indirectly defines the hydrologic response of a catchment, changes in parameters such as temperature, precipitation, evaporation are influencing the streamflow and flow regimes, substantially. Hydrologic variables (streamflow, precipitation, evapotranspiration, etc.) are particularly driven by atmospheric variables (temperature, relative humidity, pressure, perceptible water, etc.) (Sonali and Kumar 2013). Analysing trends in hydrological variables can provide valuable insights for identifying alteration in relationship between atmospheric and hydrologic variables within a region. Identification of trends in time series of hydrologic variables is of significant scientific and practical importance in order to plan and implement an efficient and sustainable management practice in a river basin. In the recent past, considerable number of studies have been focused on analysing the hydrological response to climate variability and climate change through identifying trends in hydrologic time series. Most of these studies have employed trend detection methods such as Sen's slope test, least square linear regression, Mann-Kendall test, Seasonal Mann-Kendall test, etc. All of the above mentioned techniques hold a common assumption that the hydrologic variable is stationary during observed period of record. However, substantial anthropogenic climate change and various other human disturbances have compromised the assumption of stationarity (Milly et al. 2007). Changes in basin climate are altering the mean and extremes of hydrologic variables resulting in tempering of probability density function established from instrumental records. These changes in climate are non-monotonic and non-uniform in nature, making trend detection complicated in a non-stationary environment (Franzke 2010). Therefore, identification of trends using conventional methods with stationarity assumption may result in erroneous conclusions and there needed a technique independent of such assumption.

A spectral analysis method called the wavelet transform (WT) has found its application in the analysis of non-stationary geophysical time series (Lau and Weng 1995; Lindsay et al. 1996). The WT decomposes one-dimensional non-stationary time series into multiple low and high frequency components to represent intra-annual, inter-annual and decadal fluctuations. The method has been acknowledged to be superior to other conventional signal analysis techniques, for example, Fourier transform (FT). In FT, a signal is decomposed into sine wave functions with infinite duration, whereas, WT decomposes signals using wavelet functions with limited duration and zero means. In recent times, the WT method has gained popularity for analysing the time series of geophysical variables under climate change scenario (Nalley et al. 2012, 2013). The method provide insight to the different periodic components that dominates the trend in hydrologic variables which can later be inferred to large and regional scale climate circulations. It provides a complete picture of dynamics contained in the signal being analysed. The present study aims to investigate relationship between atmospheric circulations and trends in monthly, seasonal, and annual precipitation over Upper Ganga basin, India using the discrete wavelet transform (DWT) approach. Different lower resolution (low frequency) components can be derived from a time series using the DWT approach. DWT simplifies the decomposition process as it is based on dyadic discretization (integer power of two) and generates one-dimensional signal which is easier to analyse. MK trend test is applied to the decomposed time series to assess their statistical significance. The decomposed series having MK-Z value closest to that of original time series is considered to be the dominant periodic component. Later, climate system(s) with similar periodicity as that of dominant periodic component are identified in order to investigate the relationship between the trends in precipitation and dominant climate system.

2 Study Area

The Ganga River is the longest east flowing river in India which forms at the confluence of Bhagirathi and Alaknanda River in Uttarakhand state of India within the mountain range of the Himalayas. The entire Ganga River basin is divided into three zones namely Upper Ganga basin, Middle Ganga basin and Lower Ganga basin. The present study is carried out over the Upper Ganga basin (up to Haridwar) situated in northwest Himalayan region in India. The areal extent of the region lies within $29° 38'–31° 24'$ N latitude and $78° 09'–80° 22'$ E longitude covering an area of about $22,292$ km^2 up to Haridwar. The Upper Ganga basin extends from snow/glacier covered greater Himalayas in the north to forest covered Himalayan foothills in the south. Runoff generated from snow melt and monsoon rainfall nurtures the population living in the downstream reaches of the basin. The elevation in the river basin ranges from 7799 m in the Himalayan mountain peaks to 277 m in the plains. In the western Himalayas, rainfall distribution responds to moist monsoon winds from the Bay of Bengal and moisture bearing westerlies during winters. A major fraction of rainfall in Upper Ganga basin is received from the Indian summer monsoon (ISM) extending from June to September, thus, monsoon rains have vital social and economic consequences. Location map of study area along with topographic details is shown in Fig. 1.

3 Materials and Methods

3.1 Data Collection

The daily precipitation gridded dataset developed by India Meteorological Department (IMD) (Pai et al. 2014) at $0.25° \times 0.25°$ resolution has been used to generate basin average precipitation time series for the period from 1901 to 2015. A

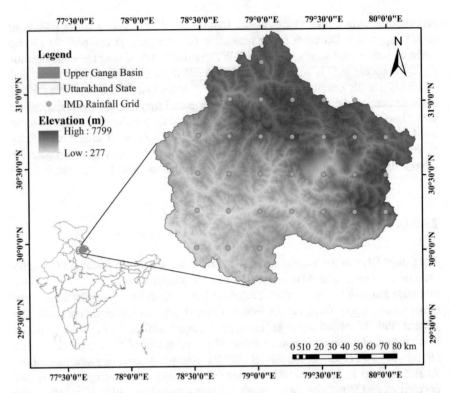

Fig. 1 Location of study area and topography of the Upper Ganga basin with IMD rainfall grid points

total of seven time series viz. monthly, seasonally-based, pre-monsoon (March–May), monsoon (June–September), post-monsoon (October–November), winter (December–February) have been generated from the available daily precipitation data to carry out various analyses. Each dataset has been tested for the presence of seasonality and autocorrelation.

3.2 Wavelet Transform

A wavelet is a small piece of wave having zero mean and finite length in space. Wavelet transform (WT) is a mathematical function which scales a signal into high frequency (low pass filter) and low frequency components (high pass filter) (Adarsh and Janga Reddy 2015). It utilises a variable-size window function which can be enlarged and shifted in the time and frequency domain (Lau and Weng 1995). High frequency components are captured by narrow window whereas low frequency components are resolved by a wide window. In the process of time series decomposition using WT, mother wavelet is shifted along the signal in multiple steps in order to generate wavelet coefficients. Wavelet coefficients store the information regarding position and extent of different events at various scales. Different dilated versions of the mother wavelet represent different scales. In signal decomposition, the WT is considered advantageous over conventional signal processing technique like Fourier transform (FT) as FT uses sine function resulting in loss of time information of the signal being processed. The WT can be applied in two ways: continuous and discrete wavelet transform. In the present study, discrete wavelet transform (DWT) is applied to decomposed precipitation time series into various periodic components.

3.2.1 Discrete Wavelet Transform

Various studies have employed DWT for trend analysis and forecasting (Karthikeyan and Kumar 2013; Nalley et al. 2013). In DWT, window function scales and translates on dyadic scales i.e. in integer power of 2 (Adarsh and Janga Reddy 2015). The mother wavelet function (ψ) in DWT is described as:

$$\psi_{(a,b)}\left(\frac{t-\gamma}{s}\right) = \frac{1}{(s_0)^{\frac{a}{2}}}\psi\left(\frac{t-by_0s_0^a}{s_0^a}\right) \tag{1}$$

where, ψ represents the discrete mother wavelet, a and b are integers representing the magnitude of enlargement (stretching) and translation (shifting) of the wavelet, respectively, γ_0 defines the location along the signal having value greater than zero and s_0 is the dilation length with value greater than 1. The wavelet coefficients of time series x_t at discrete integer time step is given as:

$$W_\psi(a,b) = \frac{1}{(2)^{a/2}}\sum_{t=0}^{N-1}x_t\psi\left(\frac{t}{2^a}-b\right) \tag{2}$$

with the values of s_0 and γ_0 are 2 and 1, respectively (Mallat 1989; Daubechies 1992).

3.3 Test for Serial Correlation

Lag-k autocorrelation coefficient (ACF) can be computed as (Partal and Kahya 2006):

$$R = \frac{\left(\frac{1}{n-1}\right)\sum_{t=1}^{n-k}[x_t - \bar{x}_t][x_{t+k} - \bar{x}_t]}{\left(\frac{1}{n}\right)\sum_{t=1}^{n}[x_t - \bar{x}_t]^2} \tag{3}$$

$$\frac{\{-1 - 1.645\sqrt{n} - 2\}}{n-1} \leq R \leq \frac{\{-1 + 1.645\sqrt{n} - 2\}}{n-1} \tag{4}$$

where, R = autocorrelation coefficient (ACF) at lag-k of the time series x_t. and $\bar{x}_t$ = mean of the data. Lag-1 autocorrelation coefficients (ACFs) are used to identify the presence of significant autocorrelation. The time series is said to be independent of serial correlation if the value of lag-1 ACF is found to be within the interval computed using Eq. (2). Autocorrelation coefficients (ACFs) are analysed at various lags to obtain correlograms using MATLAB®.

3.4 Mann-Kendall Test

The MK test (Mann 1945; Kendall 1975) is a non-parametric trend test widely employed to test the significance of trend in a time series. The test statistic S is given as:

$$S = \sum_{i=1}^{n-1} \sum_{j=i+1}^{n} sgn(x_j - x_i) \tag{5}$$

where x_j and x_i are data values in sequence i and j $(j > i)$, n is the length of data series, and $sgn(x_j - x_i)$ is sign function as:

$$sgn(x_j - x_i) = \begin{cases} 1, & for\ (x_j - x_i) > 0 \\ 0, & for\ (x_j - x_i) = 0 \\ -1, & for\ (x_j - x_i) < 1 \end{cases} \tag{6}$$

For identically distributed random variable with $n \geq 8$, test statistics closely follow normal distribution with the mean and variance given by:

$$E(S) = 0 \tag{7}$$

$$Var(S) = \frac{n(n-1)(2n+5) - \sum_{i=1}^{m} t_i(t_i - 1)(2t_i + 5)}{18} \tag{8}$$

where, t_i denotes the number of ties of extent, i and m are the number of tie groups. The standardized normal variate of test statistics Z_S is computed as:

$$Z_S = \begin{cases} \frac{S-1}{\sqrt{Var(S)}}, & for\ S > 0 \\ 0, & for\ S = 0 \\ \frac{S+1}{\sqrt{Var(S)}}, & for\ S < 0 \end{cases} \tag{9}$$

The positive values of Z_S indicate increasing trend while negative Z_S indicates decreasing trend. At a given level of significance 'α', the null hypothesis H_0 of no trend is rejected if $|Z_S| > Z_{1-\alpha/2}$, where $Z_{1-\alpha/2}$ is the value of standard normal variate corresponding to a probability of $\alpha/2$. In present study, the hypothesis is tested at 5% significance level, i.e., $\alpha = 0.05$. At 5% significance level, the null hypothesis of no trend is rejected, if $|Z_S| > 1.96$.

Hamed and Rao (1998) proposed a modified version of MK test in which effect of autocorrelation is taken into account by using a modified variance of the test statistic. The modified variance is described as:

$$V^*(S) = Var(S).\frac{n}{n_S^*} = \frac{n(n-1)(2n-5)}{18}.\frac{n}{n_S^*} \qquad (10)$$

where, $\frac{n}{n_S^*}$ represents a correlation in the data due to the presence of serial correlation. The $\frac{n}{n_S^*}$ is evaluated using:

$$\frac{n}{n_S^*} = 1 + \frac{2}{n(n-1)(n-2)}\sum_{i=1}^{n-1}(n-i)(n-i-1)(n-i-2)\rho_s(i) \qquad (11)$$

where n denotes number of observations, and $\rho_s(i)$ is the autocorrelation function values computed for the rank of the observations. The standardized test statistics Z is computed as:

$$Z = \frac{S}{[V^*(S)]^{0.5}} \qquad (12)$$

4 Methodology

Prior to decomposition using WT, various time series of precipitation are analysed for presence of autocorrelation and seasonality effect at 5% significance level. To identify trends in the time series, Mann-Kendall test is applied if no serial correlation exists and Modified Mann-Kendall test is applied if there exist significant autocorrelation. A flow chart of methodology adopted in the present study is shown in Fig. 2. Various monthly, seasonal and annual precipitation time series are analysed for inherent periodic components using DWT. All the computations were performed in MATLAB®. The first step in the methodology involves selection of an appropriate mother wavelet function, level of decomposition and border extension method. Among various families of mother wavelet, the Haar, Symlets and Daubechies families of wavelets have found their application in analysis of trends in geophysical time series (Nalley et al. 2013; Sang 2013; Adarsh and Janga Reddy 2015; Araghi et al. 2015). In particular, past studies have used Daubechies (db) family of wavelets owing to its ease of application. It provides full scaling and translation orthogonal properties with non-zero basis function over a finite interval (Ma et al. 2003; de Artigas et al. 2006).

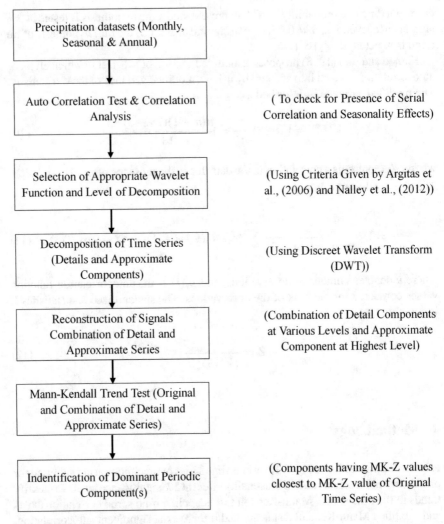

Precipitation datasets (Monthly, Seasonal & Annual)

↓

Auto Correlation Test & Correlation Analysis

(To check for Presence of Serial Correlation and Seasonality Effects)

↓

Selection of Appropriate Wavelet Function and Level of Decomposition

(Using Criteria Given by Argitas et al., (2006) and Nalley et al., (2012))

↓

Decomposition of Time Series (Details and Approximate Components)

(Using Discreet Wavelet Transform (DWT))

↓

Reconstruction of Signals Combination of Detail and Approximate Series

(Combination of Detail Components at Various Levels and Approximate Component at Highest Level)

↓

Mann-Kendall Trend Test (Original and Combination of Detail and Approximate Series)

↓

Indentification of Dominant Periodic Component(s)

(Components having MK-Z values closest to MK-Z value of Original Time Series)

Fig. 2 Flowchart of methodology

Different forms of db wavelets (db1, db2, …, db45) are available in wavelet transform package of MATLAB®. For selection of appropriate mother wavelet for signal decomposition, db1–db10 wavelets are considered in the study. To check for the border distortion effect introduced during decomposition of signal with finite length, three types of border extension are available in MATLAB® viz. periodic extension, zero padding and boundary value replication (symmetrization). Symmetrization is the default extension mode in MATLAB which ensure signal recovery by symmetric boundary replication. For identifying the adequate combination of the above mentioned three parameters, de Artigas et al. (2006) and Nalley et al. (2012) have proposed two different approaches, each of which are summarized in the following section. The method proposed by de Artigas et al. (2006) involves minimization of mean relative error (MRE) between the original and approximate (A) series corresponding to last decomposition level as:

$$MRE = \frac{1}{n} \sum_{j=1}^{n} \frac{|a_j - x_j|}{|x_j|} \tag{13}$$

where x_j is the original time series with n number of data points and a_j is the approximate component of x_j.

In the second approach proposed by Nalley et al. (2012), the combination producing minimum RE is selected, where RE is calculated as:

$$RE = \frac{|Z_{op} - Z_{or}|}{|Z_{or}|} \tag{14}$$

Z_{or} represents the MK Z-value of original time series; and Z_{op} is the MK Z-value of the approximation component of the last decomposition level of DWT. The combination of mother wavelet type, level of decomposition and border decomposition producing minimum value of MRE and RE are selected for further analyses. To calculate the number of decomposition levels, de Artigas et al. (2006) proposed the following equation:

$$L = \frac{Log\left(\frac{n}{2v-1}\right)}{Log(2)} \tag{15}$$

where n is the number of records in monthly time series, v is the number of vanishing moments and L is maximum decomposition level. In MATLAB, the number of vanishing moments for a Daubechies (db) wavelet is half of the length of its starting filter.

The *multilevel one-dimensional wavelet analysis* function in MATLAB® has been employed to decompose precipitation time series using discrete wavelet transform (DWT). The approximation (A) and detail (D) components are generated through convolving the time series with low-pass and high-pass filters, followed by a dyadic scale discretization. The detail component at each decomposition level is added with approximate component corresponding to last level of decomposition as important characteristics of original time series such as trend are stored in last the approximate component. Decomposed time series are reconstructed into one dimensional signal using *multilevel one-dimensional wavelet reconstruction* function in MATLAB. The original signal is decomposed using window of scale varying by integer powers of 2 i.e. the signal is broken down in halves, then in quarters, and it continues onward (Dong et al. 2008). The original signal (time series) is decomposed at various levels by power of 2 and at each level, approximation (A) components of previous level are decomposed. The periodic component dominating the trend in original time series is identified by comparing the MK-Z value of original series with MK-Z value of detail component at different decomposition levels. As each detail component represents a time series of defined periodicity (in integer power of 2), the component having MK-Z value closest to that of original series defines the dominance of variable with corresponding periodicity. Lastly, the existing atmospheric processes directly or indirectly controlling the precipitation patterns in southern Asia are identified.

5 Results and Discussion

5.1 Serial Correlation and Seasonality

All seven precipitation time series (monthly, seasonally-based, annual, monsoon, pre-monsoon, post-monsoon and winter) have been tested for the presence of serial correlation. The computed lag-1 autocorrelation coefficients along with their upper and lower bounds are given in Table 1. As evident from the table, significant lag-1 autocorrelation has been observed in monthly, seasonally-based and monsoon time series, whereas, annual, post-monsoon, winter and pre-monsoon time series are found to be independent of serial correlation. In general, monthly time series are expected to have stronger autocorrelation than the annual data series (Hamed and Rao 1998). Also, strong effect of seasonality has been observed in the monthly time series as repeated patterns of semi-annual and annual cycles can be seen in its correlograms (Fig. 3). Similarly, the correlograms of seasonally-based time series are presenting an evidence of annual to inter-annual cyclic patterns (Fig. 3). No clear effect of seasonality can be observed in the rest of time series. The trends in time series following strong seasonality pattern are identified using Modified Mann-Kendall test as discussed in previous section and time series independent of the effect of autocorrelation are employed with Mann-Kendall test.

Table 1 Lag-1 autocorrelation coefficients for various precipitation time series for Upper Ganga basin

Time series	ACF (lag-1)	Upper limit	Lower limit
Monthly*	**0.521**	**0.054**	**−0.054**
Seasonally-based*	**−0.353**	**0.093**	**−0.093**
Annual	−0.052	0.187	−0.187
Monsoon*	**0.333**	**0.187**	**−0.187**
Post-monsoon	0.116	0.187	−0.187
Winter	0.183	0.187	−0.187
Pre-monsoon	−0.052	0.187	−0.187

*Significant at 5% significance level

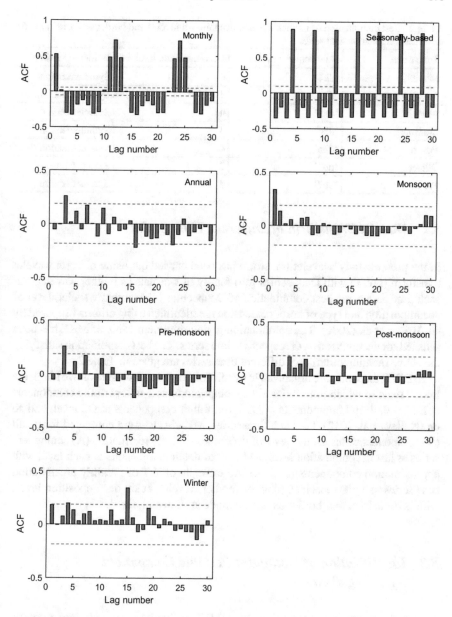

Fig. 3 Correlograms of monthly, seasonally-based, annual, monsoon, post-monsoon, pre-monsoon and winter time series

Table 2 Best combination of mother wavelet, decomposition level and border extension type for different precipitation data series

Time series	Daubechies wavelet	Decomposition level	Border extension
Monthly	'db4'	6	Symmetrization
Seasonally-based	'db4'	4	Symmetrization
Annual	'db5'	4	Symmetrization
Monsoon	'db6'	4	Symmetrization
Post-monsoon	'db5'	4	Symmetrization
Winter	'db3'	2	Zero padding
Pre-monsoon	'db7'	4	Symmetrization

5.2 Wavelet Transform of Different Time Series

In the present study, wavelet transform has been carried out using discrete wavelet transform (DWT) with Daubechies (db) family of wavelets as mother wavelet. For each time series, the best combination of Daubechies (db) mother wavelet, level of decomposition and type of border extension are identified using criteria proposed by de Artigas et al. (2006). The combination giving minimum value of MRE has been selected for decomposition of respective time series. The best combination satisfying the above mention criteria for different time series are given in Table 2.

In DWT, the scales are organized in dyadic format (integer powers of two) from the lowest scale to which time series are decomposed to as 2-unit periodic components at D1 level, 4-unit components at D2 level, 8-unit components at D3 level, and so on (Nalley et al. 2012). For each time series, MK Z-values are computed for detail (D) component series at each level of decomposition, approximation (A) component series at last decomposition level, and sum of detail components at each level with approximation components (D + A). An example of WT of monthly precipitation time series using Daubechies (db4) as mother wavelet, at six decomposition levels with 'symmetrization' border extension mode is shown in Fig. 4.

5.3 Identification of Dominant Periodic Component Affecting Trend

The modified version of Mann-Kendall (MMK) test has been applied to the monthly, seasonally-based and monsoon time series as all these are effected by strong autocorrelation. The original Mann-Kendall (MK) test has been applied to the remaining time series independent of the effect of autocorrelation. The results of both MK and MMK test are presented in Table 3. As evident from the table, monthly, annual, monsoon ad pre-monsoon precipitation is exhibiting an insignificant increasing trend at 5% significance level, whereas, precipitation time series of seasonally-based, post-monsoon

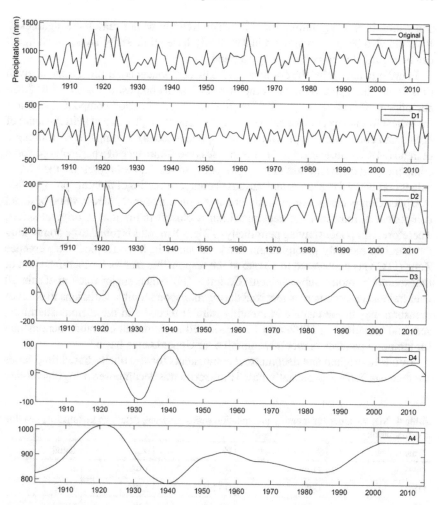

Fig. 4 Monthly precipitation time series and its decomposition using DWT with db4 wavelet into six levels

Table 3 MK Z-values for different original precipitation time series

Time series	S	Z	Z_n	Trend
Monthly	7129	0.42	1.96	+
Seasonally-based	−210	−0.06	1.96	−
Annual	209	0.5	1.96	+
Monsoon	313	0.75	1.96	+
Post monsoon	−245	−0.59	1.96	−
Winter*	**−941**	**−2.01**	**1.96**	−
Pre-monsoon	623	1.5	1.96	+

*Significant at 5% significance level

and winter is showing falling trend in which trend in winter season are significant. It should also be noted that trend in seasonally-based time series are so weak that it can be regarded as to be following no trend.

In order to determine the dominant periodic component controlling trend in given time series, the approximation component of last decomposition is added to detail component of each decomposition level as the approximation component captures the lowest frequency components of the signal. The strength of trend (MK Z-value) of each detail component (approximation component added) is compared with strength of trend (MK Z-value) in original time to assess the dominance of a particular periodic component on the overall trend in original time series. Table 4 represents the MK Z-values computed for original and various decomposed series of monthly precipitation in Upper Ganga basin. For monthly precipitation time series, MK-Z value of detail component at third level of decomposition is closest to the that of original series. Therefore, component having periodicity of 2^3 i.e. 8 months is found to be dominating the trend in monthly precipitation. Seasonal shifts in Inter-Tropical Convergence Zone (ITCZ) is considered to be one of the responsible factors for manifestation of monsoon over Indian sub-continent (Gadgil 2003). Since, major fraction of rainfall over Indian sub-continent is received during monsoon months, variations in onset of monsoon may cause change in monthly rainfall received in the region. Therefore, seasonal variations in ITCZ may be considered as dominant climatic phenomenon for temporal patterns in monthly rainfall over Upper Ganga basin.

Trends in original and decomposed components for seasonally-based time series are shown in Table 5. Seasonally-based time series was decomposed using db4 mother

Table 4 MK Z-values for monthly precipitation series (original, detail, approximation, detail + approximation)

Time series	S	Z	$|Z_n|$	Trend
Original	7129	0.42	1.96	+
D1	1884	0.11	1.96	+
D2	−2304	−0.13	1.96	−
D3	−244	−0.01	1.96	−
D4	−2830	−0.17	1.96	−
D5	−882	−0.05	1.96	−
D6	−18,980	−1.11	1.96	−
A6	−11,626	−0.68	1.96	−
A6 + D1	−17,466	−1.02	1.96	−
A6 + D2	−17,604	−1.03	1.96	−
A6 + D3[a]	**8918**	**0.52**	**1.96**	**+**
A6 + D4	−18,844	−1.10	1.96	−
A6 + D5	−18,578	−1.09	1.96	−
A6 + D6	48,496	2.84	1.96	+

[a]Dominant periodic component

Table 5 MK Z-values for seasonally-based precipitation series (original, detail, approximation, detail + approximation)

Time series	S	Z	$|Z_n|$	Trend
Original	−210	−0.06	1.96	−
D1	−154	−0.05	1.96	−
D2	−558	−0.17	1.96	−
D3	−672	−0.20	1.96	−
D4	3130	0.95	1.96	+
A4	506	0.15	1.96	+
A4 + D1[a]	**−32**	**−0.01**	**1.96**	**−**
A4 + D2	72	0.02	1.96	+
A4 + D3	−514	−0.16	1.96	−
A4 + D4	3026	0.92	1.96	+

[a]Dominant periodic component

wavelet up to four decomposition level with 'symmetrization' border extension. From the table, the seasonally based series has been observed to be following practically no trend. Subsequently, different trend can be seen in various decomposed components with no series presenting any significant trend. As evident from Table 5, MK-Z value of level-1 detail component (D1) is closest to that of original time series, therefore, level-1 periodic component with a periodicity of 2^1 seasons i.e. about 8 months may be adopted as dominant for seasonally-based precipitation. Again, shifts in ITCZ have seasonal periodicity which corresponds to that of identified dominant component. Therefore, trends in seasonally-based precipitation over Upper Ganga basin may be attributable to the seasonal variations in ITCZ.

The MK Z-value for different decomposed components of annual time series are shown in Table 6. As evident from the table, a positive trend exists in the annual precipitation in UGB. Further, trends in all detail and approximation components are observed to be positive in nature which shows that almost all physical processes with various periodicities are contributing to the overall annual time series. The level-2 detail component representing 2^2 years i.e. 4-year periodic component may be adopted as dominant periodic component for annual precipitation as MK-Z value of D2 + A component is 0.41 which is closest to that of original time series. Among various existing atmospheric process, El-Niño Southern Oscillations (ENSO) have periodicity of 4–7 years (Webster et al. 1998; Gadgil et al. 2004). In addition, trends observed in observed and decomposed time series of monsoon rainfall are same as that of annual rainfall. These results are evident as monsoon rainfall constitutes major portion of total annual rainfall received in the region. Subsequently, 4-year periodic component corresponding to D2 + A periodic component is found to the dominant periodic component for monsoon rainfall also (Table 7). Moreover, various past studies have documented the dominance of ENSO on Indian summer monsoon rainfall (Sikka 1980; Pant and Parthasarathy 1981; Rasmusson and Carpenter 1983;

Table 6 MK Z-values for annual precipitation series (original, detail, approximation, detail + approximation)

Time series	S	Z	$|Z_n|$	Trend
Original	209	0.50	1.96	+
D1	111	0.27	1.96	+
D2	−31	−0.07	1.96	−
D3	163	0.39	1.96	+
D4	439	1.06	1.96	+
A4	459	1.11	1.96	+
A4 + D1	295	0.71	1.96	+
A4 + D2[a]	**169**	**0.41**	**1.96**	+
A4 + D3	281	0.68	1.96	+
A4 + D4	693	1.67	1.96	+

[a]Dominant periodic component

Table 7 MK Z-values for monsoon precipitation series (original, detail, approximation, detail + approximation)

Time series	S	Z	$|Z_n|$	Trend
Original	313	0.75	1.96	+
D1	11	0.02	1.96	+
D2	−133	−0.32	1.96	−
D3	97	0.23	1.96	+
D4	247	0.59	1.96	+
A4	587	1.42	1.96	+
A4 + D1	5	0.01	1.96	+
A4 + D2[a]	**271**	**0.65**	**1.96**	+
A4 + D3	455	1.10	1.96	+
A4 + D4	1231	2.97	1.96	+

[a]Dominant periodic component

Webster et al. 1998; Ashok and Saji 2007). Therefore, ENSO may be adopted as the dominant climate cycle controlling the temporal patterns in monsoon and annual rainfall over Upper Ganga basin.

Post-monsoon precipitation in UGB is found to be dominated by 2-year periodic component as MK-Z value of D1 + A component is closest to that of the original time series (Table 8). Among key atmospheric processes identified to have influence on rainfall over Indian sub-continent, periodic cycle of seasonal variations in ITCZ, Indian Ocean Dipole (IOD) movement and ENSO varies between 2 and 7 years. Although, various attempts have been made by researchers to understand the physics behind rainfall occurrence over Indian sub-continent, the interaction between these

Table 8 MK Z-values for post-monsoon precipitation series (original, detail, approximation, detail + approximation)

Time series	S	Z	$|Z_n|$	Trend
Original	−245	−0.59	1.96	−
D1	151	0.36	1.96	+
D2	−223	−0.54	1.96	−
D3	−173	−0.42	1.96	−
D4	37	0.09	1.96	+
A4	−1931	−4.67	1.96	−
A4 + D1[a]	**−247**	**−0.56**	**1.96**	−
A4 + D2	−1075	−2.60	1.96	−
A4 + D3	−1119	−2.70	1.96	−
A4 + D4	−893	−2.16	1.96	−

[a]Dominant periodic component

individual process is not yet well understood. Therefore, attribution of change in precipitation patters specifically among these climate patterns is difficult. As, 2-year periodic component is found to be dominant in post-monsoon period, variability of rainfall during this period may be due to interaction between the processes with bi-annual periodicities.

As obtained for post-monsoon period, precipitation patterns in winter season are also dominated by 2-year periodic component (Table 9). Subsequently, variability and temporal in precipitation during winter season may also be attributable to climate processes with 1–2-year periodicity such as IOD, seasonal variation in ITCZ or ENSO. In addition, the original precipitation time series for winter season is exhibiting a significant decreasing trend with MK-Z value being −2.01. Lastly, MK Z-values computed for original and decomposed time series of pre-monsoon precipitation are given in Table 10. Combination of detail and approximation components at all four decomposition level has been observed to following significant increasing

Table 9 MK Z-values for winter precipitation series (original, detail, approximation, detail + approximation)

Time series	S	Z	$|Z_n|$	Trend
Original	−941	−2.01	1.96	−
D1	11	0.02	1.96	+
D2	−13	−0.03	1.96	−
A2	−985	−2.38	1.96	−
A4 + D1[a]	**−761**	**−1.84**	**1.96**	−
A4 + D2	−669	−1.61	1.96	−

[a]Dominant periodic component

Table 10 MK Z-values for pre-monsoon precipitation series (original, detail, approximation, detail + approximation)

Time series	S	Z	$|Z_n|$	Trend
Original	623	1.50	1.96	+
D1	27	0.06	1.96	+
D2	95	0.23	1.96	+
D3	−383	−0.92	1.96	−
D4	−35	−0.08	1.96	−
A4	1651	3.99	1.96	+
A4 + D1	937	2.26	1.96	+
A4 + D2[a]	**717**	**1.70**	**1.96**	+
A4 + D3	1285	3.10	1.96	+
A4 + D4	1463	3.53	1.96	+

[a]Dominant periodic component

trend. However, MK-Z value of D2 + A component (1.70) is closest to that of original time series (1.50). This indicates that a climate process with 4-year periodicity is dominating the trend observed in pre-monsoon series. Evidently, the insignificant increasing trend observed in pre-monsoon rainfall over UGB may fairly be attributable to ENSO (4–7-year periodicity).

6 Conclusion

The wavelet transform (WT) approach which is conventionally been used for signal processing has been applied to precipitation time series in Upper Ganga basin to analyse trends and dominant periodic component controlling the trends. The application of DWT to the precipitation time series is found to provide useful information regarding inherent controlling processes. The methodology presented in the present study can be applied to other hydro-meteorological variables also. In the results, monthly, annual, monsoon and pre-monsoon precipitation series are found to be exhibiting an insignificant increasing trend at 5% significance level. Whereas significant decreasing trend are observed in winter precipitation series. A fair amount of rainfall in the region is also received during winter season from retreating northeast monsoon in southern foothills of Himalayas and westerlies in northern Great Himalayas. The significant decreasing trend observed in winter precipitation reveals the drying tendency of region in western Himalayas.

The atmospheric oscillations dominating temporal trends in various seasonal and annual precipitation series are identified by comparing the dominant periodic components with periodicity of known atmospheric patterns. The temporal trend in monthly and seasonally-based precipitation series are found to be dominated by seasonal variations in Inter Tropical Convergence Zone (ITCZ). The cycles of ENSO are found to be dominating the trends observed in pre-monsoon, monsoon and annual precipitation over Upper Ganga basin. Further, temporal variations in precipitation during remaining two seasons i.e. post-monsoon and winter are found to be influenced by the combined effect of inter-annual scale climate processes namely ENSO, IOD and ITCZ.

Overall, ENSO climate circulation is having a dominant effect on most of the precipitation time series which is also supported by various past studies. The summer monsoon rainfall over Indian subcontinent is a combined results of various climate processes occurring simultaneously at different temporal scale. The interaction between these processes, however, is not well studied yet. Therefore, the relationship established between monsoon precipitation trends and various global climate circulations need to be verified by carrying out correlation studies. Further, such relationships may be helpful in identifying linkages between precipitation and different climate processes and improved precipitation forecast at regional scale.

Acknowledgements The authors are grateful to the anonymous reviewers for their useful comments and suggestions.

References

Adarsh S, Janga Reddy M (2015) Trend analysis of rainfall in four meteorological subdivisions of southern India using nonparametric methods and discrete wavelet transforms. Int J Climatol 35(6):1107–1124

Araghi A, Baygi MM, Adamowski J, Malard J, Nalley D, Hasheminia SM (2015) Using wavelet transforms to estimate surface temperature trends and dominant periodicities in Iran based on gridded reanalysis data. Atmos Res 155:52–72

Ashok K, Saji NH (2007) On the impacts of ENSO and Indian Ocean dipole events on sub-regional Indian summer monsoon rainfall. Nat Hazards 42(2):273–285

Daubechies I (1992) Ten lectures on wavelets. Society for industrial and applied mathematics

de Artigas MZ, Elias AG, de Campra PF (2006) Discrete wavelet analysis to assess long-term trends in geomagnetic activity. Phys Chem Earth, Parts A/B/C 31(1):77–80

Dong X, Nyren P, Patton B, Nyren A, Richardson J, Maresca T (2008) Wavelets for agriculture and biology: a tutorial with applications and outlook. Bioscience 58(5):445–453

Franzke C (2010) Long-range dependence and climate noise characteristics of Antarctic temperature data. J Clim 23(22):6074–6081

Gadgil S (2003) The Indian monsoon and its variability. Annu Rev Earth Planet Sci 31(1):429–467

Gadgil S, Vinayachandran PN, Francis PA, Gadgil S (2004) Extremes of the Indian summer monsoon rainfall, ENSO and equatorial Indian Ocean oscillation. Geophys Res Lett 31(12)

Hamed KH, Rao AR (1998) A modified Mann-Kendall trend test for autocorrelated data. J Hydrol 204(1–4):182–196

Karthikeyan L, Kumar DN (2013) Predictability of nonstationary time series using wavelet and EMD based ARMA models. J Hydrol 502:103–119

Kendall MG (1975) Rank correlation methods, 4th edn. 2d impression, Charles Griffin, London

Lau KM, Weng H (1995) Climate signal detection using wavelet transform: how to make a time series sing. Bull Am Meteor Soc 76(12):2391–2402

Lindsay RW, Percival DB, Rothrock D (1996) The discrete wavelet transform and the scale analysis of the surface properties of sea ice. IEEE Trans Geosci Remote Sens 34(3):771–787

Ma J, Xue J, Yang S, He Z (2003) A study of the construction and application of a Daubechies wavelet-based beam element. Finite Elem Anal Des 39(10):965–975

Mallat SG (1989) Multifrequency channel decompositions of images and wavelet models. IEEE Trans Acoust Speech Signal Process 37(12):2091–2110

Mann HB (1945) Nonparametric tests against trend. Econom: J Econom Soc, pp 245–259

Milly PCD, Julio B, Malin F, Robert M, Zbigniew W, Dennis P, Ronald J (2007) Stationarity is dead. Ground Water News Views 4(1):6–8

Nalley D, Adamowski J, Khalil B (2012) Using discrete wavelet transforms to analyze trends in streamflow and precipitation in Quebec and Ontario (1954–2008). J Hydrol 475:204–228

Nalley D, Adamowski J, Khalil B, Ozga-Zielinski B (2013) Trend detection in surface air temperature in Ontario and Quebec, Canada during 1967–2006 using the discrete wavelet transform. Atmos Res 132:375–398

Pachauri RK, Meyer L, Plattner GK, Stocker T (2015) IPCC, 2014: Climate Change 2014: Synthesis Report.In: Contribution of Working Groups I, II and III to the Fifth Assessment Report of the Intergovernmental Panel on Climate Change. IPCC

Pai DS, Sridhar L, Rajeevan M, Sreejith OP, Satbhai NS, Mukhopadyay B (2014) Development of a new high spatial resolution (0.25° × 0.25°) Long Period (1901-2010) daily gridded rainfall data set over India and its comparison with existing data sets over the region data sets of different spatial resolutions and time period. Mausam 65(1):1–18

Pant GB, Parthasarathy SB (1981) Some aspects of an association between the southern oscillation and Indian summer monsoon. Arch Meteorol, Geophys, Bioclim, Ser B 29(3):245–252

Partal T, Kahya E (2006) Trend analysis in Turkish precipitation data. Hydrol Process 20(9):2011–2026

Rasmusson EM, Carpenter TH (1983) The relationship between eastern equatorial Pacific sea surface temperatures and rainfall over India and Sri Lanka. Mon Weather Rev 111(3):517–528

Sang YF (2013) A review on the applications of wavelet transform in hydrology time series analysis. Atmos Res 122:8–15

Sikka DR (1980) Some aspects of the large scale fluctuations of summer monsoon rainfall over India in relation to fluctuations in the planetary and regional scale circulation parameters. Proc Indian Acad Sci-Earth Planet Sci 89(2):179–195

Sonali P, Kumar DN (2013) Review of trend detection methods and their application to detect temperature changes in India. J Hydrol 476:212–227

Webster PJ, Magana VO, Palmer TN, Shukla J, Tomas RA, Yanai MU, Yasunari T (1998) Monsoons: processes, predictability, and the prospects for prediction. J Geophys Res Oceans 103(C7):14451–14510

Utilizing Stream Network for Regional Flood Frequency Analysis in Ganga Basin

Hemanta Medhi and Shivam Tripathi

Abstract Regional flood frequency analysis (RFFA) is used for estimating flood quantiles at ungauged sites or sites with low record length. RFFA estimates flood quantiles at ungauged sites using regional regression models. These regression models are mostly obtained using the ordinary least square regression (OLS), weightage least square regression (WLS) or generalized least square regression (GLS) methods. Literature shows that the GLS method gives better estimates as it accounts for cross correlation in the discharge among the gauged sites, which is related to the geographical distance between the gauging sites. However, the stream network that provides information on natural or physical connectivity between the gauging sites is ignored. In this study, the GLS based regression model is tested for estimating flood quantiles at ungauged sites by using the stream network distance. This methodology is applied to 53 sub-basins in the Ganga basin, India. The comparison of the performance measures using the geographic and the stream network distance shows improvement in the flood quantile estimates for the stream distance based regression model. This study shows the usefulness of stream network information for RFFA and suggests the need for future research to efficiently incorporate stream network information in flood quantile estimation.

Keywords Runoff · Stream network · Flood frequency analysis · Regression equation

H. Medhi (✉) · S. Tripathi (✉)
Department of Civil Engineering, Indian Institute of Technology Kanpur, Kanpur
208016, India
e-mail: hemanta.medhi@gmail.com

S. Tripathi
e-mail: shiva@iitk.ac.in

M. S. Chauhan and C. S. P. Ojha (eds.), *The Ganga River Basin: A Hydrometeorological Approach*, Society of Earth Scientists Series,
https://doi.org/10.1007/978-3-030-60869-9_12

1   Introduction

Managing floods has been a challenging problem in water resources engineering. Floods may damage hydraulic structures like bridges, dams, and embankments, and wash off roads leading to loss of life and property. Hence to mitigate flood damage or to take preventive measures, one needs to know the flood magnitude in the area. This necessitates the collection of flow data in rivers and tributaries. To these collected flow data, flood frequency analysis (FFA) is applied by fitting suitable probability distributions (IACWD 1982). However, collecting flow data from a stream at every location is not possible, and in remote areas, constructing a gauging site (station) is not always feasible. Again, the gauging site of interest may have limited records of observed flow data. For basins with small record lengths, estimating the magnitude of large return period floods is a formidable task, and the uncertainties involved are enormous (Stedinger and Griffis 2008). The problem of having less or no data at a location limits the usage of at-site flood frequency analysis. An alternative method is regional flood frequency analysis (RFFA), in which the flood of a particular return period at any location is estimated by pooling information from neighboring sites.

In hydrology, RFFA has received wide attention and is popularly used to esti-mate flood quantile at ungauged sites (Ilorme and Griffis 2013; Das and Cunnane 2012; McIntyre et al. 2005). RFFA facilitates in the extrapolation of flood quantiles from sites at which records have been collected to a site where flood information is unavailable or limited. This is done by developing a regional relationship between flood statistics of a site and basin attributes like drainage area, slope, and precipi-tation (Markus et al. 2003). Thus, knowing the attributes of an ungauged basin, the flood magnitude can be estimated.

RFFA involves three major steps: (a) regionalization, which is the process of obtaining hydrologically homogeneous regions (basins with similar hydrologic behavior), (b) obtain regional regression models, in which the basin attributes and the at-site flood quantiles for each of the gauged site in the homogeneous regions are used to form a regional regression model, and (c) obtain flood quantile estimates for the ungauged sites by using regional regression model for the homogeneous region to which the ungauged site is allocated based on similarity in attribute space.

Traditionally, geographic and political boundaries were considered for identifying homogeneous regions (Rao and Srinivas 2008). However, the regions delineated on the basis of geography and political boundaries lack a scientific basis, and hence, such methods are no longer used. Methods like clustering analysis, a region of influence, and self-organizing map which use the geographical, physiographical, and meteo-rological attributes of the basin for homogeneous grouping sites are now commonly used in hydrological studies (Burn 1997; Srinivas et al. 2008; Medhi and Tripathi 2015). The average silhouette width (Rousseeuw 1987), Dunn's index (Dunn 1973), cophenetic correlation coefficient (Sokal and Rohlf 1962), Davies-Bouldin index (Davies and Bouldin 1979) are few of the many methodologies proposed in the literature to test the validity of the clustered gauging sites. However, the heterogeneity measure based on L-moments is the most widely accepted test for homogeneity of hydrological data (Hosking and Wallis 1997). Unlike other methods, it uses the hydrological responses to obtain the heterogeneity index for the homogeneous regions.

Once a homogeneous region is identified, the index flood method (IFM) or the quantile regression technique (QRT) is used to estimate the flood quantiles for a site in the homogeneous region. The index flood method (Dalrymple 1960; Robson and Reed 1999) used in RFFA for flood estimation is based on the assumptions that—(a) observations at any given site are independent and identically distributed, (b) observations at different sites are independent, and (c) sites in a homogeneous region have identical frequency distributions apart from a scale factor. The mean of the annual maximum flood is often considered the index flood, which is multiplied by an appropriate quantile from the dimensionless regional growth curve to estimate the flood quantile. For an ungauged site, the index flood is estimated by using the regression relationship between the basin attributes and the index flood of the gauged sites (Grover et al. 2002).

The Quantile regression technique is widely used in the United States to obtain regional equations (Walker and Krug 2003) for estimating flood quantiles. The ordinary least square regression (OLS) (Kroll and Stedinger 1998), weightage least square regression (WLS) (Tasker 1980), and generalized least square regression (GLS) (Griffis and Stedinger 2007) are few of the many regression methods found in the literature. Stedinger and Tasker (1985) applied these methods to synthetically generated dataset and found the GLS method, which accounts for the concurrent records and record length of the dataset (Stedinger and Tasker 1985), gives better estimates. Unlike the OLS method, where equal weightage is given to the errors, the GLS method gives different weightage depending on the length of the data, concurrent data period, the standard deviation of the data, and the cross-correlation of flow data between the gauging sites. The cross-correlation is estimated as a function of the distance between the gauging sites.

The hydrological sites are a part of the stream network, and this information can be utilized to improve the flood quantile estimates. Walker and Krug (2003) developed an ad hoc method to adjust the flood estimate at a site based on its stream distance from a neighboring gauged site. However, there is no method available in the literature that systematically utilizes stream network information for RFFA. The main objective of this study is to propose a method for incorporating stream connectivity information in regional regression models for flood estimation. The proposed method is tested by applying it to gauging sites in the Ganga basin. The chapter is structured as follows—Sect. 2 describes the study area, and Sect. 3 presents the data used in the study. The proposed methodology is described in Sect. 4, and the results are presented and discussed in Sect. 5. Section 6 concludes the study and outlines future work.

2 Study Area

The extent of the Ganga basin is shown in Fig. 1. The Ganga River travels a distance of about 2525 km. During its course, it is joined by numerous tributaries before it finally reaches the Bay of Bengal. The main flow in the Ganga River is contributed by the melting of snow and from the monsoon rainfall. The major tributaries contributing to the Ganga River are the Ramganga, Gomti, Ghaghara, Gandaki, Koshi, Mahananda, Yamuna, Tamsa, Son, and Punpun. The basin area of the Ganga basin is approximately 1,060,000 km² lying in the extent of 73° 21′ E to 90° 35′ E longitude and 22° 25′ N to 31° 28′ N latitude (Paul 2017). The Ganga basin has a very complex topography. It is bounded on the north by the mighty Himalayas and in south lays the central plain of India called the Indo-Gangetic plain. The Indo-Gangetic plain is about 1600 km long, and its width varies from 200 to 800 km. The climatic conditions are very different in different parts of the basin. The Himalayas have very low temperatures, and at the same time, the Indo-Gangetic plain has a very high temperature.

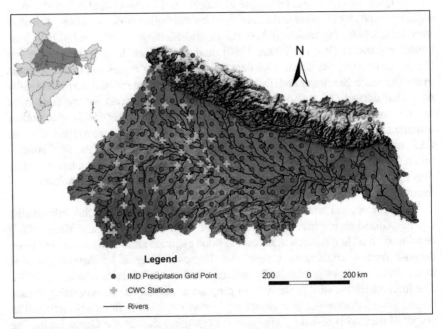

Fig. 1 Spatial distribution of the 53 CWC gauge sites (green plus) and IMD precipitation grid points (red circle) in the Ganga basin

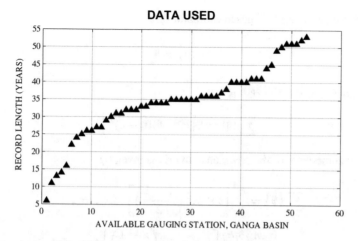

Fig. 2 Record length of the 53 gauge sites used in the study

3 Data Used

There are over 250 Central Water Commission, India (CWC) gauge sites spreading across the Ganga basin. A total of 53 sites concentrated towards the western side of the Ganga basin were available for this study. Those 53 sites are shown in Fig. 1 (green plus symbol). Daily discharge data were available for the sites and were expressed in SI units (m³/s). The figure also shows the IMD precipitation grid points (0.5° × 0.5°; red dots) where precipitation values were interpolated from the observations of the rain gauge sites (Rajeevan and Bhate 2009). 35 years (1971–2005) of gridded precipitation data is available for the study. The period of record for daily discharge data for the available 53 sites varied from 1959 to 2011. The maximum record length of the data obtained is 53 years and the minimum being 6 years. Figure 2 shows the record length of the different gauge sites in the Ganga basin. The basin drainage area for the 53 sites is delineated from the SRTM DEM 90 m resolution data and the drainage density computed using ArcGIS. The three basin attributes—drainage area, precipitation and drainage density are used to obtain the regression model.

4 Methodology Used

Let us consider i sites ($i = 1, ..., N$) and each site having n_i years of data, $X_i = \{x_{1_i}, ..., x_{n_i}\}$. The mean and standard deviation for the population are denoted by μ_i and σ_i, respectively, and for the sample are denoted by $\bar{x}_i$ and s_i respectively. Let θ denote any flood statistics like the 100 year return period. The population statistics and its estimator are denoted by, $\theta = [\theta_1, \theta_2, ..., \theta_N]$ and $\hat{\theta} = [\hat{\theta}_1, \hat{\theta}_2, ..., \hat{\theta}_N]$, respectively.

Assuming $\hat{\theta}$ to be an unbiased estimator of θ

$$E\left[\hat{\theta}\right] = \theta \tag{1}$$

The covariance matrix of $\hat{\theta}$ is

$$\sum(\hat{\theta}) = E\left[(\hat{\theta} - \theta)(\hat{\theta} - \theta)^T\right] \tag{2}$$

The ij component of the covariance matrix is given by

$$\sum_{ij}(\hat{\theta}) = \frac{\sigma_i^2}{n_i}\left[1 + K_p^2\left(\frac{\kappa - 1}{2}\right)\right] \text{ for } i = j$$

$$= \frac{\rho_{ij}m_{ij}\sigma_i\sigma_j}{n_i n_j}\left[1 + \rho_{ij}K_p^2\left(\frac{\kappa - 1}{4}\right)\right] \text{ for } i \neq j \tag{3}$$

where, K_p is the frequency factor, κ is the mode of the kurtosis of annual peak flows obtained for an individual site, n_i is the number of years of data for the ith site, σ_i is the standard deviation for the ith site, m_{ij} is the number of concurrent observations between the sites i and j, and ρ_{ij} is the cross-correlation in peak discharge at these sites.

The cross-correlation is usually obtained by fitting a suitable model between the computed correlation between the peak discharge and the geographic distance between the gauged sites. The model used in this study is

$$\rho_{ij} = a_1^{\frac{d_{ij}}{a_2 d_{ij} + 1}} \tag{4}$$

In this study, the stream network distance between the connected gauges sites is used to modify the ρ_{ij} value in Eq. (4).

An example of a linear regression model for estimating a streamflow characteristic (θ) using basin characteristics is

$$\theta_i = \alpha + \beta \ln(A_i) + \varepsilon_i \tag{5}$$

where $i = (1, 2, 3, ..., N)$ is the index for gauge site, $\theta_i =$ stream flow characteristic (dependent variable), α and $\beta =$ regression parameters, $A_i =$ basin attributes and $\varepsilon_i =$ normal and independently distributed model error with mean zero and variance γ^2.

In matrix notation Eq. (5) can be written as

$$\theta = \Xi\beta + \varepsilon \tag{6}$$

where, $\Xi = \begin{pmatrix} 1 & \ln(A_1) \\ \vdots & \vdots \\ 1 & \ln(A_N) \end{pmatrix}$, $\beta = \begin{bmatrix} \alpha \\ \beta \end{bmatrix}$, and $\varepsilon = \begin{bmatrix} \varepsilon_1 \\ \vdots \\ \varepsilon_N \end{bmatrix}$.

So using Eqs. (1) and (6), we have

$$E\left[\hat{\theta}\right] = \Xi\beta \tag{7}$$

with the covariance of $\hat{\theta}$ about $\Xi\beta$ as

$$E\left[\left(\hat{\theta} - \Xi\beta\right)\left(\hat{\theta} - \Xi\beta\right)^T\right] = \Lambda(\gamma^2) = \gamma^2 I_N + \sum\left(\hat{\theta}\right) \tag{8}$$

The model parameter $\hat{\beta}$ is estimated as follows

$$\hat{\beta} = \left(\Xi^T \sum(\hat{\theta})\, \Xi\right)^{-1} \Xi^T \sum(\hat{\theta})^{-1}\hat{\theta} \tag{9}$$

The leave-one-out cross-validation (LOOCV) method is utilized to check the performance of RFFA in estimating flood quantiles at ungauged sites. One site is left out at a time (test site), and its T-year regional flood quantile is estimated by the regional equation obtained using the remaining sites (training sites). This step is repeated until all the sites are left out once, and their regional T-year flood quantiles are estimated. The performances of the test sites are measured by absolute relative root mean square error (RRMSE) defined as

$$RRMSE = \sqrt{\frac{1}{N} \sum_{i=1}^{N} \left(\frac{\widehat{Q}_i^a - \widehat{Q}_i}{\widehat{Q}_i^a}\right)^2} \times 100 \tag{10}$$

where $\widehat{Q}_i$ and $\widehat{Q}_i^a$ are the estimated and at-site T-year flood quantiles at the site i, respectively. Smaller values of RRMSE indicate better performance. The at-site flood quantiles are estimated by fitting log-Pearson type 3 (LP3) distribution, as recommended in Bulletin 17B (IACWD 1982).

5 Results

SRTM (Shuttle Radar Topographic Mission) dataset at a 90-m spatial resolution is used to delineate the sub-basins and generate the stream networks. The location of 53 CWC gauge sites along with the delineated drainage area of the basin for Kanpur CWC site is shown in Fig. 3. The basin area is delineated using the Arc Hydro tool (Maidment 2002).

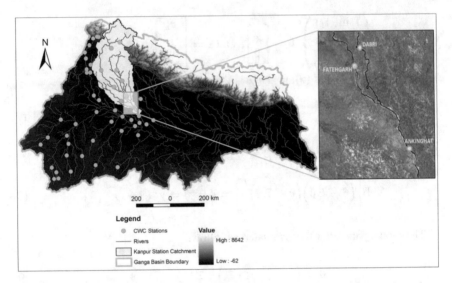

Fig. 3 Elevation profile of the study area and the location of the sub basins outlet. The inset shows three gauging sites, two pairs are connected by stream and one pair is not connected by stream

For the 53 available sites in the Ganga basin, the correlation among flow data for the concurrent years between gauging sites is calculated. The latitude and the longitude of the sites were used to find the geographic distance among each other.

A scatter plot between the geographic distance and the calculated correlation among the sites is shown in Fig. 4. A bin size of 100 is used to find the mean correlation and distance for the bin. These mean values were then used to fit the model in Eq. (4).

The fitted model using geographic distance is given by Eq. (11). The coefficient of determination (R^2) for this model is 0.82.

$$\rho_{ij} = 0.9932^{\frac{d_{ij}}{0.0004d_{ij}+1}} \tag{11}$$

An important feature among the gaging sites is network connectivity. When a river moves from upstream to downstream, all the sites that lie in its path were considered connected. Hence the distance along the path of the stream (instead of geographic distance) is more meaningful for flood estimation. The motivation behind the use of stream distance is discussed in the following paragraph with an example.

In a basin, there may be some sites that are very close to each other but may not be connected by stream network. While, there may be some sites which lie in the connected stream path, but are far apart. Figure 3 shows three sites, where Ankinghat and Dabri, and Ankinghat and Fatehgarh are connected by a stream. But sites Dabri and Fatehgarh are not connected. Correlation between annual peak flow of site Dabri and Ankinghat is 0.55, and for the sites, Fatehgarh and Ankinghat are 0.73. The correlation between the annual peak flow of Dabri and Fatehgarh is 0.38, as shown

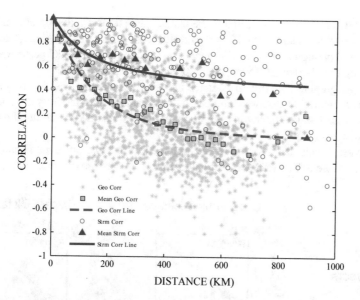

Fig. 4 Correlation in annual peak discharge among gauging sites in the study area and their geographical and stream distance. The abbreviation Geo Corr represents correlation w.r.t geographic distance while Strm Corr denotes correlation w.r.t stream distance

in Table 1. It is evident that the connected sites have high correlation even when they are far apart than the sites which are not connected but close to one another. This information is used to modify the relationship between correlation and the distance by incorporating the stream connectivity.

The correlation coefficient with the stream distances is plotted, as shown in Fig. 4. A bin size of 25 is selected for stream distance because the pairs of connected sites are less.

For the stream distance, the fitted model is given in Eq. (12).

$$\rho_{ij} = 0.9954^{\frac{d_{ij}}{0.0045d_{ij}+1}} \tag{12}$$

The model in Eq. (12) has a coefficient of determination of 0.79. The fitted models (Eqs. 11 and 12) are used to compare the performance of geographic distance and stream distance in the estimation of flood quantiles.

Table 1 Example showing the importance of stream connected information

Sites	Connectivity	Stream dist. (km)	Geographic dist. (km)	Correlation
Ankinghat/Dabri	Connected	103.99	76.84	0.552
Ankinghat/Fatehgarh	Connected	85.57	69.85	0.737
Dabri/Fatehgarh	Not connected	Nil	11.62	0.382

The GLS regression is applied to the Ganga basin to obtain the regression model between the flood quantiles and three basin attributes, namely, drainage area, drainage density and precipitation. LOOCV method is used to test the model performance in predicting the 50-year, 100-year, and 500-year return period flood.

The geographic and stream distance between the sites is used to compute the covariance matrix, as discussed in Sect. 4. The scatter plot of the 50-year, 100-year, and 500-year flood estimated by the two approaches, and the at-site flood estimate are shown in Fig. 5.

RRMSE is calculated for the results obtained for the two different distances and shown in Table 2. The RRMSE using geographic and stream distance for a 50-year

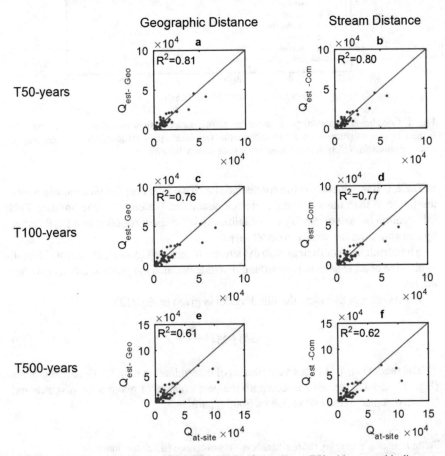

Fig. 5 Comparison of the at-site and the estimated flood quantiles. **a** T50 with geographic distance, **b** T50 with stream distance, **c** T100 with geographic distance, **d** T100 with stream distance, **e** T500 with geographic distance and **f** T500 with stream distance. The abbreviation $Q_{at\text{-}site}$ represents at-site flood quantile, $Q_{est\text{-}Geo}$ represents flood quantile estimates using geographic distance, and $Q_{ets\text{-}Str}$ represents the flood quantile estimates using stream distance for the connected sites

Table 2 RRMSE of geographic and stream distance based models in LOOCV for estimating estimating 50, 100 and 500 year return period flood quantiles

Sl. No.	Distance	T50	T100	T500
1	Geographic	94.93	93.57	94.13
2	Stream	94.18	90.91	92.95

return period is obtained to be 94.93 and 94.18, respectively. The RRMSE for 100-year and 500-year return period floods are 93.57 and 94.13, respectively, using the geographic distance, while 90.91 and 92.95, respectively, using the stream distance, respectively. For all the three return periods, the model based on stream distance performed better. The difference is highest for the 100 year return period flood estimate. This improvement in the performance measure suggests the importance of the stream distance in the estimation of the flood quantiles.

6 Conclusion

The stream distance is used in this study to modify the covariance matrix used in GLS regression, and the flood quantile estimates are compared to the results obtained using the geographic distance for 50-years, 100-years, and 500-years return period flood. Improvement in the flood quantile estimates for the stream distance based method is seen for all the return period floods. The results showed the importance of the stream network for flood estimation.

Uncertainty in the prediction of the flood estimates was not considered in this study. Bayesian statistics can be applied for quantifying the prediction uncertainties. Further, in the present study, the flood data were assumed to be stationary, and the developed models are valid only under the present climatic conditions. Future work should explore ways to refine these equations for use in climate change scenarios.

One of the drawbacks of this study is the assumption that all the available sites in the study region were homogenous. Methods are available to identify and check the homogeneity of a region in terms of flood statistics. Future work should identify and test the homogeneity of a region prior to conducting the regional analysis.

References

Burn DH (1997) Catchment similarity for regional flood frequency analysis using seasonality measures. J Hydrol 202(1–4):212–230

Dalrymple T (1960) Flood-frequency analyses, manual of hydrology: part 3 (No. 1543-A) USGPO

Das S, Cunnane C (2012) Performance of flood frequency pooling analysis in a low CV context. Hydrol Sci J 57(3):433–444

Davies DL, Bouldin DW (1979) A cluster separation measure. IEEE Trans Pattern Anal Mach Intell 2:224–227

Dunn JC (1973) A fuzzy relative of the ISODATA process and its use in detecting compact well-separated clusters. J Cybern 3:32–57

Griffis VW, Stedinger JR (2007) The use of GLS regression in regional hydrologic analyses. J Hydrol 344(1–2):82–95

Grover PL, Burn DH, Cunderlik JM (2002) A comparison of index flood estimation procedures for ungauged catchments. Can J Civ Eng 29(5):734–741

Hosking JRM, Wallis JR (1997) Regional frequency analysis: an approach based on L-moments. Cambridge University Press, Cambridge; New York

Ilorme F, Griffis VW (2013) A novel procedure for delineation of hydrologically homogeneous regions and the classification of ungauged sites for design flood estimation. J Hydrol 492:151–162

Inter-Agency Committee on Water Data (IACWD) (1982) Guidelines for determining flood flow frequency—Bulletin 17B. Hydrology Subcommittee, Washington, DC

Kroll CN, Stedinger JR (1998) Regional hydrologic analysis: ordinary and generalized least squares revisited. Water Resour Res 34(1):121–128

Maidment DR (2002) Arc hydro: GIS for water resources, 2nd edn. ESRI, New York

Markus M, Knapp HV, Tasker GD (2003) Entropy and generalized least square methods in assessment of the regional value of streamgages. J Hydrol 283(1–4):107–121

McIntyre N, Lee H, Wheater H, Young A, Wagener T (2005) Ensemble predictions of runoff in ungauged catchments. Water Resour Res 41(12)

Medhi H, Tripathi S (2015) On identifying relationships between the flood scaling exponent and basin attributes. Chaos: An Interdiscip J Nonlinear Sci 25(7):075405

Paul D (2017) Research on heavy metal pollution of river Ganga: a review. Ann Agrar Sci 15(2):278–286

Rajeevan M, Bhate J (2009) A high resolution daily gridded rainfall dataset (1971–2005) for mesoscale meteorological studies. Curr Sci, pp 558-562

Rao AR, Srinivas VV (2008) Regionalization of watersheds: an approach based on cluster analysis. Water science and technology library. Springer, Dordrecht, Netherlands

Robson A, Reed D (1999) Flood estimation handbook, volume 3: statistical procedures for flood frequency estimation. Institute of Hydrology, Wallingford

Rousseeuw PJ (1987) Silhouettes: a graphical aid to the interpretation and validation of cluster analysis. J Comput Appl Math 20:53–65

Sokal RR, Rohlf FJ (1962) The comparison of dendograms by objective methods. Taxon 11(2):33–40

Srinivas VV, Tripathi S, Rao AR, Govindaraju RS (2008) Regional flood frequency analysis by combining self-organizing feature map and fuzzy clustering. J Hydrol 348(1–2):148–166

Stedinger JR, Griffis VW (2008) Flood frequency analysis in the United States: time to update. J Hydrol Eng 13:199–204

Stedinger JR, Tasker GD (1985) Regional hydrologic analysis: 1. Ordinary, weighted, and generalized least squares compared. Water Resour Res 21(9):1421–1432

Tasker GD (1980) Hydrologic regression with weighted least squares. Water Resour Res 16(6):1107–1113

Walker JF, Krug WR (2003) Flood-frequency characteristics of Wisconsin Streams. U.S. Geological Survey, Water Resources Investigations Report 03-4250

Plotting Positions for the Generalized Extreme Value Distribution: A Critique

Sonali Swetapadma and C. S. P. Ojha

Abstract Plotting ordered ranked data is a standard graphical technique applied in many hydrologic and water resource fields that utilizes the plotting position formula. This graphical approach mainly deals with assigning a probability of occurrence to extreme events like floods, rainfall, etc. The concept of unbiasedness has inspired several researchers to develop various plotting positions, including the shape parameter of a probability distribution. The derivation of Generalized Extreme Value (GEV) distribution form the statistical theory of extreme random variables justifies its vast application in the flood frequency studies. Considering the significance of GEV distribution in the field of hydrology, in the present study, existing plotting positions for GEV are compared in terms of root mean square error (RMSE) and relative bias (RBIAS) between theoretical reduced variates of GEV distribution and those obtained from these formulas. The Sum of square error between the top three reduced variates is used as a statistical indicator to evaluate their performance in the estimation of higher quantiles. The study is carried out varying the sample size (n) over the most frequently occurred range of shape parameter (k) in hydrological applications, i.e., $-0.3 \leq k \leq 1.5$. No single plotting position is found to perform better over the entire range of shape parameter. However, the formula considering the skewness coefficient of the sample satisfies comparatively a broad range of shape parameter. The necessity of modifying these plotting positions has been discussed to address the effect of sample size and skewness coefficient.

Keywords Plotting position formula · GEV distribution · Flood frequency · Shape parameter · Coefficient of skewness

S. Swetapadma (✉) · C. S. P. Ojha
Department of Civil Engineering, IIT Roorkee, Roorkee, Uttarakhand 247667, India
e-mail: sonaliswetapadma1992@gmail.com

C. S. P. Ojha
e-mail: cspojha@gmail.com

© The Author(s), under exclusive license to Springer Nature Switzerland AG 2021 193
M. S. Chauhan and C. S. P. Ojha (eds.), *The Ganga River Basin: A Hydrometeorological Approach*, Society of Earth Scientists Series,
https://doi.org/10.1007/978-3-030-60869-9_13

1 Introduction

Plotting position plays a vital role in many hydrometeorological applications, mainly dealing with probability distribution models and regression analysis. It helps in the comparison of theoretical distributions fitted to the observed rainfall data or discharge data series and also provides a practical graphical or visual approach in analyzing the outputs from any regression analysis. Plotting the ordered data is a common graphical practice for frequency analysis in which observed data are sorted in descending order of their magnitude and traced on a probability paper based on their probability of occurrence. The cumulative probability associated with each data is predicted by using a plotting position formula and marked on a probability paper, where a straight line is fitted to these data to estimate various return period peak flow values. This graphical analysis provides a visual aid to compare the fitness of several probability distributions to a hydrologic data series, thereby deciding the best fit distribution for a particular gauging site. Even though the analytical approaches are more efficient in estimating the parameters of frequency distributions, the graphical method is still broadly used in many hydrologic applications. Also, the results of an analytical approach are usually supported graphically by plotting the sample on a probability paper.

The term plotting position was first introduced by Foster (1934), which mainly refers to the exceedance probability of an event. Since the first formula given by Hazen (1914), many formulas along with their computational methods were proposed by various researchers such as, (Weibull 1939; Beard 1943; Gumbel 1958; Blom 1958; Langbein 1960; Gringorten 1963; Benson 1962; Cunnane 1978; Xuewu et al. 1984; Arnell et al. 1986; Nguyen et al. 1989; Guo 1990a, b; Nguyen and In-na 1992); which include both empirical and numerical approaches. A comprehensive discussion on this topic was given by Cunnane (1978) and Rao and Hamed (2000). Blom (1958) derived a generalized equation for plotting position as $P = (i - \alpha)/(n - \alpha - \beta + 1)$, and Gringorten (1963) proposed the value of α to be 0.44 for extreme value type I distribution. Cunnane (1978) observed that the well-known (Weibull 1939) method was biased for non-uniform distributions, mainly with positive values of the skewness coefficient. So, he derived an unbiased plotting position as the expected values of the order statistics $E[y_i]$ for different probability distributions. Nguyen et al. (1989) developed an unbiased plotting position formula for Pearson type III distribution considering the effect of shape parameter or the corresponding skewness coefficient. Nguyen and In-na (1992) suggested a formula for Pearson type III distribution based on both historical and systematic flood records. A distribution-free plotting position given by $(i - 0.326)/(n + 0.348)$ from median order statistics of the data set was derived by Yu and Huang (2001). De (2000) proposed an unbiased formula for EV I or Gumbel distribution using exact plotting position from Gumbel order statistics. Makkonen (2006) carried out a research on plotting position in extreme value study and found that the Weibull formula is the only correct method as it predicts much shorter return periods of extreme events than other formulas. Yahaya et al. (2012) compared seventeen plotting position formulae for Gumbel distribution and found

Table 1 Commonly used plotting position formulas

Method	Probability of exceedance P_i
Weibull	$i/n + 1$
Hazen	$(i - 0.5)/n$
Blom	$(i - 0.44)/(n + 0.12)$
Gringorten	$(i - 3/8)/(n + 1/4)$

the most appropriate one for different sample lengths. Some of the frequently applied empirical plotting position formulae are listed in Table 1, where i is the rank assigned to each observation, and n is the total number of observations. Weibull is the most commonly used plotting position formula, while Blom and Gringorten have been recommended for Normal and Gumbel distribution, respectively, Mehdi and Mehdi (2011). The detailed literature survey in this area suggests that the selection of a suitable formula depends upon the primary objective of the study and the corresponding probability distribution of the data series. However, no such convincing arguments have been put forward to choose a particular plotting position formula under certain circumstances.

The concept of unbiased plotting position has influenced many hydrologists and statisticians to develop plotting positions considering the shape parameter or skewness coefficient of the underlying probability distribution. Arnell et al. (1986) proposed that the skewness coefficient should be included in the plotting position for GEV distribution. Based on this concept, (In-na and Nguyen 1989) derived a formula for GEV distribution from the descending series. Goel and De (1993) obtained an unbiased plotting position that contains the skewness coefficient based on probability-weighted moments. Similarly, (Kim et al. 2012) developed an unbiased plotting position for GEV distribution, which was found to perform better within the range of shape parameter ± 0.2. Some of such existing plotting position formulas for GEV distribution are listed in Table 2, where i is the rank assigned to each observation, n is the total number of observations, and Υ is the skewness coefficient.

Table 2 Existing plotting position formulas for GEV distribution

Method	Probability of exceedance P_i
Gringorten	$P_i = (i - 0.44)/(n + 0.12)$
Cunnane	$P_i = (i - 0.4)/(n + 0.2)$
In-na and Nguyen	$P_i = (n - i + 0.05\Upsilon + 0.65)/(n - 0.08\Upsilon + 0.38); \ -3.8 \le \Upsilon \le 3.54$
Goel and De	$P_i = (i - 0.02\Upsilon - 0.32)/(n - 0.04\Upsilon + 0.36);$ $-0.3 \le k \le 1.5$
Kim et al.	$P_i = (i - 0.32)/(n + 0.0149\Upsilon^2 - 0.1364\Upsilon + 0.3225);$ $-0.2 \le k \le 0.2$

In the present study, the primary aim is to compare existing plotting position formulae for GEV distribution and to propose the best one to be applied in frequency analysis studies. The accuracy of these formulas is assessed by comparing the error between theoretical reduced variates (TRV) and those obtained from the formula for different values of shape parameter (k) and sample size (n). A few statistical parameters such as root mean square error, relative bias, and sum of square error in the top three reduced variates are used to compare the plotting positions mentioned in Table 2. Based on the detailed comparison, the need for modification of the existing formulas has been discussed in detail.

2 Theoretical Background

2.1 Generalized Extreme Value Distribution (GEV)

Generalized extreme value is a type of continuous probability distribution derived from the extreme value theory. The cumulative distribution function (CDF) of GEV distribution introduced by Jenkinson (1955) is given below.

$$F(x) = \exp\left[-\left\{1 - \frac{k(x - u)}{\alpha}\right\}^{\frac{1}{k}}\right], k \neq 0 \tag{1}$$

$$F(x) = \exp\left[-\exp\left\{-\frac{(x - u)}{\alpha}\right\}\right], k = 0 \tag{2}$$

where u, α, and k are the location, scale, and shape parameter of GEV distribution, respectively. The domain of the random variable x is given as,

$$1 - \frac{k(x - u)}{\alpha} > 0, \text{ for } k \neq 0 \tag{3}$$

$$-\infty < x < \infty, \text{ for } k = 0 \tag{4}$$

The GEV distribution converges to Gumbel, Frechet, and Weibull distribution for zero, positive and negative values of shape parameter (k). This shape parameter (k) is related to the coefficient of skewness (Υ), as given below (NERC 1975).

$$\gamma = \mu_3/\mu_2^{\frac{3}{2}} \tag{5}$$

where μ_2 and μ_3 are the second and third central moments, respectively.

$$\mu_2 = \frac{\alpha^2}{k^2}\left[\Gamma(1 + 2k) - \Gamma^2(1 + k)\right] \tag{6}$$

Fig. 1 Variation of skewness coefficient with respect to the shape parameter of GEV distribution

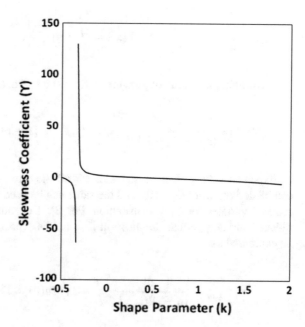

$$\mu_3 = \frac{\alpha^3}{k^3}\left[-\Gamma(1+3k) + 3\Gamma(1+k)\Gamma(1+2k) - 2\Gamma^3(1+k)\right] \tag{7}$$

where $\Gamma(.)$ represents the incomplete gamma function. The relationship between shape parameter and the skewness coefficient of GEV distribution is shown below in Fig. 1.

2.2 Theoretical Reduced Variates of GEV Distribution

The probability density function $g(x_r)$ of rth order statistics for a sorted random sample of 'n' observations such as $x_1 \leq x_2 \leq x_3 \leq x_4 \ldots \leq x_{n-1} \leq x_n$ is given by (Arnell et al. 1986)

$$g(x_r) = \frac{n!}{(n-r)!(r-1)!}F(x_r)^{r-1}\{1 - F(x_r)\}^{n-r}f(x_r) \tag{8}$$

where $f(x_r)$ and $F(x_r)$ are the probability density function (PDF) and cumulative distribution function (CDF) of the random sample, respectively. The expected value of x_r is given as,

$$E[x_r] = \int\limits_{-\infty}^{\infty} x_r g(x_r) dx \tag{9}$$

Substituting the value of g (x_r) from Eq. (8) into Eq. (9),

$$E[x_r] = \frac{n!}{(n-r)!(r-1)!} \int\limits_{-\infty}^{\infty} x_r * F(x_r)^{r-1} * \{1 - F(x_r)\}^{n-r} * f(x_r) dx \tag{10}$$

Expected values of order statistics for all the three cases of GEV distribution can be derived from Eq. (10), and the same can be used to calculate the theoretical reduced variates for GEV distribution. For EV I or Gumbel distribution (k = 0), reduced variate y_1 equals to $-\ln\{-\ln(F)\}$ and its expected value from Eq. (10) can be expressed as,

$$E[y_{1(r)}] = \frac{n!}{(n-r)!(r-1)!} \int\limits_{0}^{1} -\ln\{-\ln(F)\} * F^{r-1} * (1-F)^{n-r} dF \tag{11}$$

For EV II distribution (k < 0), reduced variate $y_2 = \{-\ln(F)\}^k$ and the expression for the expected value of order statistics are given below.

$$E[y_{2(r)}] = \frac{n!}{(n-r)!(r-1)!} \int\limits_{0}^{1} \{-\ln(F)\}^k * F^{r-1} * (1-F)^{n-r} dF \tag{12}$$

Similarly, for EV III distribution with k > 0 and reduced variate $y_3 = -\{-\ln(F)\}^k$,

$$E[y_{3(r)}] = -\frac{n!}{(n-r)!(r-1)!} \int\limits_{0}^{1} \{-\ln(F)\}^k * F^{r-1} * (1-F)^{n-r} dF \tag{13}$$

These expected values of order statistics in terms of probability-weighted moments (PWM) are given below.

$$E[y_{1(r)}] \approx \frac{n!}{(n-r)!(r-1)!} \sum_{s=0}^{n-r} \binom{n-r}{s} (-1)^s [C_E + \ln(r+s)](r+s)^{-1} \tag{14}$$

$$E[y_{2(r)}] \approx \frac{n!}{(n-r)!(r-1)!} \Gamma(1+k) \sum_{s=0}^{n-r} \binom{n-r}{s} (-1)^s (r+s)^{-(1+k)} \tag{15}$$

$$E[y_{3(r)}] \approx -\frac{n!}{(n-r)!(r-1)!}\Gamma(1+k)\sum_{s=0}^{n-r}\binom{n-r}{s}(-1)^s(r+s)^{-(1+k)} \quad (16)$$

C_E represents Euler's constant i.e. 0.5772. For the largest sample size i.e. $r = n$,

$$E[y_{1(n)}] = C_E + \ln(n) \quad (17)$$

$$E[y_{2(n)}] = n^{-k}\Gamma(1+k) \quad (18)$$

$$E[y_{3(n)}] = -n^{-k}\Gamma(1+k) \quad (19)$$

The expected values of order statistics obtained from the integral form (Eqs. 11–13) and summation form (i.e., Eqs. 14–16) are compared, as shown below in Fig. 2. It's observed that for $n < 35$, both the expressions have similar behavior, while for

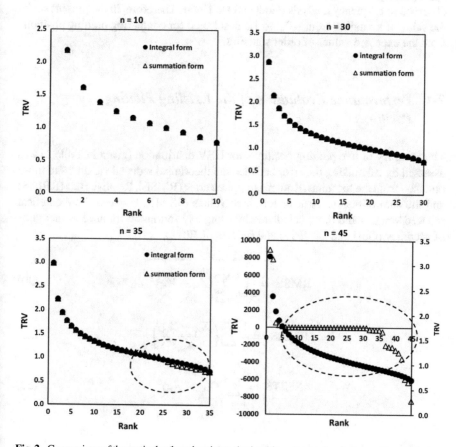

Fig. 2 Comparison of theoretical reduced variates obtained from integral and summation expression

n ≥ 35, the expressions obtained by combining the PWM concept and comparative approach are inaccurate due to rounding off errors in the middle part of the series.

2.3 Selection of a Range of Shape Parameter (K)

Estimating the shape parameter of GEV distribution is analytically more complicated than calculating the skewness coefficient of the sample. Therefore, the latter is generally used in various modified formulas of GEV distribution, as listed in Table 2. As evident from Eq. (7), for 'k' $= -1/3$, the skewness coefficient (Υ) becomes infinitely large. The coefficient of skewness shows systematic variation before $k = -1/3$, i.e., within a range of -0.49 to -0.34, as shown in Fig. 1. Also, $k \leq -1/2$, the skewness coefficient cannot be defined due to negative gamma functions. From the graph, it can be observed that from $k = -0.3$ to $k = 1.5$, the relationship between Υ and k is appropriately defined, and this range also includes most of the hydrologic conditions observed in frequency analysis (Goel and De 1993). Therefore, in the present study, the value of k ranging from -0.3 to 1.5 is selected for comparing plotting positions from the expected values of order statistics.

2.4 Performance Evaluation of the Existing Plotting Positions

The accuracy of five plotting positions for GEV distribution (given in Table 2) was assessed by calculating the error between the theoretical reduced variates and those obtained from the formulas. Root mean square error (RMSE), Relative Bias (RBIAS) and the sum of square error of top three values (SSET3) between the theoretical reduced variates and those calculated by using any formula were used as measures of accuracy (Goel and De 1993) and (Kim et al. 2012).

$$\text{RMSE} = \sqrt{\frac{1}{n} \sum_{i=1}^{n} (X_{si} - X_{oi})^2} \tag{20}$$

$$\text{RBIAS} = \frac{1}{n} \sum_{i=1}^{n} \left(\frac{X_{si} - X_{oi}}{X_{oi}} \right) \tag{21}$$

$$\text{SSET3} = \sum_{i=n-2}^{n} (X_{oi} - X_{si})^2 \tag{22}$$

where 'n' is the sample size, and X_{oi} and X_{si} represent the theoretical reduced variate of GEV distribution and reduced variate of a plotting position for i^{th} variable, respectively. Theoretical reduced variates were calculated from the expressions described in the previous section. In the present study, the shape parameter (k) values ranging from −0.3 to 0.15 were selected, and each 'k' sample length (n) is varied between 10 and 100.

3 Results and Discussion

Since it is not easy to get discharge data series from any particular river basin satisfying GEV as the best fit distribution along with covering the entire range of shape parameter or skewness coefficient, the formulas given in Table 2 were compared by applying it to synthetically generated data series of GEV distribution with varying sample length and shape parameter. Root mean square error between the theoretical reduced variates and those obtained from the plotting positions were calculated for a varying range of sample length and shape parameter and was plotted in Fig. 3. The variation of RMSE according to sample length for k = −0.3, −0.05, and 0.30 are shown here as examples. A similar pattern was observed for all other values of the shape parameter also. Each plotting position performed better as the sample size is increased at any value of shape parameter, i.e., RMSE decreased with an increase in sample lengths, and, also, the RMSEs were found to be stable within ± 10% at higher values of sample lengths. Since the formula by In-na and Nguyen was developed for skewness coefficient (Υ) ranging from −3.8 to 3.54, RMSE of the same couldn't be obtained at k = −0.3, i.e., for $\Upsilon = 13.484$. For a particular sample length (n), the calculated errors of these formulas were convexly distributed with the shape parameter (k).

No particular plotting position was found to perform better over the entire range of shape parameter considered in the present study. For $-0.25 < k \leq -0.3$, the formula proposed by Goel and De possessed the minimum error. Similarly, $-0.14 \leq k \leq -0.25$ and $-0.01 \leq k \leq -0.13$, the plotting position developed by Gringorten and In-na and Nguyen had the least RMSE, respectively. The best performing formulas were independent of sample length for the entire range of negative shape parameters, i.e., from −0.01 to −0.3. However, for the positive shape parameters, i.e., 0.01–0.3, different best plotting positions were obtained according to sample length (n). When k lied between 0.01 and 0.04, for n ≤ 30, the formula derived by In-na and Nguyen was the best, and for n > 30, the Gringorten formula had better accuracy of estimation. Likewise, within the range of 0.05–0.20, plotting positions by Kim et al. and Cunnane showed the least RMSE values for n ≤ 20 and n > 40, respectively. At the in-between sample sizes of 20 and 40, both the formulas behaved in the same manner. The RMSE of all these plotting positions at some selected values of shape parameter is shown in Fig. 4 for an illustration purpose.

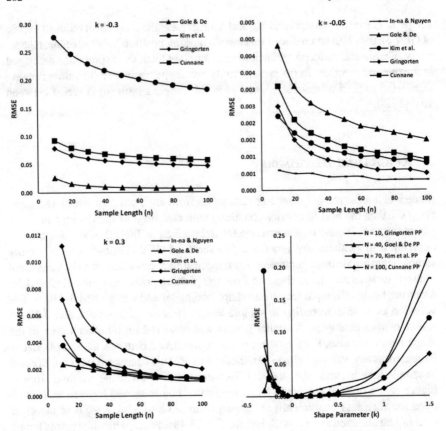

Fig. 3 Variation of RMSE of plotting positions with respect to sample length (n) and shape parameter (k)

All the plotting positions behaved differently over a different range of shape parameter (k). The RBIAS of all the formulas gradually decreased with an increase in sample size (n) for any value of k. However, for a particular sample length, the RBIAS values were concavely distributed with respect to shape parameter or the corresponding skewness coefficient. Even though Kim et al. proposed the formula for $-0.2 \leq k \leq 0.2$, it had the minimum relative bias value for -0.2 to -0.13 and also from -0.05 to 0.12. However, the remaining ranges, i.e., $(-0.13, -0.05)$ and $(0.12, 0.20)$ were satisfied by the formula suggested by In-na and Nguyen, and Gringorten respectively. The plotting position developed by Goel and De had the lowest RBIAS values when the shape parameter lied between -0.30 and -0.26 and 0.25 to 0.38 also. For the remaining values of shape parameter (k), i.e., $(-0.25, -0.21)$, $(0.21, 0.24)$, and $(0.39, 1.5)$, the formula proposed by Gringorten, Cunnane, and In-na and Nguyen showed the minimum RBIAS over the entire range of sample sizes. The relative bias of all the plotting positions for some selected values of sample sizes and shape parameters is shown in Fig. 5 as an example.

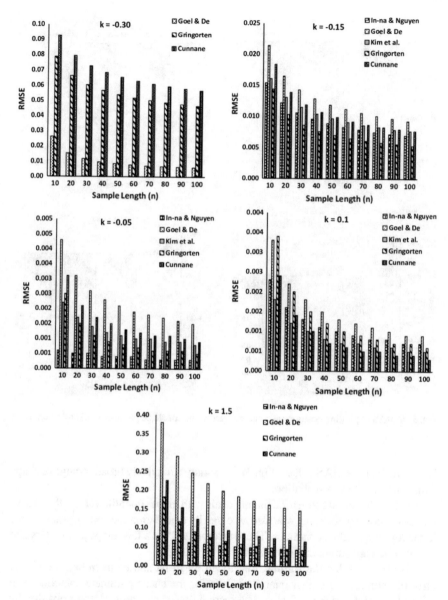

Fig. 4 RMSE of plotting positions at some selected values of shape parameter (k)

The above analysis justified that no single plotting position had the least error for the entire range of shape parameters of GEV distribution. Based on RMSE, the degree of accuracy of the best performing formulas varied according to the sample size for positively valued shape parameters. However, the plotting position developed by Kim et al. and In-na Nguyen can be selected as the top two formulas based upon

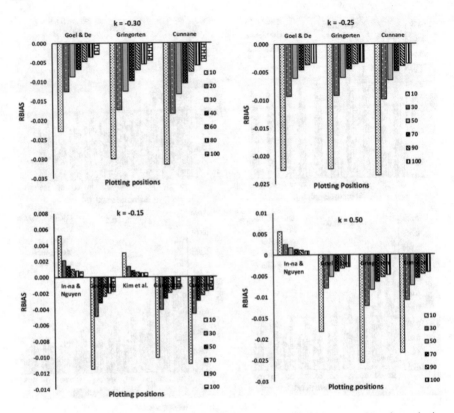

Fig. 5 RBIAS of plotting positions at some selected values of shape parameter (k) and sample size (n)

both RMSE and RBIAS values (Fig. 6) that satisfy a relatively broader range of shape parameters of GEV distribution.

Since design flood estimation mainly deals with higher quantile values, the performance of plotting positions was also assessed by calculating the sum of square error in the top three reduced variates. These error values for a few shape parameters and sample sizes are summarized in Table 3 as an example.

It is observed that the formula developed by In-na and Nguyen had the lowest error of estimates of higher quantile values for the shape parameter ranging from -0.12 to -0.03 and $0.4-1.5$ also. The second best-performing plotting position by (Goel and De 1993) showed the minimum error for $-0.25 \leq k \leq -0.13$ and also, $0.19 \leq k \leq 0.39$. Even though various formulas performed well based on RMSE and RBIAS values, they were not so good to estimate higher quantile values with a better degree of accuracy.

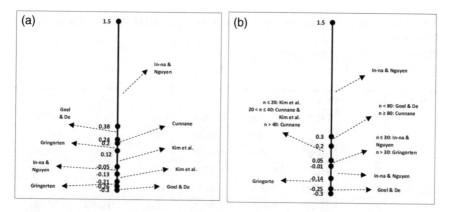

Fig. 6 Performance evaluation of plotting positions over the entire range of shape parameters based on, **a** RBIAS, and **b** RMSE

3.1 Need for Modification of Plotting Positions for GEV Distribution

Since no single formula is found to be perfect for the entire range of shape parameter considered in the study, a suitable formula should be selected depending upon the aim of the analysis (more emphasis on higher quartiles, error criteria to evaluate the degree of accuracy of quantile estimations, etc.), sample size and skewness of a particular sample, which itself is a tedious task. Therefore, the current plotting positions need to be modified to propose plotting positions performing better irrespective of sample length and shape parameter of GEV distribution. Here, the authors have highlighted one of such existing formulas derived by Goel and De (1993) as an example in this regard.

Goel and De (1993) proposed an unbiased plotting position for GEV distribution, as given in Table 2. The formulas were derived for three cases, such as exact plotting positions, plotting position for the largest in a given sample, and synthetically derived reduced variates by using both general and Cunnane form. Based on various error criteria, the formula developed by applying the Cunnane form with exact plotting positions was recommended as the final one. The general form used in that study was,

$$P_m = \frac{m + a}{n - b} \tag{23}$$

i.e. $nP_m - m = a + bP_m$ (24)

The unknown coefficients ('a' and 'b') of the above equation were calculated by regressing $nP_m - m$ over P_m for different values of sample length (n) and shape parameter (k) or skewness coefficient (ϒ). Goel and De (1993) assumed a linear

Table 3 Evaluation of plotting position formulas based on the sum of square error in the top three reduced variates

Shape parameter (k)	Plotting positions	Sample length (n)					
		10	30	50	70	90	100
−0.25	1	NA	NA	NA	NA	NA	NA
	2	0.0227	0.0358	0.0454	0.0533	0.0602	0.0633
	3	0.0172	0.0327	0.0430	0.0512	0.0584	0.0616
	4	0.0207	0.0323	0.0408	0.0479	0.0540	0.0568
	5	0.0311	0.0513	0.0655	0.0772	0.0873	0.0920
−0.15	1	0.0024	0.0033	0.0039	0.0043	0.0047	0.0048
	2	0.0042	0.0057	0.0065	0.0072	0.0078	0.0080
	3	0.0026	0.0039	0.0047	0.0052	0.0056	0.0058
	4	0.0016	0.0019	0.0021	0.0023	0.0024	0.0025
	5	0.0030	0.0039	0.0044	0.0049	0.0052	0.0054
−0.1	1	7.88E−05	1.01E−04	1.13E−04	1.21E−04	1.27E−04	1.30E−04
	2	1.10E−03	1.40E−03	1.50E−03	1.60E−03	1.70E−03	1.70E−03
	3	4.48E−04	6.16E−04	6.96E−04	7.50E−04	7.93E−04	8.11E−04
	4	2.94E−04	2.73E−04	2.85E−04	2.97E−04	3.80E−04	3.13E−04
	5	6.46E−04	7.20E−04	7.80E−04	8.26E−04	8.64E−04	8.81E−04
0.15	1	8.40E−05	6.10E−05	5.24E−05	4.75E−05	4.56E−05	4.27E−05
	2	7.85E−05	5.67E−05	4.88E−05	4.41E−05	4.24E−05	3.97E−05
	3	4.76E−05	3.65E−05	3.17E−05	2.88E−05	2.77E−05	2.59E−05
	4	1.01E−04	9.17E−05	8.26E−05	7.63E−05	7.38E−05	6.96E−05
	5	1.93E−05	1.51E−05	1.35E−05	1.25E−05	1.20E−05	1.13E−05
0.5	1	6.45E−05	2.33E−05	1.43E−05	1.03E−05	8.04E−06	7.24E−06
	2	4.76E−04	1.44E−04	8.52E−05	6.04E−05	4.68E−05	4.20E−05
	3	NA	NA	NA	NA	NA	NA
	4	0.0025	9.30E−04	5.69E−04	4.10E−04	3.20E−04	2.89E−04
	5	0.0017	6.12E−04	3.73E−04	2.68E−04	2.09E−04	1.88E−04

Note '1', '2', '3', '4', and '5' represent formula ID of the plotting position formulae given in Table 2

relationship between these coefficients and skewness (Υ) for any given sample size (n) and expressed it as,

$$a = C_1 + C_2 * \Upsilon \tag{25}$$

$$b = C_3 + C_4 * \Upsilon \tag{26}$$

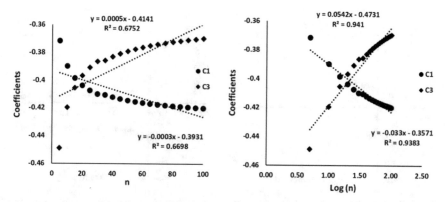

Fig. 7 Variation of the coefficients (C_1 and C_3) given by Goel and De (1993) with sample size (n)

The value of these coefficients was estimated by regressing 'a' on Υ and 'b' on Υ for varying sample length (n). C_2 and C_4 were found to be constant such as 0.02 and 0.04, respectively. However, they assumed the relationship of the other two coefficients, i.e., C_1 and C_3, with sample size (n) to be linear. By using the values of these coefficients from Table 2 of the article (Goel and De 1993), C_1 and C_3 are plotted against sample size (n), as shown below in Fig. 7.

It is evident that the relationship of these coefficients with the logarithm of sample size, i.e., log(n) has a better linear fit with $R^2 \approx 0.94$ as compared to the sample size 'n' ($R^2 \approx 0.67$). Similarly, the plotting position proposed by Kim et al. and In-na Nguyen (Table 2) considered the linear relationship of sample size (n). The use of logarithmic of sample size might reduce the error of estimation, making it independent of sample length also.

Also, the values of 'a' and 'b' listed in Table 1 of the article (Goel and De 1993) had a non-linear variation with respect to skewness coefficient (Υ), which was assumed to be linear in the derivation. These coefficients possessed a linear relationship with the shape parameter instead of the skewness coefficient (Fig. 8). Since the effect of sample length and skewness has not yet been correctly addressed in the existing formulas, there lies a chance of developing a better plotting position for GEV distribution.

4 Summary

No single plotting position was found to perform better over the entire range of frequently occurred shape parameter of GEV distribution. However, the formula derived considering the effect of skewness coefficient satisfied comparatively a broader range than other conventional ones. Therefore, the choice of a suitable formula mainly depends upon the aim of the analysis (relative emphasis on higher quantiles, error criteria to evaluate the degree of accuracy of quantile estimations,

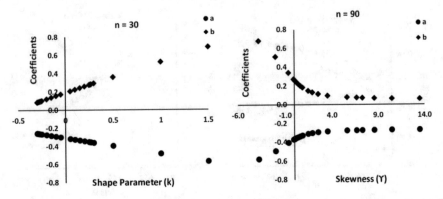

Fig. 8 Variation of coefficients 'a' and 'b' with respect to shape parameter and skewness coefficient (for n = 30 and 90)

etc.), sample size, and skewness of a particular sample. Still, there are some limitations with the existing formulas such as, for 'k' $\approx$ 0 or Υ = 1.139, these don't reduce to Gringorten formula, which has been recommended for Gumbel distribution (Mehdi and Mehdi 2011). Analysis of the derived formula by (Goel and De 1993) suggests that the effect of length and skewness of a sample has not been suitably addressed while developing the formulas. So there is a need for modifying these plotting position formulas for more accurate quantile estimation, which is an integral part of hydrometeorological applications in any study areas.

References

Arnell NW, Beran M, Hosking JRM (1986) Unbiased plotting positions for the general extreme value distribution. J Hydrol 86(1–2):59–69

Beard LR (1943) Statistic analysis in hydrology. Trans Am Soc Civ Eng 108:1110–1160

Benson M (1962) Plotting positions and economics of engineering planning. J Hydraul Div Am Soc Civ Eng 88(HY6):57–71

Blom G (1958) Statistical estimates and transformed beta-variables. Wiley, New York

Cunnane C (1978) Unbiased plotting positions—a review. J Hydrol 37:205–222

De M (2000) A new unbiased plotting position formula for Gumbel distribution. Stoch Environ Res Risk Assess 14:1–7

Foster HA (1934) Duration curves. Trans Am Soc Civ Eng 99:1213–1235

Goel NK, De M (1993) Development of unbiased plotting position formula for general extreme value distributions. Stoch Env Res Risk 7:1–13

Gringorten II (1963) A Plotting rule for extreme probability paper. J Geophys Res 68(3):813–814

Gumbel EJ (1958) Statistics of extremes. Columbia University Press, New York

Guo SL (1990a) Unbiased plotting position formulae for historical floods. J Hydrol 121:45–61

Guo SL (1990b) A discussion on unbiased plotting positions for the general extreme value distribution. J Hydrol 121:33–34

Hazen A (1914) Storage to be provided in impounding reservoirs for municipal water supply. Trans Am Soc Civ Eng 39(9):1943–2044

In-na N, Nguyen VTV (1989) An unbiased plotting position formula for the generalized extreme value distribution. J Hydrol 106:193–209

Jenkinson AF (1955) The frequency distribution of the annual maximum (or minimum) values of meteorological elements. Q J Roy Meteor Soc 81(348):158–171

Kim S, Shin H, Joo K, Heo J (2012) Development of plotting position for the general extreme value distribution. J Hydrol 475:259–269. https://doi.org/10.1016/j.jhydrol.2012.09.055

Langbein WB (1960) Plotting positions in frequency analysis. U. S Geol Surv Water-supply Pap 1543(A):48–51

Makkonen L (2006) Plotting positions in extreme value analysis. J Appl Meteorol Climatol 45:334–340

Mehdi F, Mehdi J (2011) Determiniation of plotting position formula for the Normal, Log- Normal, Pearson (III), Log-Pearson (III) and Gumbel distribution hypotheses using the probability plot correlation coefficient. World Appl Sci J 15(8):1181–1185

NERC (1975) Flood studies report. Nat Environ Res, Council, London

Nguyen V-T-V, In-na N (1992) Plotting formula for pearson type III distribution considering historical information. Environ Monit Assess 23(1–3):137–152

Nguyen V-T-V, In-na N, Bobee B (1989) New plotting-position formula for pearson type III distribution. J Hydraul Eng 115(6):709–730

Rao AR, Hamed KH (2000) Flood frequency analysis. CRC Press

Weibull W (1939) A statistical theory of strength of materials. Ingenious Vetenskaps Akad. Handl 151(15)

Xuewu J, Jing D, Shen HW, Salas JD (1984) Plotting positions for pearson type III distribution. J Hydrol 74(1–2):1–29

Yahaya AS, Nor N, Rohashikin N, Jali M, Ramli NA, Ahmad F (2012) Determination of the probability plotting position for type i extreme value distribution. J Appl Sci 12:1501–1506. https://doi.org/10.3923/jas.2012.1501.1506

Yu G, Huang C (2001) A distribution free plotting position. Stoch Env Res Risk 15:462–476

Application of Machine Learning Techniques for Clustering of Rainfall Time Series Over Ganges River Basin

Vikram Kumar, Manvendra Singh Chauhan, and Shanu Khan

Abstract The active growing population and urbanization have resulted in some changes in climatic variables which may have some influence over the rainfall. Thus, it is of great significance to identify clusters in any basin. These clusters can be based on rainfall amount. For Ganges basin, rainfall records for 20 stations located in Ganges basin are analysed using k mean clustering algorithm. Results in respect of ten clusters are presented and interpreted to identify homogeneous and non-homogeneous regions in a part of Ganga basin covering three states of Uttrakhand, U.P. and Bihar.

Keywords k-mean clustering · Ganges basin · Homogeneous region

1 Introduction

Management of water resources in the Ganges basin is largely governed by the state government, and there has been no basin-wise determination for total surface water planning and allocation. One way to manage the available water in the basin is storage of excess wet-season river flow at the upstream for use during dry season.

One of the significant problems in hydrological research is related to precise forecasting of rainfall and discharge from the rivers, apparently timely advices of extreme weather (rainfall, flood) which canprevent losses of human kind and harms initiated by natural disasters (Kumar and Sen 2018). Excess rainfall during the monsoon season leads to flooding, while low flows during the month of March–April are often

V. Kumar (✉)
Department of Civil Engineering, Gaya College of Engineering, Gaya 823003, India
e-mail: hyvikram@gmail.com

S. Khan
Department of Electrical Engineering, MNIT Jaipur, Jaipur 302017, India
e-mail: shanukhanIN@ieee.org

M. S. Chauhan
Department of Civil Engineering, Holy Mary Institute of Technology and Science, Hyderabad, Telangana 501301, India

© The Author(s), under exclusive license to Springer Nature Switzerland AG 2021
M. S. Chauhan and C. S. P. Ojha (eds.), *The Ganga River Basin: A Hydrometeorological Approach*, Society of Earth Scientists Series,
https://doi.org/10.1007/978-3-030-60869-9_14

insufficient for water supply for different purposes. This results in conflicts among the upstream and downstream users over the basin. Management of available seasonal water to meet the demands is still a major challenge in many mountainous river basins in the upstream (Kumar and Sen 2020). To overcome such issues, machine learning is becoming important (Shanu et al. 2015) in every discipline. To plan a framework for precise precipitation estimation, valuable information can be gained from differing fields, for example, climate information mining (Yang et al. 2007; Kumar and Sen 2018; Kumar et al. 2017), operational hydrology (Li and Lai 2004; Manvendra et al. 2017), ecological machine learning (Hong 2008), and statistical hydrology (Pucheta et al. 2009). The rainfall-runoff modeling and its understanding has been described quantitatively since the nineteenth century. However, it is only in the last decade or so that Artificial Neural Networks (ANNs) and other machine learning algorithms have been applied to the rainfall-runoff modelling including other hydrological studies. During heavy rainfall and failure of the instrument, the data transmission system becomes non-functional, and causes data gap in recording. Furthermore, due to lack of such data gap, it may affect the accuracy in flood forecasting and developing the model. Use of clusters can be very effective in development of ANN or other machine learning algorithms. In fact, input from clustered regions are expected to be more valuable for developing any forecasting algorithm. With this in view, in this chapter, the attempt is to perform cluster analysis to see whether any forecasting attempt to develop ANN or other tools can benefit. Hence, in this study, monthly average rainfall data of 102 years (1900–2002) for 20 different locations are used for clustering.

2 Study Area

In India, Ganga basin covers the state of Madhya Pradesh, Uttar Pradesh, Himachal Pradesh, Rajasthan, West Bengal, Uttarakhand, Bihar, Jharkhand, Chhattisgarh, Haryana, and Union Territory of Delhi spanning a region of 861,452 km^2 which is closely 26% of the aggregate land zone of the nation. The Ganga starts as Bhagirathi from the Gangotri icy masses in the Himalayas at a height of around 7010 m in Uttarkashi region of Uttarakhand and streams for an aggregate length of around 2525 km up before it enters to the Bay of Bengal (Arabian sea) (Fig. 1). In these locations, the unmanaged surface water use and the greater cost of groundwater as compared to surface water creates interlinked management issues. Dry season groundwater siphoning brings down the water table rather consistently all through the watershed, while during rainstorm, stream flow steered through trenches recharges groundwater, which is then siphoned out in the following dry season.

The Ganga basin lies in longitudes E73°2′ to E89°5′ and latitudes N21°6′ to N31°21′ having concentrated length and width of approximately 1543 and 1024 km. In view of the major population, 3% of the Indian populace (448.3 million according to 2001 statistics) and rapid growth of urbanization, the Indian government has put lot of effort in term of human resource and nation capital budgets to develop the

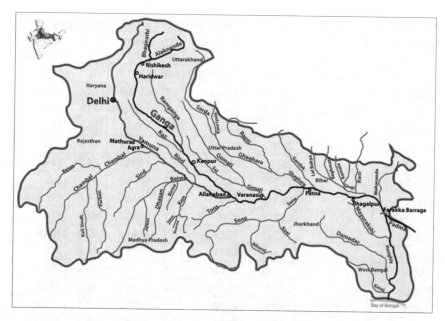

Fig. 1 Simplified map of the Ganga basin

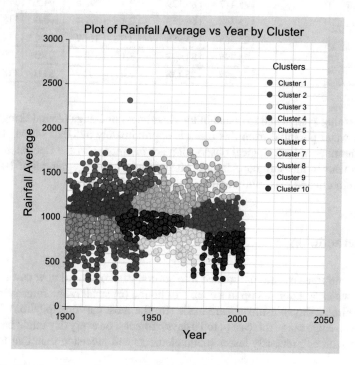

Fig. 2 Classification of annual average rainfall of all the districts

Table 1 Latitude–longitude and statistics of average monthly of 102-year rainfall data (cm)

Location	State	Latitude	Longitude	Min	Max	Mean	Standard deviation
Bhageswar	Uttarahand	29.84	79.769	6.0	322.9	101.7	119.1
Hardiwar		29.946	78.164	5.7	256.6	68.4	91.4
Agra	Uttar Pradesh	27.177	78.008	4.4	265.5	62.6	97.4
Allahabad		25.436	81.846	4.3	338.6	86.5	121.3
Azamgarh		26.074	83.186	3.1	298.4	81.4	113.7
Ballia		25.758	84.149	2.7	309.2	84.5	114.4
Barelliy		28.367	79.43	3.1	283.2	73.4	101.2
Etawa		26.812	79.005	3.0	272.6	66.5	99.2
Jaunpur		25.749	82.699	3.7	303.3	81.4	114.1
Kanpur		26.527	79.83	3.4	295.5	73.4	106.6
Lucknow		26.847	80.946	4.4	291.8	74.9	105.9
Mathura		27.492	77.674	4.6	232.7	55.4	84.4
Mirjapur		25.134	82.564	4.3	333.2	89.1	121.3
Varanasi		25.318	82.974	3.7	319.2	84.7	116.5
Arariya	Bihar	26.111	87.302	2.4	362.7	114.6	130.0
Begusarai		25.417	86.129	2.3	298.8	94.5	113.6
Bhagalpur		25.348	86.982	2.4	299.4	100.0	115.5
Buxar		25.565	83.978	2.9	314.7	85.4	114.4
Gaya		24.795	84.999	3.1	328.9	95.0	120.0
Patna		25.594	85.138	2.0	302.7	91.0	115.0

flood warning system and to maintain the cleanliness of Ganga (Aviral Ganga). The average water potential from this basin is approximately 525,020 MCM. There is a large number of hydrological observation are monitored by the Central and State Governments to measure rainfall data. Mean monthly average rainfall of 102 years with minimum and maximum rainfall is summarized in Table 1.

3 Methodology

The climate forecasts are partitioned into several categories, (a) now casting: forecasting which is extended up to a couple of hours, (b) small range forecasting (1–2 or sometime 3 days): estimates are made (for the most part precipitation) for each 24 h.(c) Average range forecast (4 to 10 days): forecast on every day might be recommended with continuously lesser subtle elements and precision than that for short range figures, (d) extended range forecast (over ten days to a month or year): there is no unique definition for long range forecasting, which may go from a month to month to an occasional forecast (Ganesh and Nikam 2013). The forecasting algorithms may utilise Classification, Prediction, Preprocessing, Clustering, Association and Visualization. As the focus of this work is on Clustering, details of the same are briefly summarized here.

3.1 Clustering

At present, clustering by K-means is one of the widely used clustering techniques. Initially, one can select k objects as the initial centroid of k class. This unsupervised learning algorithm utilizes the evaluation of

$$\sum_{j-1}^{k} \sum_{i-1}^{x} \left\| x(i)^{(j)} - c(j) \right\|^2$$

where, $\left\| x(i)^{(j)} - c(j) \right\|$ is the distance between the rainfall data (x_i) and the formed cluster center (cj), from their corresponding cluster center. For details of methodology, one can refer to Hartigan (1975). Classifying cluster as homogeneous relies on the algorithm in which it calculates the distance between the neighbor station data into different clusters. To measure the homogeneity having similar pattern follows the squared Euclidean distance approach.

The computational statistics requires the sample variance of coefficient of variation which is obtained using relation

$$V_1 = \frac{\sum_i^p ni \left(L_{cv}^i - L_{cv} \right)^2}{\sum_{i=1}^p n_i} \tag{1}$$

where n_i is the number of rainfall data points in the feature vector (rainfall gauge station) i among the available p vectors, L^i_{CV} is the coefficient of variation computed for feature vector i, and L_{CV} is the average of coefficient of variation taken over all p feature vectors.

The hypothesis of homogeneity is evaluated based on the value of H_1 computed through the below equation

$$H_1 = \frac{V_1 - \mu V_1}{\sigma_{V1}} \tag{2}$$

where μV_1 and σ_{V1} are average and standard deviation calculated from the simulated data using four parameters Kappa distribution. If the sites (rainfall gauge stations in the same cluster) are acceptable homogeneous if $H_1 < 1$, possibly homogeneous if $1 \leq H_1 < 2$ or definitely heterogeneous if $H_1 \geq 2$.

4   Results and Discussion

For all the studied 20 locations around Ganga basin, instances were collected for each month for 102 years, so total 102 instances. In the present study, the k mean Cluster analysis has been used to classify the rainfall time series into different clusters. In general, the rainfall time series constitutes the past several records which is a high dimensional data so old-fashioned simple clustering approach may not be used because the recorded rainfall time series may comprise outliers and clusters of different sizes which advocates the nearness among the pair of rainfall data in terms of nearest neighbors.

Figure 2 shows ten different rainfall-clusters of studied subregion of Ganga basin, which has been categorized into three different rainfall categories (RC1, RC2 and RC3). The analysis for clustering included by selecting the minimum three number of clusters and the optimal number of clusters by increasing the number of clusters and minimization of squared root error. In the RC1 (low range cluster), the total annual average rainfall values in mm from (586.75 to 724.24) in which cluster 1, 6 and 9 lie, RC2 (medium range cluster) consists of clusters 5, 7 and 10 whose rainfall varies from (877.3 to 1449.65) where as RC3 (high range cluster) where the annual average rainfall is more than 1449.65 and consists of clusters 2, 3, 4, 8. As well observed RC3 i.e. cluster during the monsoon season has maximum departures from the annual average as compared to RC1 and RC2 clusters. The rainfall during the non-monsoon season between October-May were relatively very less for RC1 and RC2 which accounts with the known fact that during the October-May is out-of-phase with the remaining months. Based on the analysis of vearge annual rainfall, the clusters are identified as follows:

Cluster 1: Agra, Allahabad, Azamgarh, Ballia, Bareli, Jaunpur;

Cluster 2: Mirzapur, Varanasi, Bageswar, Arariya, Begusarai, Bhagalpur, Buxar, Gaya and Patna; and

Cluster 3: Etawa, Kanpur, Lucknow, Mathura and Haridwar.

In order to understand the monsoon rainfall pattern for all the 20 districts over the Ganga basin, we acquired distinct groups of cluster for all the months. Figure 3 shows the 10 clusters segregation on the basis of monsoon rainfall which falls during the June–September over the basin.

The range of cluster in this case is different than what was discussed in case of annual rainfall behavior value; here the RC1, consists of cluster 2, 6 and 8 in which rainfall ranges from (502.8 – 617.5 mm) and generally lower than what in case of annual rainfall. Similarly, RC2 consists of clusters 4, 5 and 9 whose rainfall varies from (617.5 to 858.19) where as RC3 where the monsoonal average rainfall is more

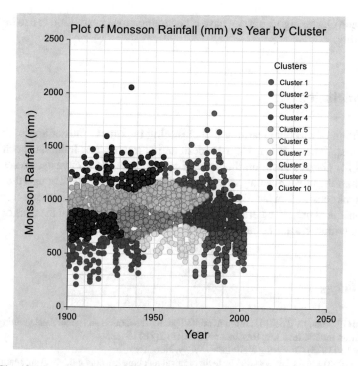

Fig. 3 Classification of monsoonal average rainfall of all the districts

than 858.19 which consists of clusters 1, 3, 7 and 10. The RC3 are generally the data of mid-July to end of August of each year whereas RC2 data are during the early June and during the September month. Based on the analysis of average monsoonal rainfall, clusters are identified as follows:

Cluster 1: Mirzapur, Varanasi, Bageswar, Arariya, Begusarai, Bhagalpur, Buxar, Gaya and Patna.

Cluster 2: Agra, Allahabad, Azamgarh, Baliya, Bareli, Jaunpur, Etawa, Kanpur, and Lucknow.

Cluster 3: Mathura and Haridwar.

According to Fig. 3, RC2 was assessed as 'acceptably homogeneous' whereas RC1 and RC3 were considered to be 'possibly homogeneous'. In the clustering of monsoon rainfall, K-means algorithm identified 3 clusters almost identical to that obtained in case of annual average rainfall. Here, there are also more stations in cluster RC3 which captures most of the spatial domain of Ganga basin. Particularly, RC3 cluster is distributed over the middle and upstream of the Ganga basin. The other two clusters have been found almost similar to those observed in case of annual rainfall. Also, homogeneity in the rainfall station by the clusters approach identified by using the k mean cluster algorithm can have other applications in addition to use in rainfall forecasting.

5 Conclusion

Application of machine learning like clustering by k mean cluster approach for 20 different stations of Bihar, Uttar-Pradesh and Uttrakhand of India has been implemented for Ganges basin on the average monthly data of 102 years. Present study recognizes the clusters along with within-cluster homogeneity. Though the k mean algorithms construct an identical number of clusters for precipitation series for annual and monsoonal period, where the formed clusters are not identical in terms of cluster size and within-cluster homogeneity.

References

Gaikwad GP, Nikam VB (2013) Different rainfall prediction models and general data mining rainfall prediction model. Int J Eng Res Technol (IJERT) 2(7):115–123

Hartigan JA (1975) Clustering algorithms. Wiley, New York

Hong W (2008) Rainfall forecasting by technological machine learning models. Appl Math Comput 200(1):41–57

Kumar V, Shanu, Jahangeer (2017) Statistical distribution of rainfall in Uttarakhand, India. Appl Water Sci 7:4765–4776

Kumar V, Sen S (2018) Evaluation of spring discharge dynamics using recession curve analysis: a case study in data-scarce region, Lesser Himalayas India. Sustain Water Res Manage 4(3):539–557. https://doi.org/10.1007/s40899-017-0138-z

Kumar V, Sen S (2020) Assessment of spring potential for sustainable agriculture: a case study in Lesser Himalayas. Appl Eng Agric 36(1):11–24. https://doi.org/10.13031/aea.13520

Li PW, Lai EST (2004) Short-range quantitative precipitation forecasting in Hong Kong. J Hydrol 288(1–2):189–209

Manvendra S, Vikram K, Atul R, Dikshit PKS, Dwivedi SB (2017) Spatial distribution of the suspended sediment of river Ganga at Varanasi, U.P., India. Int J Earth Sci Eng 10(3):533–540. https://doi.org/10.21276/ijee.2017.10.0310.

Monira SS, Faisal Zaman M, Hirose H (2012) A rainfall forecasting method using machine learning Models and its application to the Fukuoka city case. Int J Appl Math Comput Sci 22(4):841–854. https://doi.org/10.2478/v10006-012-0062-1

Pucheta JA, Cristian MRR, Martín RH, Carlos AS, Patiño HD, Benjamín RK (2009) A feed-forward neural networks-based nonlinear autoregressive model for forecasting time series, Comput y Sistemas 14(4):423–435

Shanu K, Vikram K, Atul KR (2015) An efficient algorithm for modeling and estimation of ground water level fluctuation. J Ind Water Res Soc 35(4):7–13

Shanu K, Vikram K Sandeep C (2018) A classed approach towards rainfall forecasting: machine learning network. Int J Swarm Intell 3(4):276–289. https://doi.org/10.1504/IJSI.2017.10008892

Yang Y, Lin H, Guo Z, Jiang J (2007) A data mining approach for heavy rainfall forecasting based on satellite image sequence analysis. Comput Geosci 33(1):20–30

Applicability of the InVEST Model for Estimating Water Yield in Upper Ganga Basin

Shray Pathak, C. S. P. Ojha, and R. D. Garg

Abstract The value of ecosystem services to society has increased many folds with time, along with the efforts to manage it. Integrated Valuation of Ecosystem Services and Tradeoffs (InVEST) occurs to be an effective model that assists in managing ecosystem services. However, the applicability of the InVEST model still requires testing in various geographic locations. With recent trends and observations, it is understood that there is a large Spatio-temporal variation in the hydro-meteorological parameters. Subsequently, the estimated water yield cannot be predicted better if the model assessed the single mean values of all the input hydro-meteorological parameters. Thus, a comprehensive analysis has been done in estimating water yield by mean Lumped Zhang model and spatial InVEST model. Also, the study weighed the performance of the InVEST model in estimating water yield in one of the most diverse and undulated topographic basins of India, i.e., Upper Ganga Basin. The estimated water yield values are compared with the observed in order to understand the variability of the yield models in predicting water yield in the Upper Ganga Basin.

Keywords Ecosystem services · InVEST · Water yield · Rainfall · Land use land cover · GIS

S. Pathak (✉)
School of Geographic Sciences, East China Normal University, Shanghai 200241, China
e-mail: shraypathak@gmail.com

C. S. P. Ojha · R. D. Garg
Department of Civil Engineering, Indian Institute of Technology Roorkee, Roorkee, Uttarakhand 247667, India
e-mail: cspojha@gmail.com

R. D. Garg
e-mail: rdgarg@gmail.com

© The Author(s), under exclusive license to Springer Nature Switzerland AG 2021
M. S. Chauhan and C. S. P. Ojha (eds.), *The Ganga River Basin: A Hydrometeorological Approach*, Society of Earth Scientists Series,
https://doi.org/10.1007/978-3-030-60869-9_15

1   Introduction

Ecosystem services in society should be properly managed in order to provide valuable services to civilians and humanity (Reeth 2013). One of the important ecosystem services is to provide fresh water for the welfare of society by ensuring the development of increased population, agricultural activities, industries, improvising living standards, tourism, etc. (Cudennec et al. 2007). Water yield is one of the important aspects of the attributes and processes that produce ecosystem services related to water resources (Vigerstol and Aukema 2011). It assists in estimating the water availability at the specific locations for consumptive use and in situ water supply. Climate change and uncertainties associated with hydro-meteorological parameters have been a major concern for the scientific society. One of the most developed cities, i.e., Cape Town in South Africa, is presently facing the water crisis. Although the city has vast resources and advanced technologies, due to the unavailability of the water resources for drinking and performing various non-potable works, the city has diminished its existence, and a lot of civilians are migrating from the town. Thus, it is important to conserve and protect freshwater resources in the region. However, its initial step is to analyze the available freshwater resources along with its various sources. It is essential for the qualitative assessment and visualization of the water yield to understand the interaction between the people and water (Pathak et al. 2017). Further, it helps in achieving and developing an effective strategy to manage and protect freshwater resources (Li et al. 2011). Also, it significantly helps in developing a sustainable approach to develop a regional-scale conservation policy (Ouyang et al. 2013). However, estimation of water yield is a complex process that is influenced by various driving factors such as precipitation, slope, vegetation cover, soil permeability, land use, etc. (Zhang et al. 2012; Shukla et al. 2018). Since, twentieth century a large number of water yield models were developed and employed (Abbott et al. 1986; Arnold et al. 1998). Various water yield models are applicable for different regions, along with various assumptions and expectations. In hilly regions, the undulated topography and high spatio-temporal variations made it more difficult to estimate water yield with conventional yield models (Liston and Elder, 2006). A lumped Zhang model (Zhang et al. 2004) is a worldwide adopted model that incorporates the mean value of input variables to estimate water yield at a basin scale. However, this model fails to weigh the climatic variability and parameters spatially.

Thus, a model is required, especially for hilly terrain, that has the capability to include the hydro-meteorological variations at a spatial scale and not at a basin scale (Pathak et al. 2019a). In this study, an ecosystem service model called Integrated Valuation of Ecosystem Service and Tradeoffs (InVEST) (Tallis et al. 2010), which is a spatially explicit model that capacitates all the input hydro-meteorological parameters along with the water yield output at a spatial scale. The model runs on the gridded scale, which requires relatively low data and expertise and provides results spatially. This model has the ability to include all the Spatio-temporal variations of the input and output results. The model has been applied in various locations across the globe (Chen et al. 2011; Fu et al. 2013; Marquès et al. 2013; Mdk et al.

2013). This simplified model has a long history and is based upon the Budyko theory. Even today this model receives a lot of interest in the field of hydrology and water resources management (Budyko and Ronov 1979; Zhang et al. 2001, 2004; Ojha et al. 2008; Donohue et al. 2012; Zhou et al. 2012; Xu et al. 2013; Wang and Tang 2014; Pathak et al. 2019b). The InVEST model has been applied in diverse fields with vast scientific literature available. Terrado et al. (2014) implemented the InVEST model densely populated Llobregat watershed in dry and wet extreme conditions, centering on the role of meteorological parameters. Hoyer and Chang (2014) tested the different scenarios of climate change and urbanization in Tualatin and Yamhill watersheds by applying the same water yield model. Also, the sensitivity analysis was performed for the input variables, and climatic parameters were concluded as a dominant factor. Another study concluded precipitation as the most dominant factor in estimating water yield (Hamel and Guswa 2014). The same model was applied to the two mountainous catchments in India, i.e., Tungabhadra river catchment and the Sutluj river catchment. The study emphasized the precipitation and reference evapotranspiration as the major dominating factor in estimating water yield (Goyal and Khan 2017). Liu et al. (2017) identified potential conservation sites in Southwest China by integrating biodiversity and ecosystem services through the InVEST model. Trisurat et al. (2018) quantified the water yield under near and long term climate change scenarios and further identified its impact on rice cultivation at the lower Mekong basin. Furthermore, the InVEST model is checked with conventional simplified models on various watersheds in five bioclimatic regions, along with North America (Scordo et al. 2018). In the present study, the InVEST model and Lumped Zhang model are implemented to estimate the water yield at Upper Ganga Basin. The models are further validated with the observed discharge, and a comparative analysis has been performed between the models.

2 Study Area and Data Sets

2.1 Study Area

In India, the Ganga river is amongst the top 20 rivers with respect to the water discharge. The Ganga basin can be divided into three zones, i.e., upper Ganga basin, middle Ganga basin, and lower Ganga basin. For the present study, the upper Ganga basin is selected, and its geographical coordinates vary from $29° 48'$ N–$31° 24'$ N and $77° 49'$ E–$80° 22'$ E (Fig. 1). The region covers an average area of about 22292.1 km^2, which reached up to Haridwar. In terms of topography, the orthometric elevation above mean sea level id about 275 m in plains and up to 7512 m in the Himalayan region. Approx. 433 km^2 is covered with a glacier landscape and about 288 km^2 is covered under the fluvial landscape. In upper mountainous regions, about 20% area is covered with forest, and 60% of the area is used for agricultural purposes.

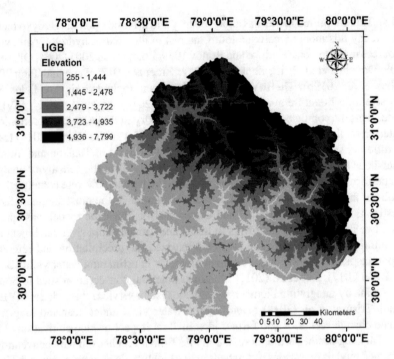

Fig. 1 Geographical representation of the upper Ganga basin

2.2 Data Sets

The basic dataset that is required as input in the water yield models is a climatic and dataset that is related to topography or landscape. The analysis has been performed for the year 2015. The following are the datasets used as an input in the water yield models.

2.3 Precipitation and Temperature

The daily time series datasets for precipitation and temperature for the year 2015 are acquired from the India Meteorological Department (IMD) at a grid size of 0.25° and 1° respectively. The daily time series for precipitation were aggregated to obtain annual time series at the spatial scale.

2.4 Land Use Land Cover and Soil Maps

For Land Use Land Cover (LULC) map, Landsat 8 Operational Land Imager (OLI) sensor satellite images were downloaded from the USGS website. Object-based supervised classification was performed in ERDAS Imagine software for LULC classification of the study region. The grid size or resolution of the raster image acquired is of 30 m resolution. Spatial maps related to the National Bureau of Soil Survey and Land Use Planning (NBSS and LUP) at 1:250,000. Digital maps were further reclassified to 30 m resolution, same as that of LULC map.

3 Methodology

3.1 Lumped Zhang Model

The mean value of the input variables, i.e., precipitation, actual evapotranspiration, potential evapotranspiration, is computed at the basin scale to compute a single mean value of the water yield for the entire basin (Zhang et al. 2004). The model is incapable of accounting for the spatial variability of all the input hydrological parameters that are associated with estimating water yield.

3.2 InVEST Model

The TheInVEST model monitors the alteration in the flows that are likely to occur in water yield due to the changes in the ecosystem of the region (Tallis et al. 2010). This model is based on the Budyko theory. It is an empirical relation that yields the ratio between actual evapotranspiration to potential evapotranspiration (Budyko 1974). The model runs in the gridded format, and all the input provided to the model are in the raster format. Thus, this model is capable of identifying and implementing the hydro-meteorological spatial variability and obtains the spatial water yield with better accuracy. The model is easier to implement with the use of Remote Sensing (RS) and Geographic Information Systems (GIS) techniques. The water yield $Y(x)$ is estimated annually at each pixel by the equation, as represented below.

$$Y(x) = \left(1 - \frac{AET(x)}{P(x)}\right) \times P(x) \qquad (1)$$

where $P(x)$ represents the annual precipitation at each pixel x, and $AET(x)$ is the actual annual evapotranspiration at each pixel x. The expression of the Budyko curve is applied in the InVEST model as given by Fu (1981) and later by Zhang et al. (2004).

The parameter index to dryness represents the ratio of the annual mean of potential evapotranspiration to precipitation. The index is used to estimate the mean annual actual evapotranspiration and is represented by Eq. (2).

$$\frac{AET(x)}{P(x)} = 1 + \frac{PET(x)}{P(x)} - \left[1 + \frac{PET(x)}{P(x)}\right]^{\left(\frac{1}{\omega}\right)} \qquad (2)$$

where, $\omega(x)$ is a non-physical parameter, which affects the natural climatic soil properties and $PET(x)$ is the annual potential evapotranspiration per pixel x (mm), which is estimated as follows.

$$PET(x) = Kc(x) \times ETo(x) \qquad (3)$$

where, $Kc(x)$ known as vegetation evapotranspiration coefficient which is directly influenced by the changes in the LULC characteristics for each pixel (Allen et al. 1998), $ETo(x)$ represents the annual reference evapotranspiration at each pixel x, which is identified based on the evapotranspiration of grass (alfalfa) grown at that particular site as shown in Eq. (6), $\omega(x)$ is an empirical parameter and mathematically it is represented by Donohue et al. [9] as shown in the Eq. (4).

$$\omega(x) = z \times \frac{AWC(x)}{P(x)} + 1.25 \qquad (4)$$

It is understood from the Eq. (4), that the minimum value of the parameter ω (x), i.e., for the bare soil will be 1.25 where the root depth is zero (Donohue et al. 2012). The parameter z is known as the seasonality factor, which represents the nature of local hydrogeological parameters, and its value varies from 1–30. Further, the parameter denotes the volumetric available plant water content, which is represented by the expression shown in Eq. (5) for each pixel at a given depth (mm).

$$AWC(x) = \text{Min.}(\text{Restricting layer depth, root depth}) \times PAWC \qquad (5)$$

Plant Available Water Content (PAWC) is the difference in the field capacity and the wilting point (McKenzie et al. 2003). It computed by the Soil Plant Air Water (SPAW) software and is solely depends upon the soil characteristics. The root depth refers to the 95% of the depth where root biomass occurs.

In the study, the modified Hargreaves technique is implemented to estimate the spatially gridded reference evapotranspiration, which is mathematically described in Eq. (6).

$$ET_o = 0.0013 \times 0.408 \times RA \times (T_{avg} + 17.0) \times (TD - 0.0123 \times P)^{0.76} \qquad (6)$$

where, T_{avg} is the average daily temperature (°C) defined as the average of the mean daily maximum and minimum temperature, ET_0 is reference evapotranspiration, RA is extraterrestrial radiation expressed in $[MJm^{-2}d^{-1}]$, and TD (°C) is the temperature range computed as the difference between mean daily maximum and minimum temperature. The extraterrestrial parameter radiation (RA) is further quantified by the expression represented in Eq. (7)

$$RA = \frac{24(60)}{\pi} \times G_{sc} \times d_r \times [w_s \sin(\varphi)\sin(\delta) + \cos(\varphi)\cos(\delta)\sin(w_s)] \quad (7)$$

where, d_r is the inverse relative distance Earth-Sun, RA is extraterrestrial radiation $[MJm^{-2}d^{-1}]$, G_{sc} is solar constant equals to 0.0820 MJm^{-2} min^{-1}, δ is the solar declination (rad), w_s is the sunset hour angle (rad), and φ is latitude (rad).

4 Results

The above methodology had been applied in the gridded format along with all the input variables that were considered at a pixel level, in Arc-GIS software.

For obtaining the spatial reference evapotranspiration (ET_0) for the upper Ganga basin, a modified Hargreaves method was applied at each pixel. ET_0 is the function of precipitation, RA, T_{avg}, TD, which was computed by considering the Eq. (6), as shown in Fig. 2. Initially, the spatial mean monthly values were computed for the study basin, which was further summed up to obtain the mean annual evapotranspiration.

Further, as per Eq. (3), the annual potential evapotranspiration (PET) is the multiplicative of the parameter ET_0 and the parameter K_c. The values of the parameter K_c depend upon the LULC and is taken from Table 1 in Shukla et al. (2018). Thus, spatial PET is obtained by applying Eq. (3) in the raster calculator toolbar in Arc-GIS, as represented in Fig. 3.

Likewise, by applying Eq. (1) in the gridded format, the spatial annual water yield was obtained for the upper Ganga basin as represented in Fig. 4.

For the same study basin, the Lumped Zhang model was applied to estimate the mean water yield. The mean estimated annual water yield for the Rishikesh sub-basin was computed as 1189.72 mm. The pour point was taken at the Rishikesh, as the observed data was available from the Uttarakhand irrigation department.

5 Validation of Results

The estimated values of water yield, PET, and AET were computed at the gridded scale. Further, these input and output variables were obtained with the available global datasets. Model simulated AET values were obtained from the Global Land Data Assimilation System (GLSAS) ET datasets from Noah model outputs. Subsequently, for the PET, global datasets obtained from the Climate Research Unit's (CRU),

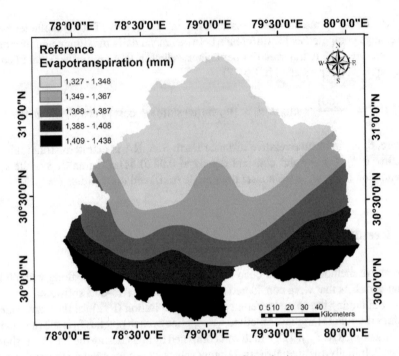

Fig. 2 Spatial reference evapotranspiration for the upper Ganga basin

Table 1 Observed and Estimated value of various InVEST model parameters

Parameters (mm)	Observed/derived (mm)		Estimated (mm)
AET	716.83		744.34
PET	1156.69		1550.42
InVEST water yield	2835.81	1928.35 (removing 32% snow melt contribution)	1720.16
Lumped zhang water yield	1928.35		1189.72

which is available at 0.5° resolution. Both the global model outputs from the above-referenced models are used for validating the estimated spatial PET and AET, as represented in Fig. 5a, b.

For validating the estimated spatial water yield, the Rishikesh pour point was considered as shown in Fig. 6. At this particular gauging station, the discharge data were available from the Uttarakhand Irrigation department.

It was observed that the contribution of the snowmelt at the Rishikesh discharge point was computed as 32% of the total runoff (Maurya et al. 2011). Thus, it is cleared from Table 1 that the estimated mean AET and PET are in good agreement with the derived values of the global models. Also, after removing 32% flow from the observed value at Rishikesh gauging station, the value seems to be closer to the estimated value by the InVEST water yield model.

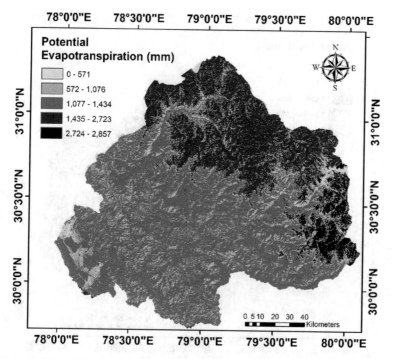

Fig. 3 Spatial potential evapotranspiration for the upper Ganga basin

6 Summary and Conclusions

In the present study, an attempt has been made to understand the applicability of the InVEST model in the highly undulated and diverse river basin in India, i.e., Upper Ganga basin. The InVEST model is based upon the Budyko theory and establishes the input and output variables in the spatial or raster format. With the effect of Climate change and uncertainties in the climatic parameters, it is difficult to understand the behavior of the basin by involving computation at a basin scale. The same has been understood by the study. The Lumped Zhang model, which works at a basin-scale and an InVEST model which works at a pixel scale, is implemented to estimate the water yield at the upper Ganga basin. The spatially estimated input variables, i.e., AET and PET, are validated with the derived global models, and these variables show good agreement. Similarly, the annual water yield estimated by the InVEST model provides better results if 32% of the snow contribution is removed from the observed runoff. However, the Lumped Zhang model fails to provide a better estimate of water yield for the upper Ganga basin. Thus, it is recommended that for such an undulated and bigger river basin, the spatial level estimation and computation techniques must be adopted in order to better understand the variability of input and output hydro-climatic variables involved in water resources management.

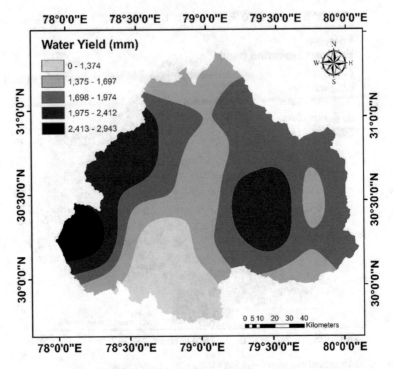

Fig. 4 Spatial annual water yield for the upper Ganga basin

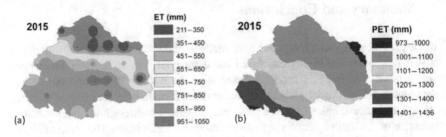

Fig. 5 a Spatial AET obtained from GLDAS Noah output datasets, **b** Spatial PET obtained from CRU datasets

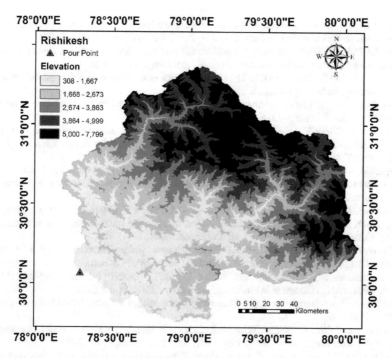

Fig. 6 Geographical representation of Rishikesh sub-basin

References

Abbott MB, Bathurst JC, Cunge JA, O'Connell PE, Rasmussen J (1986) An introduction to the European hydrological system—systeme hydrologique Europeen, "SHE", 1: History and philosophy of a physically-based, distributed modelling system. J Hydrol 87:45–59

Allen RG, Pereira LS, Raes D, Smith M (1998) Crop evapotranspiration-guidelines for computing crop water requirements-FAO irrigation and drainage paper 56. FAO, Rome 300(9):D05109

Arnold JG, Srinivasan R, Muttiah RS, Williams JR (1998) Large area hydrologic modeling and assessment part I: model development. J Am Water Resour As 34:73–89

Budyko MI (1974) Climate and Life. Academic Press, New York, USA, pp 1–507

Budyko MI, Ronov AB (1979) Evolution of chemical composition of the atmosphere during the Phanerozoic. Geokhimiya 5:643–653

Chen L, Xie G, Zhang C, Pei S, Fan N, Ge L, Zhang C (2011) Modelling ecosystem water supply services across the Lancang river basin. J Res Ecol 2:322–327

Cudennec C, Leduc C, Koutsoyiannis D (2007) Dryland hydrology in Mediterranean regions—a review. Hydrolog Sci J 52:1077–1087

Donohue RJ, Roderick ML, McVicar TR (2012) Roots, storms and soil pores: incorporating key ecohydrological processes into Budyko's hydrological model. J Hydrol 436:35–50

Fu BP (1981) On the calculation of the evaporation from land surface. Sci Atmos Sin 5(1):23–31

Fu B, Xu P, Wang Y, Peng Y, Ren J (2013) Spatial pattern of water retention in Dujiangyan County. Acta Ecol Sinica 33:789–797

Goyal MK, Khan M (2017) Assessment of spatially explicit annual water-balance model for Sutlej river basin in eastern Himalayas and Tungabhadra river basin in peninsular India. Hydrol Res 48(2):542–558

Hamel P, Guswa AJ (2014) Uncertainty analysis of a spatially-explicit annual water-balance model: case study of the Cape Fear catchment. NC Hydrol Earth Syst Sci 11:11001–11036

Hoyer R, Chang H (2014) Assessment of freshwater ecosystem services in the Tualatin and Yamhill basins under climate change and urbanization. Appl Geogr 53:402–416

Li S, Liu J, Zhang C, Zhao Z (2011) The research trends of ecosystem services and the paradigm in geography. Acta Geogr Sinica 66:1618–1630

Liston GE, Elder K (2006) A distributed snow-evolution modeling system (SnowModel). J Hydrometeorol 7(6):1259–1276

Liu S, Yin Y, Cheng F, Hou X, Dong S, Wu X (2017) Spatio-temporal variations of conservation hotspots based on ecosystem services in Xishuangbanna, Southwest China. PloS one 12(12):e0189368

Marquès M, Bangash RF, Kumar V, Sharp R, Schuhmacher M (2013) The impact of climate change on water provision under a low flow regime: a case study of the ecosystems services in the Francoli river basin. J Hazard Mater 263:224–232

Maurya AS, Shah M, Deshpande RD, Bhardwaj RM, Prasad A, Gupta SK (2011) Hydrograph separation and precipitation source identification using stable water isotopes and conductivity: River Ganga at Himalayan foothills. Hydrol Process 25:1521–1530

McKenzie NJ, Gallant J, Gregory L (2003) Estimating water storage capacities in soil at catchment scales. CRC for catchment hydrology. Australia, University of Canberra

Mdk L, Matlock MD, Cummings EC, Nalley LL (2013) Quantifying and mapping multiple ecosystem services change in West Africa. Agr Ecosyst Environ 165:6–18

Ojha CSP, Bhunya P, Berndtsson R (2008) Engineering hydrology, 1st edn. Oxford University Press, UK, p 459

Ouyang Z, Zhu C, Yang G, Weihua XU, Zheng H, Zhang Y, Xiao Y (2013) Gross ecosystem product: concept, accounting framework and case study. Acta Ecol Sinica 33:6747–6761

Pathak S, Ojha CSP, Zevenbergen C, Garg RD (2017) Ranking of storm water harvesting sites using heuristic and non-heuristic weighing approaches. Water 9(9):710

Pathak S, Ojha CSP, Shukla AK, Garg RD (2019a) Assessment of annual water-balance model for diverse Indian watersheds. J Sustain Water Built Env 5(3):04019002, 1–18

Pathak S, Garg RD, Jato-Espino D, Lakshmi V, Ojha CSP (2019b) Evaluating hotspots for stormwater harvesting through participatory sensing. J Environ Manage 242:351–361

Reeth WV (2013) Ecosystem service indicators: are we measuring what we want to manage? In: Ecosystem services, pp 41–61

Scordo F, Lavender TM, Seitz C, Perillo VL, Rusak JA, Piccolo M, Perillo GM (2018) Modeling water yield: assessing the role of site and region-specific attributes in determining model performance of the InVEST seasonal water yield model. Water 10(11):1496, 1–42

Shukla AK, Pathak S, Pal L, Ojha CSP, Mijic A, Garg RD (2018) Spatio-temporal assessment of annual water balance models for upper Ganga basin. Hydrol Earth Syst Sc 22(9):5357–5371

Tallis HT, Ricketts T, Nelson E, Ennaanay D, Wolny S, Olwero N, Vigerstol K, Pennington D, Mendoza G, Aukema J, Foster J (2010) InVEST 1.004 beta user's guide. The Natural Capital Project, Stanford University, USA

Terrado M, Acuña V, Ennaanay D, Tallis H, Sabater S (2014) Impact of climate extremes on hydrological ecosystem services in a heavily humanized Mediterranean basin. Ecol Indic 37:199–209

Trisurat Y, Aekakkararungroj A, Ma HO, Johnston JM (2018) Basin-wide impacts of climate change on ecosystem services in the Lower Mekong basin. Ecol Res 33(1):73–86

Vigerstol KL, Aukema JE (2011) A comparison of tools for modeling freshwater ecosystem services. J Environ Manage 92:2403–2409

Wang D, Tang Y (2014) A one-parameter Budyko model for water balance captures emergent behavior in darwinian hydrologic models. Geophys Res Lett 41(13):4569–4577

Xu X, Liu W, Scanlon BR, Zhang L, Pan M (2013) Local and global factors controlling water-energy balances within the Budyko framework. Geophys Res Lett 40(23):6123–6129

Zhang L, Dawes WR, Walker GR (2001) Response of mean annual evapotranspiration to vegetation changes at catchment scale. Water Resour Res 37(3):701–708

Zhang L, Hickel K, Dawes WR, Chiew FH, Western AW, Briggs PR (2004) A rational function approach for estimating mean annual evapotranspiration. Water Resour Res 40(2):W02502, 1–14

Zhang C, Li W, Zhang B, Liu M (2012) Water yield of Xitiaoxi river basin based on InVEST modeling. J Res Ecol 3:50–54

Zhou X, Zhang Y, Wang Y, Zhang H, Vaze J, Zhang L, Yang Y, Zhou Y (2012) Benchmarking global land surface models against the observed mean annual runoff from 150 large basins. J Hydrol 470:269–279

Current Trends and Projections of Water Resources Under Climate Change in Ganga River Basin

Jew Das and Manish Kumar Goyal

Abstract Climate change has a direct and indirect impact on agriculture, surface, and sub-surface water resources, human systems, and food security. The Ganga River Basin supports more than 600 million people in India. The water resources of the Ganga basin are used for irrigation, navigation, drinking purpose, and hydropower generation. In this sense, due to the dual impact of climate change and population growth, the river basin is going to be affected in different ways. Therefore, assessment and understanding of climate change impact of hydro-climatological components over the basin are indispensable. The present chapter reviews the literature regarding the climate change impact analysis over the Ganga River basin. In addition, the precipitation and temperature extreme indices are analyzed using the trend analysis. The daily precipitation and temperature gridded data are procured from the India Meteorological Department (IMD) during 1901–2015 and 1951–2015, respectively. The trend analysis is carried out using the Mann–Kendall test, and the spatial plots for the trend of extreme indices over the Ganga Basin are plotted in the geographical information system environment. In addition, current and future scenarios of precipitation, temperature, streamflow, drought, and snow cover are presented. The last part of the chapter deals with adaptation strategies and future scopes. The findings and discussion would help in formulating the effective adaptation strategies over the Ganga Basin.

Keywords Ganga river · Trend analysis · Future projection · Mann kendall test · IMD

J. Das (✉) · M. K. Goyal
Discipline of Civil Engineering, Indian Institute of Technology, Madhya Pradesh, Indore 453552, India
e-mail: jewdas05@gmail.com

M. K. Goyal
e-mail: vipmkgoyal@gmail.com

© The Author(s), under exclusive license to Springer Nature Switzerland AG 2021
M. S. Chauhan and C. S. P. Ojha (eds.), *The Ganga River Basin: A Hydrometeorological Approach*, Society of Earth Scientists Series,
https://doi.org/10.1007/978-3-030-60869-9_16

233

1 Introduction

Water plays a significant role in differentiating our planet as compared to all others, and the availability of freshwater resources is adequate enough to meet all the current and foreseeable demands on earth (Cosgrove and Loucks 2015). However, the heterogeneity in its Spatio-temporal distribution makes the available freshwater difficult to fulfill the demand, and as a result, there are many regions where the available freshwater resources are inadequate to meet the environmental, economic development, and domestic need. Water, an essential natural resource (Vörösmarty et al. 2010), is very much susceptible to the altered climate, which is exerting immense pressure on water resources and increasing the need for vulnerability assessment (Anandhi and Kannan 2018). It is expected that the world population will exceed 9 billion by 2050 and is not going to stabilize in the twenty-first century (Gerland et al. 2014), demanding 70–100% more food production (Tscharntke et al. 2012). The changing climate is one of the major threats to the water resources, food security, human systems of 1.2 billion population in the twenty-first century of India and the country avails only 4% and 9% of the total water resources and arable land of the globe, respectively (Goyal and Surampalli 2018).In this sense, understanding the water resources system and its exposure to the adverse natural and anthropogenic stressors are crucial to secure and enhance the food production to fulfill the demand of the growing population (Anandhi et al. 2016).

The landmass of India is blanketed with multifaceted geography, and the impacts of changing climate on water resources differ considerably among different regions and cannot be generalized. Additionally, the reliability of the historical climate conditions is highly criticized in the planning and management of future water resources since climate change imposes high uncertainty, which is well outside the historical climate settings. In recent decades, the devastating impact of climate change shows its cruelty all over the country. For instance, an economic loss of $100 billion was caused due to the drought during 2016, and the effect was felt over ten states accounting for 330 million people (ASSOCHAM Report 2016). According to the Planning Commission (2011), about 13.78% of the total area in India comes under flood disasters. From 1953 to 2000, 33 million people were affected due to flooding (Kumar et al. 2005). Moreover, in the global scenario, the statistics are even more threatening, about 4 billion people live at least one month in a year under severe water scarcity condition and around half a billion population face water scarcity throughout the year (Mekonnen and Hoekstra 2016). Not only the surface water resources but also the variability of groundwater resources affects the food and water security significantly. A recent study by Asoka et al. (2017) advocated that depletion of groundwater storage at the rate of 2 cm/year is observed in northern India and an increase at the rate of 1–2 cm/year is noticed in southern India during 2002–2013 as a result of pumping and precipitation patterns.

The water resources systems are transformed as a consequence of urbanizations, land cover changes, industrialization, and several man-made interventions (Vörösmarty et al. 2010), and this, in turn, altering the quality and quantity of freshwater sources.In addition, climate change will aggravate the distribution and availability of water resources, which in turn will affect the per capita water availability. Thus, quantification of water availability under the foreseen scenarios is crucial for the effective management of water resources in water-stressed countries like India. Considering the anthropogenic perturbations, Lenton et al. (2008) marked different "tipping elements" over the globe based on a critical threshold known as "tipping point" to define large-scale components of the earth system particularly sensitive to the climate change and analyzed the uncertainty associated with the underlying physical mechanisms. The ongoing climate change has the strongest and significant impact on natural systems (Stone et al. 2013). The variability in the precipitation pattern or melting of snow and ice is modifying the hydrological systems as well as affecting the quality and quantity of water resources.

In the present paper, we present an extensive analysis based on the investigation carried out by the research community, primarily focusing on the observed trends, projections based on the climate models under different scenarios, and advanced methodologies to assess the water resources under a stressed climate. This study provides a clear picture of research methodologies and an overview of the significant outcomes to understand the influence of climate change on water resources on the largest river basin India, i.e., the Ganga River basin. In addition, the present study enables the research community to develop new methodologies or improve the existing techniques for robust assessment of the impact of climate change and allows water managers and policymakers to devise proper adaptation strategies to mitigate the adverse consequences.

2 Study Area

The Ganga River basin blankets 26% of the total landmass of India, which is about 950,000 km^2, and the basin is highly populated with 530 persons/km^2. The basin lies between 73° 02′ E to 89° 50′ E longitude and 21° 60′ N to 31° 21′ N latitude. The holy river drains into 11 states of the country, having maximum width and length of ~1024 km and ~1543 km, respectively. The water resources potential of the Ganga basin is about 5,25,020 million cubic meters (Ministry of Water Resources 2014). The climate of the basin is mainly tropical and subtropical in nature with temperature 40 and 10 °C during summer and wintertime, respectively (Syed et al. 2014). The annual mean precipitation over the basin is about 2000 mm, with the majority part of precipitation occurs during the monsoon season (i.e., June to October). The major portion of the land cover (about 65.57%) over the basin is covered with agriculture, which is 40% of the total area of India. It is observed that there is a significant alteration of the land use pattern (Kothyari and Singh 1996), surface temperature (Goswami et al. 2006), and precipitation (Hoyos and Webster 2007) over the last few decades. Figure 1 depicts the location and elevation map of the river basin.

3   Trends in Precipitation and Projections

In this section, we provide a brief discussion about the precipitation and its extremes over India, focusing mainly on the largest river basin of India, the Ganga River basin. In the beginning, we provide a brief summary of precipitation over India, and then we discuss the precipitation variability over the Ganga basin.

About 70 to 90% of total annual precipitation over India results from the monsoon season, i.e., during June to September (Lacombe and McCartney 2014), and there is no significant trend is observed for annual precipitation during 1871 to 2005 considering 306 meteorological stations all over India (Goyal and Surampalli 2018). However, it is believed that anthropogenic climatic disturbances are likely to alter the monsoon precipitation (Turner and Annamalai 2012), which will impact the economic development and food security of the country. With this background, and insights investigation of rainfall trends during the historical periods is inevitable for both water resources and agricultural planning and management. A comprehensive analysis of past studies regarding trend analysis of precipitation at different spatial scales (i.e., from individual stations to country scales) is carried out by Lacombe and McCartney (2014). The analysis reveals the contrasting results among the researchers and the heterogeneity can be attributed to the varying in the time period, the various definition of seasons and extreme events, different statistical tests, trend analysis, slope estimation, and higher multidecadal variability (Goswami et al. 2006; Rajeevan et al. 2008; Krishnamurthy et al. 2009; Ghosh et al. 2012).

However, in some of the studies in trend analysis in different parts of India include, Kumar et al. (2010), who analyzed the precipitation trend at different temporal scales using 135 years of data (1870–2005) and found decreasing trend during annual and monsoon period, while at the national level they observed an increasing trend for pre-monsoon, post-monsoon, and winter rainfall; Mondal et al. (2015), who investigated the long-term trend using different techniques such as Mann-Kendall (MK) test and Sen's slope, stated a net deficit in precipitation over the country. To estimate the trends in the hydrological time series, the commonly used parametric and non-parametric statistical techniques are ordinary least square linear regression, the MK test, and Sen's slope estimation. Moreover, these methods are also modified over time for robust assessment viz., Lacombe and McCartney (2014) modified the MK-test by incorporating the scaling effect to discriminate unidirectional trend and multi-decadal cyclic phenomena. The details of these tests are widely available in the literature and hence are not discussed here.

Here, we also analyze the trends in the precipitation extreme indices as suggested by Expert Team on Climate Change Detection and Indices (ETCCDI) during 1901–2015. The analysis is carried out over the Ganga River basin. The total time period is divided into 2 groups, i.e. 1901–1950 and 1951–2015. The precipitation dataset is procured from the India Meteorological Department (IMD) at a grid resolution of $0.25° \times 0.25°$. The definitions of the precipitation extreme indices are presented in Table 1.

Table 1 Definitions and units of the the precipitation extreme indices

Sl. No	Index	Definition	Unit
1	R99p	Annual total precipitation from days >99th percentile	mm
2	R95p	Annual total precipitation from days >95th percentile	mm
3	R90p	Annual total precipitation from days >90th percentile	mm
4	RX1day	Monthly maximum one-day precipitation	mm
5	RX5day	Monthly maximum five-day consecutive precipitation	mm
6	R10	Annual count if precipitation $\geq$10 mm	Days
7	R20	Annual count if precipitation $\geq$20 mm	Days
8	CDD	Maximum number of consecutive days if precipitation <1 mm	Days
9	CWD	Maximum number of consecutive days if precipitation $\geq$1 mm	Days
10	PRCPTOT	Total annual precipitation if precipitation $\geq$1 mm	mm

Figure 2 presents the trend analysis of R99p, R95p, and R90p for both the periods (i.e., 1901–1950 and 1951–2015). The spatial mapping of the trend for R99p, R95p, and R90p during 1901–1950 exhibits similar distribution. However, the area under the significantly decreasing trend for R90p increases during 1951–2015 as compared to 1901–1950 over the northern part of the basin. In addition, mostly, the significantly decreasing trend is observed over the northern part of the basin for R99p and R95p during 1951–2015.

Figure 3 presents spatial plots of trend analysis of RX1day, Rx5day, R10, and R20 over the basin. It can be noted from Fig. 3 that there is no significantly decreasing trend is observed (except a few small patches) during 1901–1950 for all.

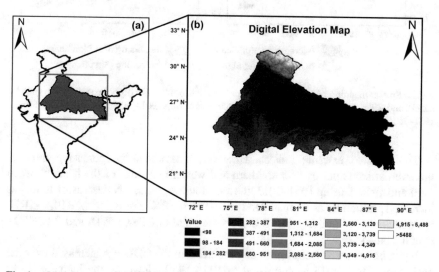

Fig. 1 **a** Location map of the Ganga basin superimposed over India map; **b** Digital Elevation Map (DEM) of the Ganga River basin

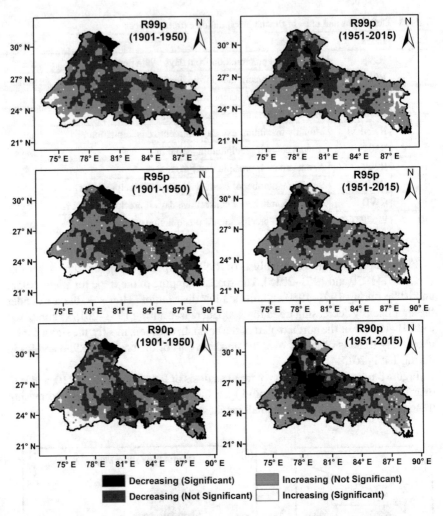

Fig. 2 Spatial mapping of trends over Ganga River basin for R99p (upper panel), R95p (middle panel), and R90p (lower panel). The left panel presents during 1901–1950 and right panel presents 1951–2015

The selected extreme indices. However, a significantly increasing trend is observed at some places over southern and western portions of the basin between 1901 and 1950. During 1951–2015, the area under significantly decreasing trend has increased profoundly as compared to 1901–1950 for RX1day, RX5day, R10, and R20 extreme indices. Similarly, Fig. 4 depicts the trend of CDD, CWD, and PRCPTOT over the basin.

It can be observed from Fig. 4 that the trend in the CDD has increased over the basin from 1951–2015 as compared to 1901–1950. However, the increase is not significant. Whereas, in the case of CWD and PRCPTOT, the spatial variability of

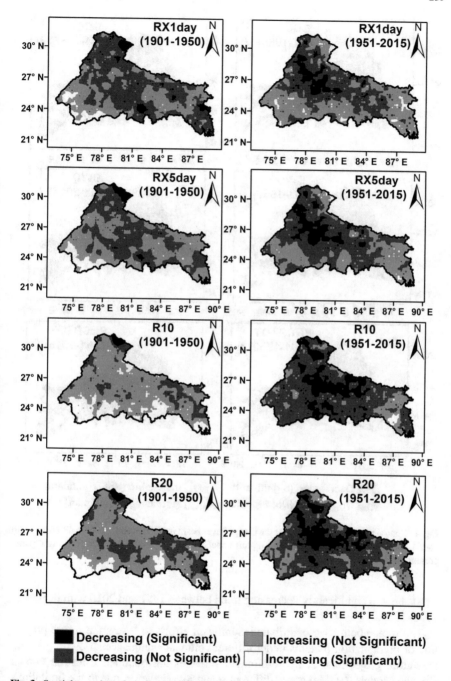

Fig. 3 Spatial mapping of trends over Ganga River basin for RX1day, RX5day, R10, and R20. The left panel presents during 1901–1950 and right panel presents 1951–2015

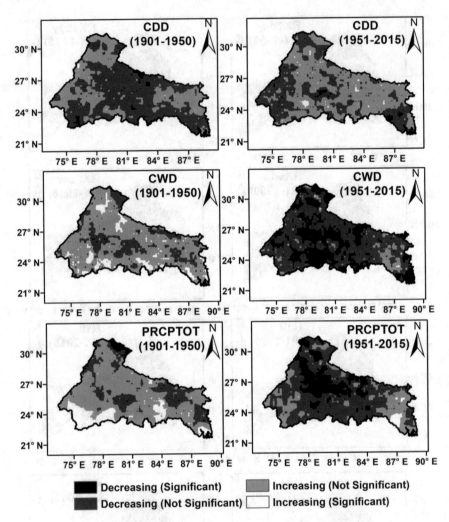

Fig. 4 Spatial mapping of trends over the Ganga River basin for CDD (upper panel), CWD (middle panel), and PRCPTOT (lower panel). The left panel presents during 1901–1950, and right panel present 1951–2015

trend shows a significantly decreasing trend between 1951 and 2015 with respect to 1901–1950.

The researchers have contributed significant efforts to assess the future projection of precipitation patterns under different climate change scenarios. In general, the future projections of hydro-climatic variables are obtained from the Global Climate Models (GCMs) under different scenarios. However, due to the limitations (e.g., incomplete understating of forcing-feedback mechanism, difficulties in cloud modeling, and difficulties in estimation of latent heat flux), the precipitation is not well modeled by the GCMs (Das et al. 2018). Therefore, the precipitation is down-scaled to the area of interest by incorporating the other atmospheric variables that are well simulated by the GCMs. To downscale, researches have used different down-scaling techniques, and statistical and dynamic downscaling techniques are the most common techniques. Readers are advised to follow Xu (1999) to understand the underlined steps and procedures involved in the downscaling techniques. However, we present synthesized results and methodologies from the past investigations (but are not limited to) over different parts of the Ganga River basin in Table 2.

Summarising the various studies, it can be established that Ganga Basin has experienced a localized trend for the historical trends of precipitation during last50years. The future projections imply that the monsoon precipitation is likely to increase. However, the magnitude of the projected changes varies with different GCMs, scenarios, and study domain.

4 Trends in Temperature and Projections

According to IMD (2016), a warming trend is prevailing throughout the country, with 2016 as the warmest year on record. During a span of 17 years (i.e., 2000–2016), 5 warmest years are recorded indicating the very high frequency of occurrence. In addition, 2001–2010 is the warmest decade for the country (Pai et al. 2013). Kothawale and Rupa Kumar (2005)stated the warming of 0.22 °C/decade between 1971 and 2003. A detailed analysis of maximum and minimum temperature over India is carried by Srivastava et al. (2017) and the critical findings of the investigation include (i) since the 1980s, annual mean temperature has shown a sharp increase, i.e., 0.2 °C/decade; (ii) the rate of warming in maximum and minimum temperatures are recorded as 1 and 0.18 °C, respectively per 100 years; (iii) highest increase in trend is noticed during post-monsoon and winter seasons than the monsoon season; (iv) The warming is mostly limited to the central, north, and eastern/north-eastern parts of India. Sonali et al. (2017) studied the long term (1901–2000) maximum and minimum temperature over India and advocated that a significant increase in the trends for maximum and minimum temperatures are observed during winter and post-monsoon, respectively. Goyal and Surampalli (2018)have analyzed the trend in the maximum, and minimum temperature incorporating the very high resolution (0.25° × 0.25°)gridded data and stated that the mean maximum and minimum temperature during the recent years (2000–2014) is significantly larger than the long-term mean, i.e., during 1951–2014. The rise in the temperature in the urban areas is more than the rural areas due to the Urban Heat Island (UHI) effect. The increase in temperature also leads to intensifying the extreme weather and climate events, such as heatwaves, and the heat waves and drought are very much concurrent in nature. More frequent heatwaves are likely to increase the mortality rate and heat stress (Murari et al. 2015). India has witnessed a relatively high frequency of heatwaves in many parts of the country (Sharma and Mujumdar 2017), and significantly increasing trends are noticed during 1961–2010 (Pai et al. 2013). The central and north-western parts of

Table 2 Historical and projected trends in rainfall over Ganga River basin

Region	Methodology	Period	Outcome	References
Upper and middle parts of Ganga Basin	Moving average, robust locally weighted regression, Mann-Kendall (MK) test, and homogeneity test	1901–1989	Decrease in the total monsoon rainfall and number of rainy days during monsoon season	(Kothyari et al. 1997)
Ganga Basin	Mann-Kendall (MK) test	90 to 100 years of data (historical)	Annual precipitation over the basin has remained stable	(Singh et al. 2008)
Ganga Basin	Mann-Kendall (MK) test and Sen's slope estimator	1951–2004	No significant trend is observed for annual precipitation	(Kumar and Jain 2011)
Ganga Basin	Linear Trend	1951–2014	Report decreasing trend during south-west monsoon precipitation	(Deshpande et al. 2016)
Ganga Basin	Mann-Kendall (MK) test and Theil–Sen's estimator	1901–2015	Reveal the mixed pattern of trends due to inherent spatial heterogeneity of rainfall	(Bisht et al. 2018)
Ganga Basin	Using five different GCMs	2046–2065	Increase upstream precipitation by 8% under A1B scenario	(Immerzeel et al. 2010)
Ganga Basin	Statistical Downscaling Model (SDSM)	2000–2100	Overall increase in the monsoon rainfall by 12.5% and 10% with respect to baseline under A1B and A2 scenarios, respectively	(Pervez and Henebry 2014)
Ganga Basin	Using four different regional climate models (RCMs)	2021–2050	Increase in heavy precipitation extremes during monsoon season	(Mittal et al. 2014)

(continued)

Table 2 (continued)

Region	Methodology	Period	Outcome	References
Ganga Basin	Using five different GCMs	2015–2099	Mean precipitation is projected to increase by 19.8%	(Masood et al. 2015)
Ganga Basin	Using downscaled outputs from 37 GCMs; Mann Kendall's test statistic and Thiel Sen's Slope estimator	Twenty-first century	Increase in the monsoon and annual precipitation under high emission scenarios	(Arora et al. 2017)
Ganga Basin	Using ensemble of RCM	2000–2050	Increase in annual mean rainfall	(Moors et al. 2011)

the country are witnessed an increase in the duration and frequency of heatwaves (Rohini et al. 2016).

Here, we analyze the four different temperature extremes, namely, TXx (maximum value of daily maximum temperature), TNx(maximum value of daily minimum temperature), TXn (minimum value of daily maximum temperature), and TNn (minimum temperature of daily minimum temperature) over Ganga River basin. The units of all the selected temperature extreme indices are in °C. The temperature dataset is collected from IMD at $0.5° \times 0.5°$ grid resolution from 1951–2015. For analysis purposes, the total length is divided into two parts, i.e., 1951–1980 and 1981–2015. The MK test of these extremes reports that there is a significantly increasing trend all over the basin for TXx during both the time periods. Moreover, during 1981–2015 the trend analysis of TNx, TXn, and TNn at each grid point over the Ganga basin reveals a significantly increasing trend. However, during 1951–1980 (Fig. 5), a mixed trend is observed over the basin for TXn. The significantly increasing trend is observed over most parts of the basin for TNx and over the western part of the basin for TNn.

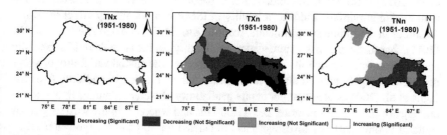

Fig. 5 Spatial mapping of trends over Ganga River basin for TNx (left), TXn (middle), and TNn (right) during 1951–1980

Both temperature and precipitation play an important role in water resources, and the variability of both the variables is interconnected. According to the Clausius-Clapeyron relationships, the water holding capacity of the air is increased by 7% due to a one-degree rise in the temperature (Held and Soden 2006). Hence, attempts have been made to project the changes in the temperature profile under different climate scenarios. However, we present synthesized results and methodologies from the past investigations (but are not limited to) over different parts of the Ganga River basin in Table 3.

Taken together, the analysis from various studies suggests that over the last 40 years, there has been an increase in the temperature profile (Nepal and Shrestha 2015). Moreover, the future projections indicate a gradual increase in temperature under the scenario of changing climate. In general, downscaling of precipitation and temperature is known as direct downscaling as these hydroclimatic variables are projected based on the transfer function obtained as a result of the association of other climatic variables known as predictors. In most of the downscaling studies, statistical downscaling is preferred to dynamic downscaling because statistical downscaling consumes less time and cost than the dynamic downscaling. It is worth mentioning that the statistical downscaling is performed based on some assumptions viz., the relationships obtained through the transfer functions during the historical period valid for the future, the selected predictors capture the climate signals completely, and spatial variability remains constant under different climate scenarios. As a result of climate change and the variability in precipitation and temperature, the final consequences come in the form of flood and drought. Therefore, it is inevitable to quantify foreseen water availability and is discussed in the following sections.

5 Streamflow and Projections

The complex nature of the Indian Summer Monsoon Rainfall (ISMR) with Spatio-temporal variability and strong sensitivity of the Asian monsoon to the global warming leads to aggravating the intensity and total monsoon precipitation (Kitoh et al. 2013). Moreover, the Indian River basins receive more than 80% of the total monsoon rainfall, and the variability in ISMR can potentially affect the water resources in the Indian subcontinent. According to UNFAO (2013) and India-WRIS (2011), based on the per capita water availability, India is ranked 132, and the annual freshwater availability is 1.91 km^3. Moreover, the spatial variability of water resources is highly uneven due to its multifaceted geography. In addition, Amarasinghe et al. (2005) stated that per capita availability of water resources in the northeast part of the country is more, i.e., 17,000 m^3 and whereas, in the western part of India, it is only 240 m^3. Investigations are carried out to examine the water availability in different river basins. Examples include Singh et al. (2014), who analyzed the long-term (1970–2010) trend using daily streamflow data over the Sutlej River basin observed a declining trend over the duration. The decrease in streamflow over the basin has significant implications on the hydropower generation and irrigation;

Table 3 Historical and projected trends in temperature over the Ganga River basin

Region	Methodology	Period	Outcome	References
Upper and middle parts of Ganga Basin	Moving average, robust locally weighted regression, Mann-Kendall (MK) test, and homogeneity test	1901–1989	An increasing trend in annual maximum temperature	(Kothyari et al. 1997)
Upper Ganga Canal	Mann-Kendall (MK) test and Sen's slope estimator	1901–2002	Annual mean, maximum and minimum temperatures have increased by 0.60 °C, 0.60 °C, and 0.62 °C, respectively	(Mishra et al. 2014)
Ganga Basin	Linear Trend	1951–2013	Number of hot days are around 45 days with maximum temperature of 40 °C	(Deshpande et al. 2016)
Upper Ganga Basin	Mann–Kendall (MK) test and change point detection	Long-term series	Rise in winter temperature by 1 °C with 1988 as a change year	(Sharma and Ojha 2018)
Ganga Basin	Using an ensemble of RCM	2000–2050	Mean annual temperature is likely to increase in the range 1–4 °C under SREB A1B	(Moors et al. 2011)
Ganga Basin	Mann-Kendall (MK) test and Sen's slope estimator	2010–2099	Projected future temperature is likely to increase under SRES-A2 and RCP6.0 scenarios	(Arora et al. 2015)
Ganga Basin	Using five different GCMs	2015–2099	Air temperature is likely to increase by 1 °C in the near future and about 4.1 °C in far future	(Masood et al. 2015)
Ganga Basin	Dynamic downscaling of HadCM3 GCM	End of the twenty-first century	Increase in the temperature 1.5–2.7 °C by the end of the middle of the century and 2.6–4.8 °C by the end of the century	(Caesar et al. 2015)

Bhutiyani et al. (2008), who analyzed the change in the streamflow over the northwestern Himalaya River basins noticed a decrease in streamflow in Beas River and Ravi River. However, winter streamflow in Chenab River showed an increasing trend, and the reason could be attributed to the snow and glacier melting during 1961–1995; Raju and Nandagiri (2017) reported no significant trend in monthly streamflow over Upper Cauvery basin during 1981–2010; While examining the streamflow variability at different outlets of Mahanadi River stated a decreasing trend at the rate of 3388 mm^3/decade during 1972–2007 (Panda et al. 2013) at Tikerpara outlet and 4.53%/year increase in annual streamflow at Mundali outlet during 1972–2003 (Dadhwal et al. 2010).

In this sense, the projection of future streamflow availability is of paramount importance in the changing climate scenarios. The downscaling of streamflow under future climate scenarios includes two types, i.e., direct and indirect downscaling. Most of the streamflow downscaling is carried out using indirect downscaling, which includes hydro-climatic variables that are downscaled to the regional climate conditions, and then the downscaled variables are given as an input to the hydrological model to generate the streamflow. In addition, in some cases, direct downscaling is carried out because a large number of the ensemble can be achieved at low computational cost (Das and Umamahesh 2018).

Here, we present synthesized results and methodologies from the past investigations (but are not limited to) over different parts of the Ganga River basin in Table 4.

6 Drought and Climate Change

Drought in India affects a large number of population and a paramount source of concern as it adversely impacts the agricultural, social, economic status of the country. The erratic nature of the precipitation and increasing temperature significantly drive the drought condition in India. As discussed, the heat waves resulting from the increase in temperature and drought conditions are found to be linked with positive feedback mechanism based on the past heat waves and temperature extremes (Fischer et al. 2007; Diffenbaugh et al. 2007). Moreover, the combination of low precipitation and high temperature (heat waves) can be induced by significant negative consequences than individual occurrence (Sharma and Mujumdar 2017). More than 40% of the total area in India was affected due to the severe drought hit India during the last five decades Kaur (2009). Shewale and Kumar (2005) stated that India Meteorological Department (IMD) identifies most drought-affected zones in India with 31 drought events during the last 130 years (1875–2004) and the zones are located in the arid western meteorological subdivision. The northeast part of India is demarcated as lowest drought prone area (Shewale and Kumar 2005). A long-term analysis of drought trends from 1901–2004 indicated the increasing trend in drought frequency and severity over India (Mallya et al. 2015). They also stated that regional drought over the agriculturally dominated area, which increases food security and socioeconomic vulnerability. Zhang et al.(2017) advocated that there is a significant increasing trend in meteorological, vegetation, and short duration droughts during the last 30 years. A district-wise analysis of drought variability by Pai et al. (2017) during 1901–2010 showed the occurrence of moderate droughts are equally likely all over the country, whereas more intense drought frequency is observed over many

Table 4 Historical and projected trends in Streamflow over Ganga Basin

Region	Methodology	Period	Outcome	References
Kosi Basin (sub-basin of Ganga)	Trend analysis	1947–1994	A decrease in streamflow and the decrease in more significant during low flows	(Sharma et al. 2000)
Upper Ganga Basin	Reconstruction from slack water and floodplain deposits	1000 years of historical flood	About 25 floods occurred over the last 1000 years	(Wasson et al. 2013)
Gomti River Basin (sub-basin of Ganga)	Mann–Kendall and Sen's slope	1982–2012	A decrease in the annual streamflow	(Abeysingha et al. 2016)
Ganga River basin	19 GCMs with GRiveT model	Up to the end of the twenty-first century	Increase in the discharge by +18.0% in future under SRES A1B scenario	(Nohara et al. 2006)
Ganga River basin	Glacio-hydrological model	Up to the end of the twenty-first century	Increase in the river flow under RCP4.5 and 8.5	(Immerzeel et al. 2013)
Upper Ganga Basin	Variable Infiltration Capacity (VIC) model	2010–2100	The sensitivity of the streamflow is high in urban areas than cropland areas	(Chawla and Mujumdar 2015)
Ganga River basin	H08 Hydrologic Model	2015–2099	Runoff is projected to increase by 33.1%	(Masood et al. 2015)

(continued)

Table 4 (continued)

Region	Methodology	Period	Outcome	References
Ganga River basin	Nine regional hydrologic models with 5 GCMs	1971–2099	Annual streamflow volume is likely to increase by 26% under RCP8.5 and 16% under RCP2.6	(Eisner et al. 2017)
Ganga River basin	Nine hydrological models and 5 GCMs	2006–2099	Increase in the mean and low flows	(Krysanova et al. 2017)
Ganga River basin	INCA-N model	1971–2099	Increase in the monsoon flows	(Whitehead et al. 2015)

districts from northwest, north and central India. Higher intensity severe drought conditions are noticed over the north, northwest, and southwest coast of peninsular India. In a recent study, Goyal and Surampalli (2018) analyzed the drought conditions (i.e., average drought duration and average drought severity) across 566 meteorological stations over India during 1901–2002. They reported a declining trend of drought properties over the crop dominated area and increasing trend over the forest of north-eastern and Western Ghats. It should be worth mentioning that they have considered only the meteorological drought for the analysis. Sarma and Singh (2019) studied the Spatio-temporal variability of meteorological drought over the Gomti basin (Sub-basin of Ganga River) during 1971–2007 and reported increase in the frequency of severe and extreme drought over the upper region of the basin Jhaet al. (2019). reported that about one fourth vegetated area of Ganga River basin is prone to vegetation drought due to lower soil moisture content.

The researchers also attempted to assess the variability in the drought under the future climatic scenarios. The examples include Salvi and Ghosh (2016), who analyzed meteorological drought condition for the twenty-first century under different emission scenarios and five GCMs, reported increase in the occurrence of extreme dry spells over central, southeast, eastern, and some parts of north-eastern India; Jenkins and Warren (2015) analyzed the drought properties under the stabilized scenarios using 3 GCMs and stated decrease in the key drought indicators viz., frequency, duration, intensity, and magnitude during 2003–2050; Ojha et al. (2013) stated drought events are likely to increase in the peninsular, west-central and central northeast regions of India during 2050–2099. They carried out the analysis using 17 GCMs; Hirabayashi et al. (2008) analyzed the drought over Ganga Basin up to the end of the twenty-first century and advocated that drought is likely to increase over the basin.

Although there is no precise definition of the drought, the simplest way to monitor the drought conditions is through different drought indices (Li et al. 2015). In addition, drought indices provide a quantitative judgment regarding the onset and termination of a drought event, as well as the severity of the event (Tabari et al. 2013). Based on the applicability, different drought indices were developed in past few decades, the indices include but are not limited to Standardised Precipitation Index (SPI) (Mckee et al. 1993), Palmer Drought Severity Index (PDSI) (Palmer 1965), Standardised Runoff Index (SRI) (Shukla and Wood 2008), Standardised Precipitation Evapotranspiration Index (SPEI) (Vicente-Serrano et al. 2010), Standardised Hydrological Index (SHI) (Sharma and Panu 2010), Agriculture Standardised Precipitation Index (aSPI) (Tigkas et al. 2018).However, the distribution fitted to the hydro climatological variables is based on the stationary assumption, i.e., parameters of the distribution do not change with time. With the increasing pace of changing climate, the stationary assumption is questionable (Milly et al. 2008). Therefore, the non-stationary approach becomes more popular nowadays, which incorporates the different climate signals in the distribution. As a result, the parameters of the distributions vary with climate change, which is more realistic in nature under the climate change scenarios. A very limited number of studies (Wang et al. 2015; Li et al. 2015; Das et al. 2020; Salvi and Ghosh 2016) are carried out to formulate the non-stationary drought index leaving a lot of scopes to design and formulate the drought events under non-stationary conditions.

7 Glacier and Climate Change

The water resources of the Himalayan river basin, such as the Ganges, are significantly influenced by the changes in the glacier and its extent, and over these river basins, India has 9,040 glaciers blanketing about 18,528 km^2 (Sharma et al. 2013). Pratap et al. (2016) studied the mass balance of glaciers over the Himalayan region for the last four decades and reported a net loss in glaciers. While analyzing the retreat of glaciers, Kulkarni et al. (2011) and Bolch et al. (2012) evaluated the annual rate of glacial shrinkage is about 0.2 to 0.7% in the period of 45 years (1960–2004). Kulkarni et al. (2007) investigated 466 glaciers in Himalayan basins using a remote sensing dataset. The examination showed a reduction of 21% in the glacial area, and the mean area was reduced from 1.4 km^2 to 0.32 km^2. Moreover, they emphasized that due to global warming, the glacial fragmentation and retreat will have a profound impact on the water resources over Himalayan river basins. Xu et al. (2009) advocated that the variability of water resources, biodiversity, shifting of ecosystem boundaries are affected due to the cascading effects of glacial retreat and increase in temperature. A recent study by Singh and Goyal (2018) compared the satellite image during 2014 and the surveyed toposheet of the Zemu glacier during 1935 and indicated changes in the glacial covering of Zemu glacier. Similarly, Bhambri et al. (2012) studied the frontal recession of Gangotri glacier (origin of the Ganga River) during 1965–2006 and stated a retreat of the glacier up to 819 $\pm$ 14 m. A detailed analysis of more than 80 different glaciers and the retreat rate obtained by different researchers are reported by Kulkarni et al. (2017) and readers are advised to follow the same. The summary report regarding the analysis of variability of glaciers through numerous investigations reports (i) the estimated glacial areal extend is $24,697 \pm 3260$ km^2; (ii) loss of glacial area for 40 years from 1960 is $12.6 \pm 7.5\%$; (iii) high uncertainty is associated in retreat estimation as a result of large variability in individual retreat; (iv) during 2000 to 2011 the mean snow cover area is varied between ~0.3 and 0.03 million km^2; (v) the average volume of snow for Ganga, Indus, and Brahmaputra basins are estimated as 9.3, 54.5, and 14.5 billion m^3 respectively.

The study of glacier and snow is generally carried out by the remote sensing techniques, which helps to monitor various parameters of snow and glaciers (e.g., glacial mass balance, the areal extent of glacier and snow cover, seasonal change-ability of albedo, and debris cover over the glacier. The satellite techniques have improved significantly that enable the user to measure the parameters of the snow and glacier using both satellite and field measurement (Kulkarni et al. 2017). In the investigation of the glacier, the satellite sensor characteristics viz., reflectance, spatial resolution, temporal coverage, and emission characteristics in various wavelengths are extensively used. The change in the glacial can lead to cause the variability in the streamflow pattern over the Himalayan rivers and tributaries. However, the scarcity of meteorological data, an improper estimate of changes over the Himalayan cryosphere, makes it more challenging to assess the contribution from changes in glaciers and snow cover to the streamflow in Himalayan river basins.

8 Adaptation Strategy and Future Focuses

According to IPCC (2001), the definition of adaptation referred to adjustment in the socio-economic and ecological system in response to the present and projected climate change effects in order to alleviate the adverse impact and take advantage of new opportunities. Adaptation of the effort made by the society to prepare for the negative or positive effects of climate change. These adjustments can be protective or opportunistic. Adaptation can include actions by individuals and communities, from a farmer planting more drought-resistant crops to a city, ensuring that new coastal infrastructure can accommodate future sea-level rise. Adger et al. (2005) argued that the implementation of a successful adaptation strategy involves certain criteria such as effectiveness in adaptation, the efficiency of adaptation, equity, and legitimacy in adaptation. However, the priority or weight for each criterion has emerged from soci-etal awareness, preparedness, and action. India, being the second largest populated country in the globe, has an agrarian economy. Climate change has severe threat on the water resources in terms of drought, flood, water quality, groundwater recharge etc. Despite of implementation of different policies by the government at various scales, water laws continues to remain inadequate and inconsistent in twenty-first century (Kumar and Bharat 2014). The reason is attributed to the various elements at local, regional, national, and international levels and their implementation issues.

To combat the impact of climate change on water resources, development in the technical aspects, improvement in the understanding and computational facilities, the estimation in the probable variability in the foreseeable future need considerable research activities for robust adaptation strategies and policies.Moreover, the multi-disciplinary research will assist in making scientific decisions on water and climate change policy (Goyal and Surampalli 2018). To improve the adaptation strategy and policy, the researchers should make significant improvement in climate change assessment and identify the suitable and robust approaches by involving key stake-holder organisations and local communities. Against this understanding, significant steps should be taken in order to access the quality data, upgradation of existing data, uncertainty analysis,and development of multicriteria based management strategy, vulnerability assessment, development of early warning system and maximize the network of data assimilation. Most importantly, the communication between the researchers and policy makers should be improved for a climate resilient nation or world as climate change impacts are highly underutilized in decision and policy making.

More precisely, in future the emphasis should divert towards the knowledge to reduce uncertainty. The use of long-range observational data of hydro-climatological variables in high mountain areas to improve the climate model outputs and the climate projections through improved historical long-range data and changes in glacier cover. Moreover, the understanding towards the glacier dynamics and permafrost melt will provide insight towards the climate change impact on the mountain areas and the rivers originated from them.

References

Abeysingha NS, Singh M, Sehgal VK, Khanna M, Pathak H (2016) Analysis of trends in streamflow and its linkages with rainfall and anthropogenic factors in Gomti River basin of North India. Theor Appl Climatol 123:785–799. https://doi.org/10.1007/s00704-015-1390-5

Adger WN, Arnell NW, Tompkins EL (2005) Successful adaptation to climate change across scales. Glob Environ Chang 15:77–86. https://doi.org/10.1016/j.gloenvcha.2004.12.005

Amarasinghe UA, Sharma BR, Aloysius N, Scott C, Smakhtin V, de Fraiture C (2005) Spatial variation in water supply and demand across river basins of India. Colombo, Sri Lanka

Anandhi A, Kannan N (2018) Vulnerability assessment of water resources—translating a theoretical concept to an operational framework using systems thinking approach in a changing climate: case study in Ogallala Aquifer. J Hydrol 557:460–474. https://doi.org/10.1016/j.jhydrol.2017.11.032

Anandhi A, Steiner JL, Bailey N (2016) A system's approach to assess the exposure of agricultural production to climate change and variability. Clim Change 136:647–659. https://doi.org/10.1007/s10584-016-1636-y

Arora H, Ojha CSP, Buytaert W, Kaushika GS, Sharma C (2017) Spatio-temporal trends in observed and downscaled precipitation over Ganga Basin. Hydrol Earth Syst Sci Discuss 1–19. https://doi.org/10.5194/hess-2017-388

Arora H, Sharma C, Ojha CSP, Kashyap D, Chaubey J, Neema M (2015) Spatio-temporal trend analysis of hydro-meteorological variables over Ganga Basin: a comparison between CMIP3 and CMIP5 Data. In: HYDRO 2015 International

Asoka A, Gleeson T, Wada Y, Mishra V (2017) Relative contribution of monsoon precipitation and pumping to changes in groundwater storage in India. Nat Geosci 10:109–117. https://doi.org/10.1038/ngeo2869

ASSOCHAM Report (2016) Drought situation to cost Rs 6.5 lakh crore to economy. The Associated Chambers of Commerce & Industry of India. Accessed May 16, 2018. https://www.assocham.org/newsdetail.php?id=5678

Bhambri R, Bolch T, Chaujar RK (2012) Frontal recession of Gangotri Glacier, Garhwal Himalayas, from 1965 to 2006, measured through highresolution remote sensing data. Curr Sci 102:489–494

Bhutiyani MR, Kale VS, Pawar NJ (2008) Changing streamflow patterns in the rivers of northwestern Himalaya: implications of global warming in the 20th century. Curr Sci 95:618–626

Bisht DS, Chatterjee C, Raghuwanshi NS, Sridhar V (2018) Spatio-temporal trends of rainfall across Indian river basins. Theor Appl Climatol 132:419–436. https://doi.org/10.1007/s00704-017-2095-8

Bolch T, Kulkarni A, Kaab A, Huggel C, Paul F, Cogley JG, Frey H, Kargel JS, Fujita K, Scheel M, Bajracharya S, Stoffel M (2012) The state and fate of Himalayan Glaciers. Science (80-) 336:310–314 . https://doi.org/10.1126/science.1215828

Caesar J, Janes T, Lindsay A, Bhaskaran B (2015) Temperature and precipitation projections over Bangladesh and the upstream Ganges, Brahmaputra and Meghna systems. Environ Sci Process Impacts 17:1047–1056. https://doi.org/10.1039/C4EM00650J

Chawla I, Mujumdar PP (2015) Isolating the impacts of land use and climate change on streamflow. Hydrol Earth Syst Sci 19:3633–3651. https://doi.org/10.5194/hess-19-3633-2015

Cosgrove WJ, Loucks DP (2015) Water management: current and future challenges and research directions. Water Resour Res 51:4823–4839. https://doi.org/10.1002/2014WR016869

Dadhwal VK, Aggarwal SP, Mishra N (2010) Hydrological simulation of Mahanadi river basin and impact of land use/land cover change on surface runoff using a macro scale hydrological model. In: InVol. 37 of Proceedings, International Society for Photogrammetry and Remote Sensing (ISPRS) TC VII Symposium–100 Years. ISPRS, Vienna, Austria, pp 165–170

Das J, Umamahesh NV (2018) Assessment and evaluation of potential climate change impact on monsoon flows using machine learning technique over Wainganga River basin, India. Hydrol Sci J 63:1020–1046. https://doi.org/10.1080/02626667.2018.1469757

Das J, Treesa A, Umamahesh NV (2018) Modelling impacts of climate change on a river basin: analysis of uncertainty using REA & possibilistic approach. Water Resour Manag. https://doi.org/10.1007/s11269-018-2046-x

Das J, Jha S, Goyal MK (2020) Non-stationary and copula-based approach to assess the drought characteristics encompassing climate indices over the Himalayan states in India. J Hydrol 580:124356. https://doi.org/10.1016/j.jhydrol.2019.124356

Deshpande NR, Kothawale DR, Kulkarni A (2016) Changes in climate extremes over major river basins of India. Int J Climatol 36:4548–4559. https://doi.org/10.1002/joc.4651

Diffenbaugh NS, Pal JS, Giorgi F, Gao X (2007) Heat stress intensification in the Mediterranean climate change hotspot. Geophys Res Lett 34:L11706. https://doi.org/10.1029/2007GL030000

Eisner S, Flörke M, Chamorro A, Daggupati P, Donnelly C, Huang J, Hundecha Y, Koch H, Kalugin A, Krylenko I, Mishra V, Piniewski M, Samaniego L, Seidou O, Wallner M, Krysanova V (2017) An ensemble analysis of climate change impacts on streamflow seasonality across 11 large river basins. Clim Change 141:401–417. https://doi.org/10.1007/s10584-016-1844-5

Fischer EM, Seneviratne SI, Lüthi D, Schär C (2007) Contribution of land-atmosphere coupling to recent European summer heat waves. Geophys Res Lett 34:L06707. https://doi.org/10.1029/2006GL029068

Gerland P, Raftery AE, ev ikova H, Li N, Gu D, Spoorenberg T, Alkema L, Fosdick BK, Chunn J, Lalic N, Bay G, Buettner T, Heilig GK, Wilmoth J (2014) World population stabilization unlikely this century. Science 346:234–237. https://doi.org/10.1126/science.1257469

Ghosh S, Das D, Kao SC, Ganguly AR (2012) Lack of uniform trends but increasing spatial variability in observed Indian rainfall extremes. Nat Clim Chang 2:86–91. https://doi.org/10.1038/nclimate1327

Goswami BN, Venugopal V, Sengupta D, Madhusoodanan MS, Xavier PK (2006) Increasing trend of extreme rain events over India in a warming environment. Science 314:1442–1445. https://doi.org/10.1126/science.1132027

Goyal MK, Surampalli RY (2018) Impact of climate change on water resources in India. J Environ Eng 144:04018054. https://doi.org/10.1061/(ASCE)EE.1943-7870.0001394

Held IM, Soden BJ (2006) Robust responses of the hydrological cycle to global warming. J Clim 19:5686–5699. https://doi.org/10.1175/JCLI3990.1

Hirabayashi Y, Kanae S, Emori S, Oki T, Kimoto M (2008) Global projections of changing risks of floods and droughts in a changing climate. Hydrol Sci J 53:754–772. https://doi.org/10.1623/hysj.53.4.754

Hoyos CD, Webster PJ (2007) The role of intraseasonal variability in the nature of Asian Monsoon precipitation. J Clim 20:4402–4424. https://doi.org/10.1175/JCLI4252.1

IMD (2016) Statement on climate of India during 2016. Earth System Science Organization, Ministry of Earth Sciences, India Meteorological Dept

Immerzeel WW, Pellicciotti F, Bierkens MFP (2013) Rising river flows throughout the twenty-first century in two Himalayan glacierized watersheds. Nat Geosci 6:742–745. https://doi.org/10.1038/ngeo1896

Immerzeel WW, van Beek LPH, Bierkens MFP (2010) Climate change will affect the Asian water towers. Science 328:1382–1385. https://doi.org/10.1126/science.1183188

India-WRIS (2011) India's water health

IPCC (2001) Climate change 2001: impacts, adaptation and vulnerability. Summary for Policy Makers, Geneva

Jenkins K, Warren R (2015) Quantifying the impact of climate change on drought regimes using the standardised precipitation index. Theor Appl Climatol 120:41–54. https://doi.org/10.1007/s00704-014-1143-x

Jha S, Das J, Sharma A, Hazra B, Goyal MK (2019) Probabilistic evaluation of vegetation drought likelihood and its implications to resilience across India. Glob Planet Change 176:23–35. https://doi.org/10.1016/j.gloplacha.2019.01.014

Kaur S (2009) Drought (India). The University of Tokyo Editoria. Accessed May 26, 2018. https://www.editoria.u-tokyo.ac.jp/projects/awci/5th/file/pdf/091216_awci/4.3-3-1_CR_India1.pdf

Kitoh A, Endo H, Krishna Kumar K, Cavalcanti IFA, Goswami P, Zhou T (2013) Monsoons in a changing world: a regional perspective in a global context. J Geophys Res Atmos 118:3053–3065. https://doi.org/10.1002/jgrd.50258

Kothawale DR, Rupa Kumar K (2005) On the recent changes in surface temperature trends over India. Geophys Res Lett 32:L18714. https://doi.org/10.1029/2005GL023528

Kothyari UC, Singh VP (1996) Rainfall and temperature trends in India. Hydrol Process 10:357–372

Kothyari UC, Singh VP, Aravamuthan V (1997) An investigation of changes in rainfall and temperature regimes of the Ganga Basin in India. Water Resour Manag 11:17–34

Krishnamurthy CKB, Lall U, Kwon HH (2009) Changing frequency and intensity of rainfall extremes over India from 1951 to 2003. J Clim 22:4737–4746. https://doi.org/10.1175/2009JCLI2896.1

Krysanova V, Vetter T, Eisner S, Huang S, Pechlivanidis I, Strauch M, Gelfan A, Kumar R, Aich V, Arheimer B, Chamorro A, van Griensven A, Kundu D, Lobanova A, Mishra V, Plötner S, Reinhardt J, Seidou O, Wang X, Wortmann M, Zeng X, Hattermann FF (2017) Intercomparison of regional-scale hydrological models and climate change impacts projected for 12 large river basins worldwide—a synthesis. Environ Res Lett 12:105002. https://doi.org/10.1088/1748-9326/aa8359

Kulkarni AV, Bahuguna IM, Rathore BP, Singh SK, Randhawa SS, Sood RK, Dhar S (2007) Glacial retreat in Himalaya using Indian remote sensing satellite data. Curr Sci 92:69–74

Kulkarni AV, Nayak S, Pratibha S (2017) Variability of glaciers and snow cover.In: Rajeevan M, Nayak S (eds) Observed climate variability and change over the Indian region. Springer, Singapore, pp 193–219

Kulkarni AV, Rathore BP, Singh SK, Bahuguna IM (2011) Understanding changes in the Himalayan cryosphere using remote sensing techniques. Int J Remote Sens 32:601–615. https://doi.org/10. 1080/01431161.2010.517802

Kumar R, Singh RD, Sharma KD (2005) Water resources of India. Curr Sci 89:794–811

Kumar V, Jain SK (2011) Trends in rainfall amount and number of rainy days in river basins of India (1951–2004). Hydrol Res 42:290–306. https://doi.org/10.2166/nh.2011.067

Kumar V, Jain SK, Singh Y (2010) Analysis of long-term rainfall trends in India. Hydrol Sci J 55:484–496. https://doi.org/10.1080/02626667.2010.481373

Kumar VS, Bharat GK (2014) Perspectives on a water resource policy for India. Discussion Paper 1. Energy and Resources Institute, New Delhi, India

Lacombe G, McCartney M (2014) Uncovering consistencies in Indian rainfall trends observed over the last half century. Clim Change 123:287–299. https://doi.org/10.1007/s10584-013-1036-5

Lenton TM, Held H, Kriegler E, Hall JW, Lucht W, Rahmstorf S, Schellnhuber HJ (2008) Tipping elements in the Earth's climate system. Proc Natl Acad Sci 105:1786–1793. https://doi.org/10. 1073/pnas.0705414105

Li JZ, Wang YX, Li SF, Hu R (2015) A nonstationary standardized precipitation index incorporating climate indices as covariates. J Geophys Res 120:12082–12095. https://doi.org/10.1002/2015JD 023920

Mallya G, Mishra V, Niyogi D, Tripathi S, Govindaraju RS (2015) Trends and variability of droughts over the Indian monsoon region. Weather Clim Extrem 12:43–68. https://doi.org/10.1016/j.wace. 2016.01.002

Masood MJF, Yeh P, Hanasaki N, Takeuchi K (2015) Model study of the impacts of future climate change on the hydrology of Ganges-Brahmaputra-Meghna basin. Hydrol Earth Syst Sci 19:747–770. https://doi.org/10.5194/hess-19-747-2015

Mckee TB, Doesken NJ, Kleist J (1993) The relationship of drought frequency and duration to time scales. In: AMS 8th Conference on Appl Climatol, pp 179–184. https://doi.org/citeulike-article-id:10490403

Mekonnen MM, Hoekstra AY (2016) Four billion people facing severe water scarcity. Sci Adv 2:e1500323–e1500323. https://doi.org/10.1126/sciadv.1500323

Milly PCD, Betancourt J, Falkenmark M, Hirsch RM, Kundzewicz ZW, Lettenmaier DP, Stouffer RJ (2008) Stationarity Is dead: whither water management? Science (80-) 319:573–574. https://doi.org/10.1126/science.1151915

Mishra N, Khare D, Shukla R, Kumar K (2014) Trend analysis of air temperature time series by Mann Kendall test—a case study of upper Ganga Canal Command (1901–2002). Br J Appl Sci Technol 4:4066–4082. https://doi.org/10.9734/BJAST/2014/8650

Mittal N, Mishra A, Singh R, Kumar P (2014) Assessing future changes in seasonal climatic extremes in the Ganges river basin using an ensemble of regional climate models. Clim Change 123:273–286. https://doi.org/10.1007/s10584-014-1056-9

Mondal A, Khare D, Kundu S (2015) Spatial and temporal analysis of rainfall and temperature trend of India. Theor Appl Climatol 122:143–158. https://doi.org/10.1007/s00704-014-1283-z

Moors EJ, Groot A, Biemans H, van Scheltinga CT, Siderius C, Stoffel M, Huggel C, Wiltshire A, Mathison C, Ridley J, Jacob D, Kumar P, Bhadwal S, Gosain A, Collins DN (2011) Adaptation to changing water resources in the Ganges basin, Northern India. Environ Sci Policy 14:758–769. https://doi.org/10.1016/j.envsci.2011.03.005

MOWR (Ministry of Water Resources) (2014) Ganga Basin

Murari KK, Ghosh S, Patwardhan A, Daly E, Salvi K (2015) Intensification of future severe heat waves in India and their effect on heat stress and mortality. Reg Environ Chang 15:569–579. https://doi.org/10.1007/s10113-014-0660-6

Nepal S, Shrestha AB (2015) Impact of climate change on the hydrological regime of the Indus, Ganges and Brahmaputra river basins: a review of the literature. Int J Water Resour Dev 31:201–218. https://doi.org/10.1080/07900627.2015.1030494

Nohara D, Kitoh A, Hosaka M, Oki T (2006) Impact of climate change on river discharge projected by multimodel ensemble. J Hydrometeorol 7:1076–1089. https://doi.org/10.1175/JHM531.1

Ojha R, Nagesh Kumar D, Sharma A, Mehrotra R (2013) Assessing severe drought and wet events over India in a future climate using a nested bias-correction approach. J Hydrol Eng 18:760–772. https://doi.org/10.1061/(ASCE)HE.1943-5584.0000585

Pai DS, Guhathakurta P, Kulkarni A, Rajeevan MN (2017) Variability of meteorological droughts over India. In: Rajeevan M, Nayak S (eds) Observed climate variability and change over the Indian region. Springer, Singapore, pp 73–87

Pai DS, Nair SA, Ramanathan AN (2013) Long term climatology and trends of heat waves over India during the recent 50 years. Mausam 64:585–604

Palmer WC (1965) Meteorological drought. U.S. Weather Bur. Res. Pap. No. 45 58

Panda DK, Kumar A, Ghosh S, Mohanty RK (2013) Streamflow trends in the Mahanadi River basin (India): linkages to tropical climate variability. J Hydrol 495:135–149. https://doi.org/10.1016/j.jhydrol.2013.04.054

Pervez MS, Henebry GM (2014) Projections of the Ganges-Brahmaputra precipitation—downscaled from GCM predictors. J Hydrol 517:120–134. https://doi.org/10.1016/j.jhydrol.2014.05.016

Planning Commission 2011 Report of working group on flood management and region specific issues for XII plan. New Delhi, India

Pratap B, Dobhal DP, Bhambri R, Mehta M, Tewari VC (2016) Four decades of glacier mass balance observations in the Indian Himalaya. Reg Environ Chang 16:643–658. https://doi.org/10.1007/s10113-015-0791-4

Rajeevan M, Bhate J, Jaswal AK (2008) Analysis of variability and trends of extreme rainfall events over India using 104 years of gridded daily rainfall data. Geophys Res Lett 35:L18707. https://doi.org/10.1029/2008GL035143

Raju BCK, Nandagiri L (2017) Analysis of historical trends in hydrometeorological variables in the upper Cauvery Basin, Karnataka, India. Curr Sci 112:577–587. https://doi.org/10.18520/cs/v112/i03/577-587

Rohini P, Rajeevan M, Srivastava AK (2016) On the variability and increasing trends of heat waves over India. Sci Rep 6:26153. https://doi.org/10.1038/srep26153

Salvi K, Ghosh S (2016) Projections of extreme dry and wet spells in the 21st century India using stationary and non-stationary standardized precipitation indices. Clim Change 139:667–681. https://doi.org/10.1007/s10584-016-1824-9

Sarma R, Singh DK (2019) Spatio-temporal analysis of drought and aridity in Gomti basin. Curr Sci 116:919–925

Sharma AK, Singh SK, Kulkarni AV, Ajai (2013) Glacier inventory in Indus, Ganga and Brahmaputra basins of the Himalaya. Natl Acad Sci Lett 36:497–505. https://doi.org/10.1007/s40009-013-0167-6

Sharma C, Ojha CSP (2018) Detection of changes in temperature in Upper Ganga River Basin. AGU Poster 35:35556. https://doi.org/10.1002/essoar.10500106.1

Sharma KP, Moore B, Vorosmarty CJ (2000) Anthropogenic, climatic, and hydrologic trends in the Kosi Basin, Himalaya. Clim Change 47:141–165

Sharma S, Mujumdar P (2017) Increasing frequency and spatial extent of concurrent meteorological droughts and heatwaves in India. Sci Rep 7:15582. https://doi.org/10.1038/s41598-017-15896-3

Sharma TC, Panu US (2010) Analytical procedures for weekly hydrological droughts: a case of Canadian rivers. Hydrol Sci J 55:79–92. https://doi.org/10.1080/02626660903526318

Shewale MP, Kumar S (2005) Climatological features of drought incidences in India. Pune, India

Shukla S, Wood AW (2008) Use of a standardized runoff index for characterizing hydrologic drought. Geophys Res Lett 35:L02405. https://doi.org/10.1029/2007GL032487

Singh D, Gupta RD, Jain S (2014) Study of long-term trend in river discharge of Sutlej River (N-W Himalayan Region). Geogr Environ Sustain 07:87–95. https://doi.org/10.24057/2071-9388-2014-7-3-87-96

Singh P, Kumar V, Thomas T, Arora M (2008) Changes in rainfall and relative humidity in river basins in northwest and central India. Hydrol Process 22:2982–2992. https://doi.org/10.1002/hyp.6871

Singh V, Goyal MK (2018) An improved coupled framework for Glacier classification: an integration of optical and thermal infrared remote-sensing bands. Int J Remote Sens 1–29. https://doi.org/10.1080/01431161.2018.1468104

Sonali P, Kumar DN, Nanjundiah RS (2017) Intercomparison of CMIP5 and CMIP3 simulations of the 20th century maximum and minimum temperatures over India and detection of climatic trends. Theor Appl Climatol 128:465–489. https://doi.org/10.1007/s00704-015-1716-3

Srivastava AK, Kothawale DR, Rajeevan MN (2017) Variability and long-term changes in surface air temperatures over the Indian Subcontinent.In: Rajeevan M, Nayak S (eds) Observed climate variability and change over the Indian region. Springer, Singapore, pp 17–35

Stone D, Auffhammer M, Carey M, Hansen G, Huggel C, Cramer W, Lobell D, Molau U, Solow A, Tibig L, Yohe G (2013) The challenge to detect and attribute effects of climate change on human and natural systems. Clim Change 121:381–395. https://doi.org/10.1007/s10584-013-0873-6

Syed TH, Webster PJ, Famiglietti JS (2014) Assessing variability of evapotranspiration over the Ganga river basin using water balance computations. Water Resour Res 50:2551–2565. https://doi.org/10.1002/2013WR013518

Tabari H, Nikbakht J, Talaee PH (2013) Hydrological drought assessment in Northwestern Iran based on Streamflow Drought Index (SDI). Water Resour Manag 27:137–151. https://doi.org/10.1007/s11269-012-0173-3

Tigkas D, Vangelis H, Tsakiris G (2018) Drought characterisation based on an agriculture-oriented standardised precipitation index. Theor Appl Climatol 135:1435–1447. https://doi.org/10.1007/s00704-018-2451-3

Tscharntke T, Clough Y, Wanger TC, Jackson L, Motzke I, Perfecto I, Vandermeer J, Whitbread A (2012) Global food security, biodiversity conservation and the future of agricultural intensification. Biol Conserv 151:53–59. https://doi.org/10.1016/j.biocon.2012.01.068

Turner AG, Annamalai H (2012) Climate change and the South Asian summer monsoon. Nat Clim Chang 2:587–595. https://doi.org/10.1038/nclimate1495

UNFAO (2013) AQUASTAT database. Rome, Italy

Vicente-Serrano SM, Beguería S, López-Moreno JI (2010) A multiscalar drought index sensitive to global warming: the standardized precipitation evapotranspiration index. J Clim 23:1696–1718. https://doi.org/10.1175/2009JCLI2909.1

Vörösmarty CJ, McIntyre PB, Gessner MO, Dudgeon D, Prusevich A, Green P, Glidden S, Bunn SE, Sullivan CA, Liermann CR, Davies PM (2010) Global threats to human water security and river biodiversity. Nature 467:555–561. https://doi.org/10.1038/nature09440

Wang Y, Li J, Feng P, Hu R (2015) A time-dependent drought index for non-stationary precipitation series. Water Resour Manag 29:5631–5647. https://doi.org/10.1007/s11269-015-1138-0

Wasson RJ, Sundriyal YP, Chaudhary S, Jaiswal MK, Morthekai P, Sati SP, Juyal N (2013) A 1000-year history of large floods in the Upper Ganga catchment, central Himalaya, India. Quat Sci Rev 77:156–166. https://doi.org/10.1016/j.quascirev.2013.07.022

Whitehead PG, Barbour E, Futter MN, Sarkar S, Rodda H, Caesar J, Butterfield D, Jin L, Sinha R, Nicholls R, Salehin M (2015) Impacts of climate change and socio-economic scenarios on flow and water quality of the Ganges, Brahmaputra and Meghna (GBM) river systems: low flow and flood statistics. Environ Sci Process Impacts 17:1057–1069. https://doi.org/10.1039/C4EM00619D

Xu C (1999) Climate change and hydrologic models: a review of existing gaps and recent research developments. Water Resour Manag 13:369–382. https://doi.org/10.1023/A:1008190900459

Xu J, Grumbine RE, Shrestha A, Eriksson M, Yang X, Wang Y, Wilkes A (2009) The melting Himalayas: cascading effects of climate change on water, biodiversity, and livelihoods. Conserv Biol 23:520–530. https://doi.org/10.1111/j.1523-1739.2009.01237.x

Zhang X, Obringer R, Wei C, Chen N, Niyogi D (2017) Droughts in India from 1981 to 2013 and implications to wheat production. Sci Rep 7:44552. https://doi.org/10.1038/srep44552

Detection of Changes in Twentieth Century Precipitation in the Ganga River Basin

Chetan Sharma and C. S. P. Ojha

Abstract Analysis of changes in 20th-century seasonal precipitation in the Ganga river basin is presented in this paper. High resolution gridded precipitation data were used for this purpose. A decrease in monsoon, as well as winter precipitation, was observed in the basin. No changes were observed in the pre- and post-monsoon seasons, however increasing trend in the precipitation was observed during the post-monsoon season, but these were found insignificant using the Mann-Kendal test at 10% significance level. Trend free pre-whitening (TFPW) method was used to remove the effect of significant lag1 autocorrelation from the data. The change points were estimated using Mann–Whitney–Pettit (MWP) method. It was observed that the year of change was earlier for locations where relatively higher decreasing trends were found. The change point in the winter season was found earlier than for the monsoon season. Overall it was observed that precipitation started decreasing significantly after the year 1960 in the basin.

Keywords Ganga river basin · Precipitation · TFPW · Mann–kendall · Climate change

1 Introduction

Intergovernmental Panel on Climate Change (IPCC), in its fifth assessment report (AR5) indicated the effect of global warming on rising sea level, increase in the frequency of extreme events, etc. (IPCC 2013). It was also reported that the changes in the climate state of different variables started from mid of the twentieth century. Effect of climate change on different hydrometeorological variables is reported in

C. Sharma (✉)
Department of Civil Engineering, Madhav Institute of Technology and Science, Gwalior, MP, India
e-mail: chetan.cvl@gmail.com

C. S. P. Ojha
Department of Civil Engineering, Indian Institute of Technology Roorkee, Uttarakhand, India
e-mail: cspojha@gmail.com

© The Author(s), under exclusive license to Springer Nature Switzerland AG 2021
M. S. Chauhan and C. S. P. Ojha (eds.), *The Ganga River Basin: A Hydrometeorological Approach*, Society of Earth Scientists Series,
https://doi.org/10.1007/978-3-030-60869-9_17

257

different studies(Barnett et al. 2005, 2008; Choi et al. 2009; Ojha et al. 2015; Mariotti et al. 2015; Sharma and Ojha 2019a, 2021).

The climate change studies require the estimation of trends and time of change using different statistical methods (Kundzewicz and Robson, 2004; Reeves et al. 2007; Khaliq et al. 2009; Sharma and Ojha 2019b). The parametric, non-parametric, and Bayesian methods are widely used to estimate trends, but the non-parametric methods are preferred as these are assumed to be free from the assumption of the parent distribution of data. Sen's nonparametric estimate (Sen 1968) along with Mann-Kendal (MK) (Mann 1945; Kendall 1975) method is widely used to find the magnitude, the direction of trend and estimation of the significance of the trend respectively (Modarres and Sarhadi 2009; Caloiero et al. 2011; Westra et al. 2013; Akinsanola and Ogunjobi 2017; Sharma and Ojha 2018a, b). The presence of serial correlation affects the power of the MK test (Yue et al. 2002). The pre-Whitening procedure was adopted by different researchers to remove the effect of autocorrelation on the significance of trends. Yue et al. (2002) presented a trend free pre-whitening method (TFPW) which removes the significant lag1 autocorrelation from the data to minimize the effect of autocorrelation on different statistical test. TFPW method have been adopted to estimate the trend of rainfall, runoff and other variables (Olguin and Sandoval 2017; Chu et al. 2018; Glas et al. 2019).

Study of hydrometeorological variables in different river basins of the world have been reported in many articles (Debeer et al. 2015; Gelfan et al. 2017; Schumacher et al. 2018; Dada et al. 2018; Abe et al. 2018; Lobanova et al. 2018; Liu et al. 2018; Sharma et al. 2019). Kothyari et al. (1997) studied the changes in precipitation and temperature in the GRB and found an increasing trend in temperature and decrease in precipitation after the year 1960. Basistha et al. (2009) studied the changes in precipitation patterns during different seasons in Uttarakhand, India, which comes under the Ganga river basin (GRB). It was found that overall the precipitation started decreasing after the year 1964. Only a few of the studies are available in the Himalayan region due to the non-availability of the data (Singh et al. 1995; Borgaonkar and Pant 2001). The changes in seasonal precipitation patterns in the GRB and time of change have been assessed in this study using sufficiently long high-resolution gridded precipitation data.

2   Study Area and Data

The Ganga river is considered one of the holiest rivers in India, and its basin is the home of more than 500 million population in India, Nepal, Tibet, and Bangladesh (Bharati et al. 2016). Ganga river basin (GRB) is one of the most populous river basins in the world where almost half of the Indian population lives (World Bank 2011). The catchment area of Ganga river basin is around 1,087,300 km^2 which lies between latitudes 22°30′ N to 31°30′ N and longitude 73°30′E to 89°00′ E (NIH 2015). The confluence of river Alaknanda and Bhagirathi at Devprayag in Uttarakhand district is the beginning of the river Ganga. Gangotri glacier is the primary source of the Ganga river and also originating place of river Bhagirathi (Krishna-Murti 1991). Some of the highest peaks of the world, i.e., Mt. Everest (8848 m), Kanchenjunga (8598 m), Makalu (8481 m), Dhaulagiri (8172 m), etc. lies in the Ganga river basin (Yang and Zheng 2004). The Ganga river basin is topographically complex and experiences a diversified climate. The northwest regions of the Ganga river basin experience lesser precipitation while the coastal areas receive very high rainfall. The study area, i.e., the GRB, is shown in Fig. 1.

Mitchell and Jones (2005) updated high resolution gridded climatological dataset developed by (New et al. 1999, 2000) covering the landmasses except for Antarctica. They filled the missing data using the mean of 1961–1999 values at each grid cell. The Climatic Research Unit (CRU) updated Mitchell and Jones (2005) in the year 2009 and updated regularly 2009 onwards using a semi-automatic approach. High resolution monthly average precipitation and temperature data at the grid interval of 0.5° × 0.5° provided by CRU was used in this study (data available at https://www.cru.uea.ac.uk/data). Each dataset was pre-screened; grid points with erroneous/missing values for a sufficiently long length was not considered in the study.

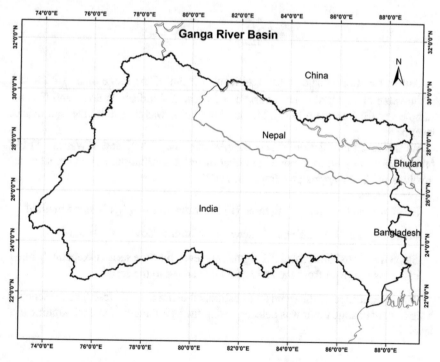

Fig. 1 Ganga River Basin (Basin boundary is shown by dark black polygon)

3   Methodology

The Indian Meteorological Department (IMD) divided the water year into four seasons, i.e., Monsoon (June–September), Post-monsoon (October–November), Winter (December-February) and Pre-Monsoon (March–May)(IMD-FAQ). The monthly total precipitation values obtained from CRU were changed to seasonal total, as suggested by IMD, after prescreening of the data for errors.

The presence of serial correlation affects the performance of statistical tests like MK and MWP. So, it is needed to find the presence of significant autocorrelation before trend or change point estimation. Autocorrelation at lag k for the time series $x_1, x_2, ...x_n$, where n represents the number of data points in the time series is given by

$$r_k = \frac{c_k}{\sigma^2} \tag{1}$$

where,

$$c_k = \frac{1}{n} \sum_{i=1}^{n-k} (x_i - \mu)(x_{i+k} - \mu) \tag{2}$$

Autocorrelation is said to be significant if it out of the range of $\pm\frac{k}{\sqrt{n}}$, where k is the value of standard normal variate at a defined significance level, and N is the sample size (Shanmugam 1997). The autocorrelation was checked at the significance level of 5%.

As the serial correlation affects the performance of MK and MWP, the TFPW procedure is adopted to remove the effect of serial correlation. Following steps are involved in TFPW approach (Yue et al. 2002).

Step1 for the time series, the trend (ß) is estimated using Sen's slope method.

Setp2 lag1 autocorrelation r_1 is determined after removing the trend ß.

Step3 if r_1 is insignificant at 5% significant level, the time series is kept unchanged, else r_1 is removed from the series, and ß is added to the series.

The significance of the trends was estimated using the MK test. The presence of a significant change point was detected using the MWP method at a 5% significance level.

4   Results and Discussions

- **Autocorrelation**

The significant autocorrelation at a 5% significance level for different seasons in the Ganga river basin is shown in Fig. 2. Around 24% of the grid points indicate the presence of significant lag1 autocorrelation considering monsoon total precipitation. Most of these grid points are located in the western and eastern regions in the basin. Similarly, ~24% grid points show the presence of autocorrelation in post-monsoon season; however, these grid points are located primarily from north-west to central south regions. Very few of the locations in the upper reaches of the Ganga river basin show the presence of significant lag1 autocorrelation for winter and pre-monsoon seasons (not shown by the figure). The results indicate that the presence of autocorrelation cannot be neglected for the estimation of trends or to find a change point.

- **Trends and Significance**

The grid points where significant autocorrelation was found, the trends were computed after removing the effect of autocorrelation using TFPW approach. The trends and its significance for monsoon season in the Ganga river basin are shown in Fig. 3a, b, respectively. The critical values of Z at 1, 5 and 10% significance level are 2.576, 1.96, and 1.645, respectively. Most of the locations in the basin indicated falling trends in the monsoon season up to 1.21 mm/year. A majority of the falling trends were found significant at a significance level of 10% ore lower, which indicates that the monsoon precipitation in these regions has decreased significantly. However, in some western regions and near the Bay of Bengal, a small increase in precipitation was found, but these were not significant even at the significance level of 10%.

The east and west regions showed a minor increase in the post-monsoon precipitation, while the south and north regions indicated a minor decrease in the precipitation. However, only a few of the decreasing trends were found significant at a 10% significance level. Fukushima et al. (2019) also studied the long-term changes of post-monsoon precipitation in India and found no significant changes (Fig. 4).

A majority of the region showed decreases in the winter precipitation, except some northern and central southern regions where a small rise in the winter precipitation was observed. Most of the decreasing trends were found significant at 10% or lower significance level. However, all of the increasing trends were found insignificant at a 10% significant level. This indicates that the winter precipitation in the GRB has somewhat reduced. The reduction in the winter precipitation in the glacier dominant regions at higher Himalayas may deteriorate the conditions of glaciers,which are already facing a threat to existence due to global warming (Sharma and Ojha 2021; Maurer et al. 2019) (Fig. 5).

The majority of the region indicated an increase in pre-monsoon precipitation. The increase in the precipitation was found up to 0.81 mm/year in some regions; however, almost all of the trend was found insignificant at a significant level of 10%.

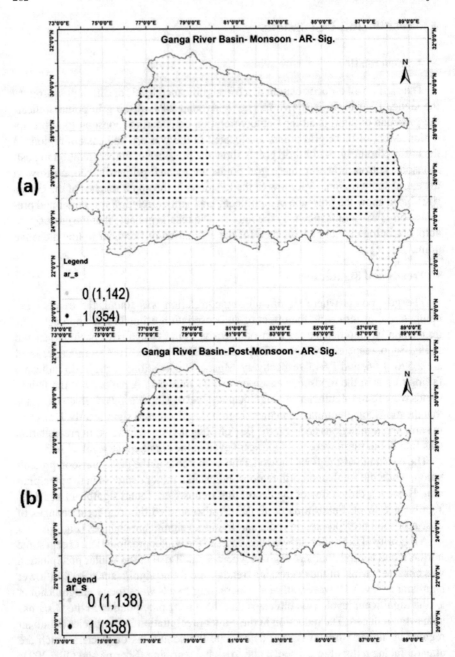

Fig. 2 Presence of autocorrelation and 5% significance level in GRB considering, **a** Monsoon season and, **b** Post-Monsoon. The significant autocorrelation is shown by black dots

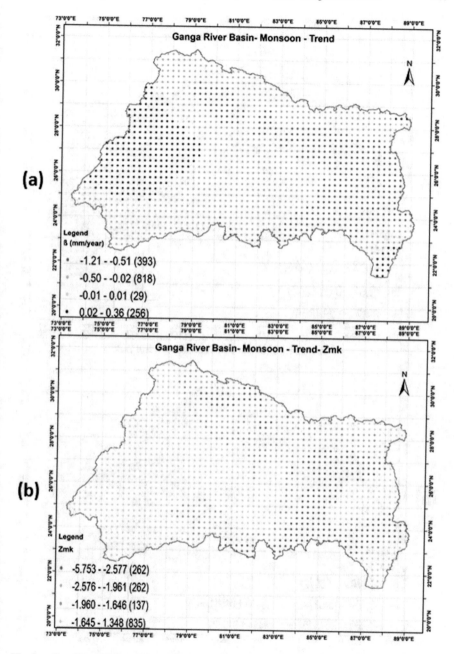

Fig. 3 **a** Trend ß and, **b** significance in GRB in monsoon season

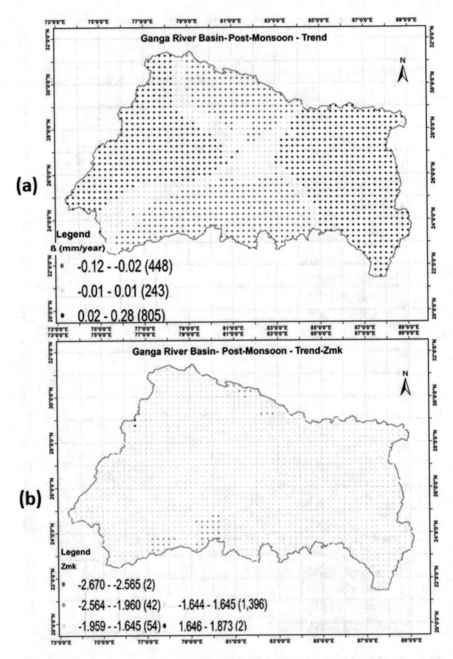

Fig. 4 **a** Trend ß and, **b** significance in GRB in Post-monsoon season

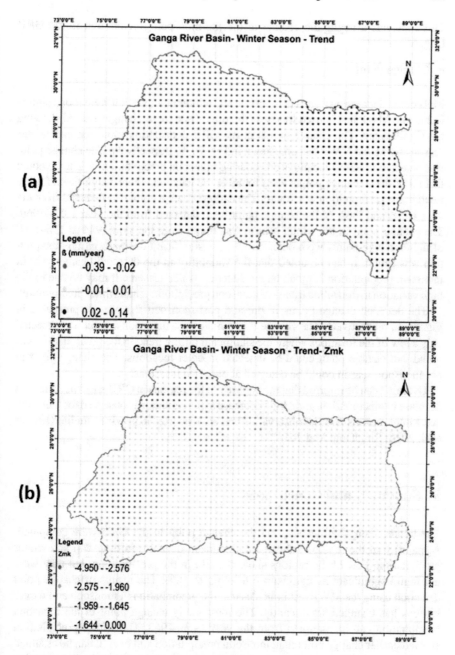

Fig. 5 **a** Trend ß and, **b** significance in GRB in winter season

This indicates a negligible change in the pre-monsoon precipitation in the GRB (Fig. 6).

- **Change Point**

Most of the decreasing trends in monsoon and winter precipitation were found significant. The MWP test is used to assess the time when these changes started showing the significance. The change points in the GRB considering monsoon and winter seasons using the MWP test are shown in Fig. 7a, b, respectively. The change point was detected in ~34% locations considering monsoon precipitation. It is to be noted that all of these change points were found at the location where a significant falling trend in the precipitation was found. The earliest change point was detected from the year 1951, and most of the change point were detected from the year 1956–1960. The change point in the winter season was found from the year 1944 to 1964. All of these change points were detected where a significant decrease in the precipitation was found. It can be noted that the variability in the detected change year in monsoon precipitation is much larger than the winter seasons. It may be due to the high variation in the falling trends of monsoon precipitation than winter precipitation.

The detected change points in pre and post-monsoon seasons are presented in Fig. 8a, b, respectively. The change point in the post-monsoon season was detected only a few of the locations where the significant falling trend in the precipitation was found. No change point could be detected in other locations. The change point in pre-monsoon season could be detected at only two locations.

Overall it can be inferred that the majority of the regions in GRB are experiencing a decrease in precipitation, and a high decrease in precipitation is observed for monsoon and winter seasons. Similar observation was also reported in previous studies (Kumar et al. 2010; Singh and Mal 2014).

5   Summary and Conclusions

The trends, their significance, and the change point in the GRB were computed. A majority of the regions indicated a significant decrease in monsoon and winter precipitation. Few of the regions showed a rise in the precipitation, but they were not found significant at significance level up to 10%. The results of change point detection using the MWP test indicates that the precipitation in monsoon and winter seasons has changed significantly. The most likely change point in the monsoon precipitation can be assumed from the year 1956–1960. The winter precipitation showed earlier change point than monsoon precipitation and overall can be assumed from the year 1950–1954. As both monsoon and winter precipitation is decreasing significantly, it can be concluded that the precipitation pattern in these seasons started to change from mid of the twentieth century. Only a few of the post and pre-monsoon data showed a significant change point. Overall it can be said that the monsoon and winter precipitation, which contribute the majority of the total annual rainfall in

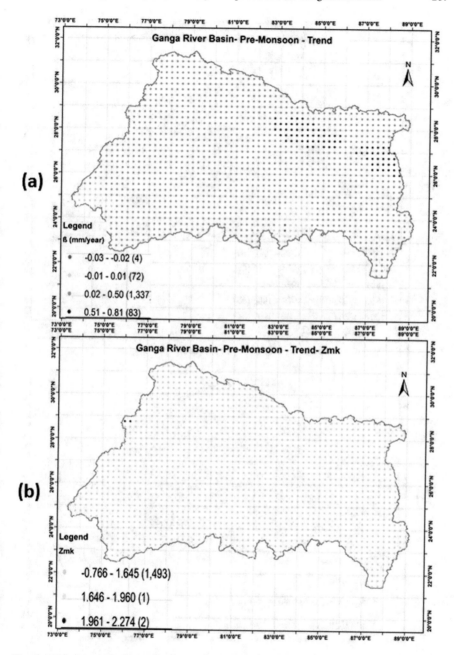

Fig. 6 **a** Trend ß and **b** significance in GRB in Pre-monsoon season

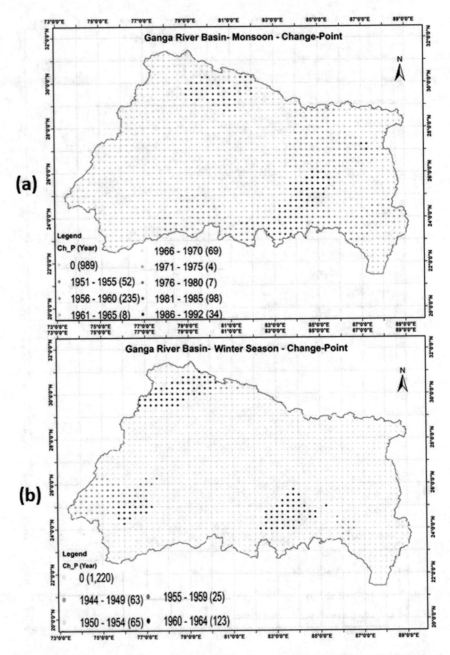

Fig. 7 Changepoint detection in GRB using MWP test considering, **a** Monsoon and, **b** Winter seasons

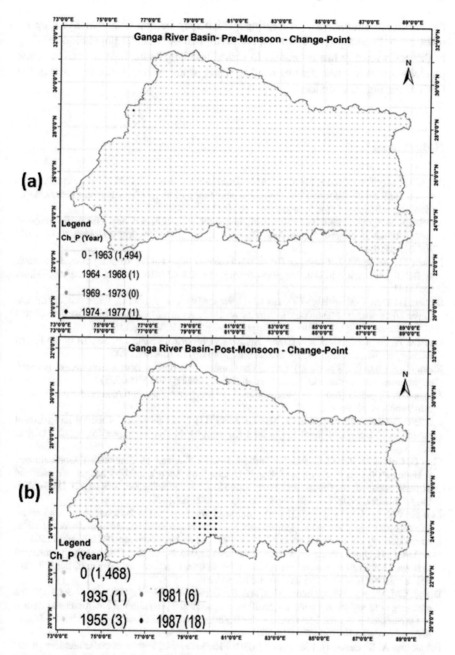

Fig. 8 Change point detection in GRB using MWP test considering, **a** Pre-Monsoon, **b** Post-Monsoon

the basin, are significantly decreasing, which may be due to global warming on other natural/anthropogenic factors. As more than 40% of the population residing in the basin already has no access to clean/sufficient water, further decline in the precipitation may cause serious trouble. Thus, water resource management in the GRB needs urgent attention.

References

Abe CA, Lobo FDL, Dibike YB, Costa MPF, Dos Santos V, Novo EMLM (2018) Modelling the effects of historical and future land cover changes on the hydrology of an Amazonian basin. Water (Switzerland) 932(1–932):19. https://doi.org/10.3390/w10070932

Akinsanola AA, Ogunjobi KO (2017) Recent homogeneity analysis and long-term spatio-temporal rainfall trends in Nigeria. Theor Appl Climatol 128(275–18):289. https://doi.org/10.1007/s00704-015-1701-x

Barnett TP, Adam JC, Lettenmaier DP (2005) Potential impacts of a warming climate on water availability in snow-dominated regions. Nature 438(303–438):309. https://doi.org/10.1038/nature04141

Barnett TP, Piere DW, Hidalgo HG, Bonfils C, Santer BD, Das T, Bala G, Wood AW, Mirin TA, Cayan DR, Dettinger MD (2008) Human-induced changes in the hydrology of the western United States. Science 319(1080–319):1083. https://doi.org/10.1126/science.1152538

Basistha A, Arya DS, Goel NK (2009) Analysis of historical changes in rainfall in the Indian Himalayas. Int J Climatol 29555–29572. https://doi.org/10.1002/joc.1706

Bharati L, Sharma BR, Smakhtin V (2016) The Ganges River Basin : status and challenges in water, environment and livelihoods, 1st edn. Routledge, London (9781138900325)

Borgaonkar H, Pant G (2001) Long-term climate variability over monsoon Asia as revealed by some proxy sources. Mausam 1:9–22

Caloiero T, Coscarelli R, Ferrari E, Mancini M (2011) Trend detection of annual and seasonal rainfall in Calabria (Southern Italy). Int J Climatol 31(44–31):56. https://doi.org/10.1002/joc.2055

Choi G, Collins D, Ren G, Trewin B, Baldi M, Fukuda Y, Afzaal M, Pianmana T, Gomboluudev P, Huong PTT, Lias N, Kwon W-T, Boo K-O, Cha Y-M, Zhou Y (2009) Changes in means and extreme events of temperature and precipitation in the Asia-Pacific Network region, 1955–2007. Int. J. Climatol. 29:1906–1929. https://doi.org/10.1002/joc.1979

Chu H, Wei J, Li J, Li T (2018) Investigation of the relationship between runoff and atmospheric oscillations, sea surface temperature, and local-scale climate variables in the Yellow River headwaters region. Hydrol Process 32(1434–1432):1448. https://doi.org/10.1002/hyp.11502

Dada OA, Li G, Qiao L, Bello YAA, Anifowose YAA (2018) Recent Niger Delta shoreline response to Niger River hydrology: conflict between forces of nature and humans. J Afr Earth Sci 139:222–231. https://doi.org/10.1016/j.jafrearsci.2017.12.023

Debeer CM, Wheater HS, Quinton WL, Carey SK, Stewart RE, MacKay MD, Marsh P (2015) The changing cold regions network: observation, diagnosis and prediction of environmental change in the Saskatchewan and Mackenzie River Basins, Canada. Sci China Earth Sci 58(46–58):60. https://doi.org/10.1007/s11430-014-5001-6

Fukushima A, Kanamori H, Matsumoto J (2019) Regionality of long-term trends and interannual variation of seasonal precipitation over India. Prog Earth Planet Sci 6(1–6):20. https://doi.org/10.1186/s40645-019-0255-4

Gelfan A, Gustafsson D, Motovilov Y, Arheimer B, Kalugin A, Krylenko I, Lavrenov A (2017) Climate change impact on the water regime of two great Arctic rivers: modeling and uncertainty issues. Climatic Change 141:499–515. https://doi.org/10.1007/s10584-016-1710-5

Glas R, Burns D, Lautz L (2019) Historical changes in New York State streamflow: Attribution of temporal shifts and spatial patterns from 1961 to 2016. J Hydrol 574:308–323. https://doi.org/10.1016/j.jhydrol.2019.04.060

IMD Frequently Asked Questions (FAQs) on Monsoon. https://webcache.googleusercontent.com/search?q=cache:1ojFVsj_Su0J:www.imd.gov.in/section/nhac/monsoonfaq.pdf+&cd=15&hl=en&ct=clnk&gl=in. Accessed 1 Aug 2019

IPCC (2013) Summary for policymakers. In: Climate Change 2013: the physical science basis. Contribution of Working Group I to the Fifth Assessment Report of the Intergovernmental Panel on Climate Change (Stocker, T.F., D. Qin, G.-K. Plattner, M. Tignor, S.K. Allen, J. Boschung, A. Nauels, Y. Xia). Cambridge, United Kingdom and New York, NY, USA

Kendall MG (1975) Rank correlation methods, 4th edn. Charless Griffin, London (0195208374)

Khaliq MN, Ouarda TBMJ, Gachon P, Sushama l, Hilaire AS (2009) Identification of hydrological trends in the presence of serial and cross correlations: a review of selected methods and their application to annual flow regimes of Canadian rivers. J Hydrol 368:117–130. https://doi.org/10.1016/j.jhydrol.2009.01.035

Kothyari UC, Singh VP, Aravamuthan V (1997) An investigation of changes in rainfall and temperature regimes of the Ganga basin in India. Water Resour Manag 11:17–34. https://doi.org/10.1023/A:1017936123283

Krishna-Murti CR (1991) Ganga, a scientific study. Northern Book Centre, India (9788172110215)

Kumar V, Jain SK, Singh Y (2010) Analysis of long-term rainfall trends in India. Hydrol Sci J 55:484–496. https://doi.org/10.1080/02626667.2010.481373

Kundzewicz ZW, Robson AJ (2004) Change detection in hydrological records—a review of the methodology. Hydrol Sci J 49:7–20. https://doi.org/10.1623/hysj.49.1.7.53993

Liu W, Wang L, Sun F, Li Z, Wang H, Liu J, Yang T, Zhou J, Qoi J (2018) Snow hydrology in the upper Yellow River basin under climate change: a land-surface modeling perspective. J Geophys Res Atmos 123:676–691. https://doi.org/10.1029/2018JD028984

Lobanova A, Liersch S, Nunes JP, Didovets I, Stagl J, Huang S, Koch H, López MDR, Rocío M, Cathrine F, Hattermann F, Krysanova V (2018) Hydrological impacts of moderate and high-end climate change across European river basins. J Hydrol Reg Stud 18:15–30. https://doi.org/10.1016/j.ejrh.2018.05.003

Mann HB (1945) Nonparametric tests against trend. Econometrica 13:245–259

Mariotti A, Pan Y, Zeng N, Alessandri A (2015) Long-term climate change in the Mediterranean region in the midst of decadal variability. Clim Dyn 44:1437–1456. https://doi.org/10.1007/s00382-015-2487-3

Maurer JM, Schaefer JM, Rupper S, Corley A (2019) Acceleration of ice loss across the Himalayas over the past 40 years. Sci Adv 5:1–12. https://doi.org/10.1126/sciadv.aav7266

Mitchell TD, Jones PD (2005) An improved method of constructing a database of monthly climate observations and associated high-resolution grids. Int J Climatol 25:693–712. https://doi.org/10.1002/joc.1181

Modarres R, Sarhadi A (2009) Rainfall trends analysis of Iran in the last half of the twentieth century. J Geophys Res Atmos 114:1–9. https://doi.org/10.1029/2008JD010707

New M, Hulme M, Jones P (1999) Representing twentieth-century space-time climate variability. Part I: development of a 1961–90 mean monthly terrestrial climatology. J Clim 12:829–856. https://doi.org/10.1175/1520-0442(1999)012<0829:RTCSTC>2.0.CO;2

New M, Hulme M, Jones P (2000) Representing twentieth-century space-time climate variability. Part II: development of 1901–96 monthly grids of terrestrial surface climate. J Clim 13:2217–2238. https://doi.org/10.1175/1520-0442(2000)013<2217:RTCSTC>2.0.CO;2

NIH (2015) National Institute of Hydrology Website. https://www.nih.ernet.in/rbis/basinmaps/ganga_about.htm. Accessed 12 May 2015

Ojha CSP, Sharma C, Upreti H, Arora H, Neema M (2015) Climate change and sustainable water management. In: IOE graduate conference, Kathmandu, pp 6–19

Olguin GA, Sandoval CE (2017) Modes of variability of annual and seasonal rainfall in Mexico. JAWRA J Am Water Resour Assoc 53:144–157. https://doi.org/10.1111/1752-1688.12488

Reeves J, Chen J, Wang XL, Lund R, Lu Q (2007) A review and comparison of changepoint detection techniques for climate data. J Appl Meteorol Climatol 46:900–915. https://doi.org/10.1175/JAM 2493.1

Schumacher M, Forootan E, Vandijk AIJM, Müller Schmied H, Crosbie RS, Kusche J, Döll P (2018) Improving drought simulations within the Murray-Darling basin by combined calibration/assimilation of GRACE data into the WaterGAP global hydrology model. Remote Sens Environ 204:212–228. https://doi.org/10.1016/j.rse.2017.10.029

Sen PK (1968) Estimates of the regression coefficient based on Kendall's Tau. J Am Stat Assoc 57:269–306

Shanmugam R (1997) Technometrics 39:426–426. https://doi.org/10.2307/1271510

Sharma C, Ojha CSP (2018a) Spatio-temporal variability of snow cover of Yamunotri catchment, India. In: IGARSS 2018—2018 IEEE international geoscience and remote sensing symposium. IEEE, pp 5192–5194. https://doi.org/10.1109/IGARSS.2018.8517408

Sharma C Ojha CSP (2018b) Detection of changes in temperature in Upper Ganga River Basin. In: 2018 AGU Fall Meeting Washington, DC. 10.1002/essoar.10500106.1

Sharma C, Ojha CSP (2019a) Statistical parameters of hydrometeorological variables: standard deviation, SNR, Skewness and Kurtosis. In: Lecture Notes in Civil Engineering. Springer, Singapore, pp 59–70. https://doi.org/10.1007/978-981-13-8181-2_5

Sharma C, Ojha CSP (2019) Changes of annual precipitation and probability distributions for different climate types of the world. Water (Switzerland) 11:1–22. https://doi.org/10.3390/w11 102092

Sharma C, Ojha CSP (2021) Climate change detection in Upper Ganga River Basin. In: Pandey A., Mishra S., Kansal M., Singh R., Singh V. (eds) Climate Impacts on Water Resources in India. Water Science and Technology Library, vol 95. Springer, Cham. https://doi.org/10.1007/978-3-030-51427-3_24

Sharma C, Ojha CSP, Shukla AK, Pham QB, Linh NTT, Fai CM, Loc HH, Dung TD (2019) Modified approach to reduce GCM bias in downscaled precipitation: a study in Ganga River Basin. Water (Switzerland) 11:1–33. https://doi.org/10.3390/w11102097

Singh P, Ramasastri KS, Kumar N (1995) Topographical influence on precipitation distribution in different ranges of Western Himalayas. Hydrol Res 26:259–284. https://doi.org/10.2166/nh. 1995.0015

Singh RB, Mal S (2014) Trends and variability of monsoon and other rainfall seasons in Western Himalaya, India. Atmos Sci Lett 15:218–226. https://doi.org/10.1002/asl2.494

Westra S, Alexander LV, Zwiers FW (2013) Global increasing trends in annual maximum daily precipitation. J Clim 26:3904–3918. https://doi.org/10.1175/JCLI-D-12-00502.1

World Bank (2011) India—National Ganga River Basin Project (English). World Bank, Washington, DC, 1–175. https://documents.worldbank.org/curated/en/710791468269101705/India-National-Ganga-River-Basin-Project

Yang Q, Zheng D (2004) Tibetan geography. China Intercontinental Press, China (9787508506654)

Yue S, Pilon P, Phinney B, Cavadias G (2002) The influence of autocorrelation on the ability to detect trend in hydrological series. Hydrol Process 16:1807–1816. https://doi.org/10.1002/hyp. 1095

Study of Twenty-first Century Precipitation and Temperature Trends Over Ganga River Basin

Chetan Sharma and C. S. P. Ojha

Abstract Analysis of 21st-century projections of precipitation and temperature in the Ganga river basin (GRB) is presented in this study. A statistical downscaling method was used to downscale the coarse-scale GCM variable at the local station. 21st-century future projections of temperature, annual and seasonal precipitation were estimated considering future climate scenarios, RCP4.5, and RCP8.5. 21st-century projections of seasonal precipitation indicate a significant increase in monsoon precipitation while a decrease in winter season precipitation. Annual average temperature may rise up to 9.3 °C considering RCP8.5, which may lead to drying of the glaciers much sooner. The annual average temperature in high altitude regions may rise much higher than the plains.

Keywords Ganga river basin · Precipitation · Temperature · Twenty-first century · GCM · River basin · Season

1 Introduction

Intergovernmental Panel on Climate Change (IPCC) is the leading agency in the world to assess the effects of climate change, generate awareness, and suggest adaption and mitigation measures. IPCC fifth assessment report (AR5) presented numerous studies and research confirming the effect of global warming on rising sea levels, melting of glaciers, and increase in the frequency of extreme events, i.e., floods and droughts, etc. (IPCC 2013). A number of studies are available indicating

C. Sharma (✉)
Department of Civil Engineering, Madhav Institute of Technology and Science, Gwalior, MP, India
e-mail: chetan.cvl@gmail.com

C. S. P. Ojha
Department of Civil Engineering, Indian Institute of Technology Roorkee, Roorkee, Uttarakhand 247667, India
e-mail: cspojha@gmail.com

© The Author(s), under exclusive license to Springer Nature Switzerland AG 2021
M. S. Chauhan and C. S. P. Ojha (eds.), *The Ganga River Basin: A Hydrometeorological Approach*, Society of Earth Scientists Series,
https://doi.org/10.1007/978-3-030-60869-9_18

273

the effect of climate change on different hydrometeorological variables (Barnett et al. 2005, 2008; Mariotti et al. 2015; Sharma and Ojha 2021, 2019; Sharma et al. 2019).

A good amount of literature is available, focusing on the review of the different trend detection methods (Kundzewicz and Robson 2004; Reeves et al. 2007; Khaliq et al. 2009). These studies indicate that non-parametric methods are preferred because of the distribution free assumption of the parent distribution of data. Sen's nonparametric estimate (Sen 1968) is the widely used method to find the magnitude and direction of trend (Jain and Kumar 2012; Gocic and Trajkovic 2013; Kundu et al. 2014; Bera 2017). However, the significance of the trends cannot be directly obtained from the Sen's method, so Mann-Kendal (Mann 1945; Kendall 1975) method is generally used to find the significance of trends (Modarres and Sarhadi 2009; Caloiero et al. 2011; Westra et al. 2013; Akinsanola and Ogunjobi 2017; Sharma and Ojha 2020, 2021). Yue et al. (2002) found that the presence of serial correlation affects the power of the Mann-Kendal test (MK). Hamed and Ramachandra Rao (1998) proposed a modified Mann-Kendal test (MMK) for estimation of trends in autocorrelated time series, which changes the variance of the MK test as per the autocorrelation structure of the data. Sen's method was used to find the magnitude and direction of the trends, and the MMK method was used to detect the significance in this study.

Changes in hydrometeorological variables across the globe have been widely reported in previous studies(Gelfan et al. 2017; Schumacher et al. 2018; Abe et al. 2018; Lobanova et al. 2018; Liu et al. 2018). Krysanova et al. (2017) compared historical changes in five river discharge variables and projected changes in precipitation and temperature in twelve big river basins, i.e., Rhine and Tagus in Europe, Niger, and the Blue Nile in Africa, Ganges, Lena, Upper Yellow, and Upper Yangtze in Asia, Upper Mississippi, MacKenzie and Upper Amazon in America, and Darling in Australia. However, the study was focused on the comparison of different discharge properties of the year 2071–2099 with the historical data for the year 1981–2010. Sharma and Ojha (2019) studied the changes in annual precipitation in different climate types of the world. However, changes in different seasons were not reported in the study. The twenty-first century projected temperature and precipitation trends in the GRB are reported in this chapter. Temperature and precipitation are considered as this one of the most important hydrometeorological variables required for efficient water management in a basin.

2 Study Area and Data

The Ganga river originates from Gangotri glacier in the state of Uttarakhand. It traverses through different topographies and meets the Bay of Bengal after completing a journey of around 2758 km (Verma et al. 2014). The Brahmaputra river originates on the northern slopes of Himalaya and travels around 2260 km in China, India, and meets the Ganga river near the Bay of Bengal in Bangladesh. The Meghna river originates in East Indian mountains and also meets the river Ganga,

Fig. 1 GRB overlaid on Google Earth map (Basin boundary is shown by dark black polygon)

the Brahmaputra near Bay of Bengal (Whitehead et al. 2018). The topography of GRB is very complex, which includes regions of higher Himalayas where elevation goes up to 8848 m to the Gangetic plains, i.e., Sundarbans in Bangladesh. The study area, i.e., the GRB, is shown in Fig. 1.

High resolution monthly average precipitation and temperature data at the grid interval of $0.5° \times 0.5°$ provided by the climatic research unit (CRU) are used in this study (data available at https://www.cru.uea.ac.uk/data). Each dataset was pre-screened; grid points with erroneous/missing values for a sufficiently long length was not considered in the study.

The National Centers for Environmental Prediction (NCEP) and the National Center for Atmospheric Research (NCAR) jointly developed global reanalysis data using observations and mathematical numerical model which simulates the recent climate. So NCEP/NCAR reanalysis data (Kalnay et al. 1996, data available at https://www.esrl.noaa.gov/psd/data/gridded/data.ncep.reanalysis.derived.html) for the year 1948–2017 was used to develop large scale atmospheric predictors and local scale predictand relationship. According to the previous researches precipitation is found to be affected by temperature, specific humidity, mean sea level pressure, geopotential height and zonal and meridional wind (Garreaud 2007; Mondal and Mujumdar 2012; Goyal and Ojha 2012), so these atmospheric variables were used in this study to downscale precipitation. Based on the available literature (Bettolli and Penalba 2018; Tomozeiu et al. 2018; Ali et al. 2019), the predictor variables used to the downscale near-surface average temperature in GRB are given in Tables 1 and 2 respectively.

Coupled Model Intercomparison Project (CMIP) was established by the Working Group on Coupled Modelling (WGCM) under the World Climate Research Program (WCRP) to study the GCM output. Arora et al. (2017) computed the GCM skill score to find the best representative GCM to downscale precipitation in GRB. A low value of skill score defines a high degree of similarity of the GCM downscaled product with observed data.GFDL_ESM2M GCM model ranked 7th on the skill score and historical, and 21st-century data is also available on WCRP. So, GFDL_ESM2M

Table 1 Predictor variables used to downscale precipitation

S. No	Predictor variable	Pressure level (hPa)
1	Air temperature	Near-surface, 200, 500, 850 and 1000
2	Geopotential height	200, 500, 850 and 1000
3	Specific humidity	850 and 1000
4	Zonal Wind component	200, 500, 850 and 1000
5	Meridional Wind component	200, 500, 850 and 1000
6	Mean Sea level pressure	–

Table 2 Predictor variables used to downscale near surface air temperature

S. No	Predictor variable	Pressure level (hPa)
1	Air temperature	Near-surface, 850
2	Geopotential height	200, 500
3	Specific humidity	850 and 1000
4	Zonal Wind component	850
5	Meridional Wind component	850
6	Mean Sea level pressure	–

GCM was selected in this study. As GCM downscaled precipitation has the lowest similarity with observed data and temperature have highest (Johnson and Sharma 2009), the GFDL_ESM2M GCM model, which best describes the precipitation, was also used to the downscale temperature.

3 Methodology

The spatial resolution of reanalysis, observed, and GCM data was different so, coarse-scale atmospheric data was re-gridded at observed grids, i.e., grid interval of 0.25° latitude by 0.25° longitude using spline interpolation. Statistical downscaling is a procedure to downscale GCM output at a regional scale using statistical methods. A statistical relationship between large scale atmospheric variables (predictor) and local scale variable (predictand) was developed. The developed relationship was used to downscale precipitation and temperature at the local scale.

This established predictor-predictand relationship was used to downscale precipitation and temperature using GCM followed by bias correction. Bias correction is a procedure to correct the bias in GCM output from reference data. Different bias correction methods and their comparison have been reported in different studies (Pan et al. 2001; Flanner et al. 2009; Li et al. 2010; Ajaaj et al. 2016; Nguyen et al. 2016; Joseph et al. 2018). Equidistant CDF matching (EDCDFm) bias correction method

(Li et al. 2010) was applied to correct the bias to downscale Hist, and future emission scenarios (RCP4.5 and RCP8.5) GCM.

The bias-corrected variable was used for further analysis. Monthly total observed and downscaled precipitation were converted to monsoon (June–September), Winter (December-February), and annual total precipitation. Weighted average monthly temperature (Tmp) was calculated, giving weightage to a number of days in a month. Sen's slope method was used to find the magnitude and direction of trends in observed and GCM downscaled data. Another important step of climate change study is to find a significance of trends. As most of the climate data is serially correlated (Sonali and Nagesh Kumar 2013; Mullick et al. 2019), the Mann-Kendal test (MK) could not be used. So, the MMK test was used to find the significance of trends.

4   Results and Discussions

- **Annual Average Temperature**

The 21st-century projections of annual average temperature in GRB considering RC4.5 and RCP8.5 are shown in Figs. 2 and 3, respectively. The annual average temperature will rise in all of the GRB considering the RCP4.5 emission scenario. A minimum rate of increase in temperature in the twenty-first century was found in Southern regions (~0.004 to 0.017 °C/year). A little higher rate of increase in temperature will be experienced in the mid-latitude regions. The highest rate of increase in temperature was observed in twenty-first century projections in the Himalayan regions of GRB. More than 95% of regions showed significant trends at a 5% significance level.

The twenty-first century temperature trends under the RCP8.5 emission scenario show high, increasing trends across the basin. The highest increase (up to 0.098 °C/year) was found in the high-altitude Himalayan regions while the lowest rate of increase in temperatures was found in the plains. All of the increasing trends were found significant at less than 1% significance level.

The annual average temperature in the GRB will increase in the twenty-first century, considering different RCP emission scenarios. Siderius et al. (2013) computed the contribution of snowmelt in the annual discharge in different reaches in GRB. It was observed that snowmelt contributes 10% to 30% of annual streamflow in the upper reaches of GRB (Collins 2013). The annual average temperature in the initial reaches of the GRB is well below zero degree Celsius, which keeps the glaciers live. A minimum and maximum increase of 0.43 and 9.3 °C in the twenty-first century will lead to completely dried up glaciers and high snowmelt contribution in the initial periods which may cause flooding. The upper reaches of the Ganga river may be completely dried up in the coming decades during the non-monsoon season due to the unavailability of snowmelt contribution in the upper reaches.

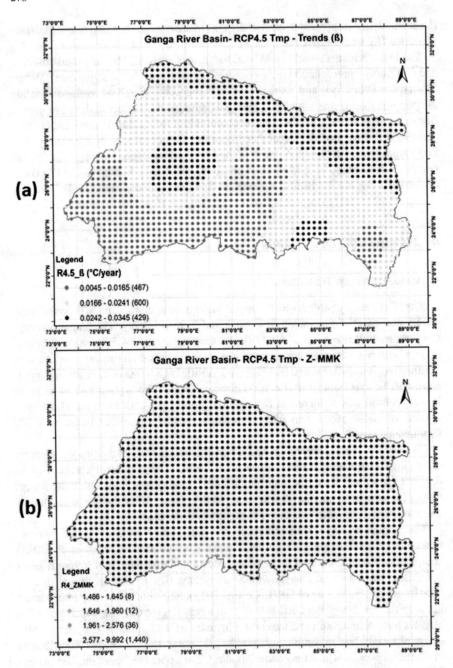

Fig. 2 Twenty-first century projections of annual average temperature in GRB considering RCP4.5 **a** trends, **b** MMK test statistic Z_{MMK} for significance of trends

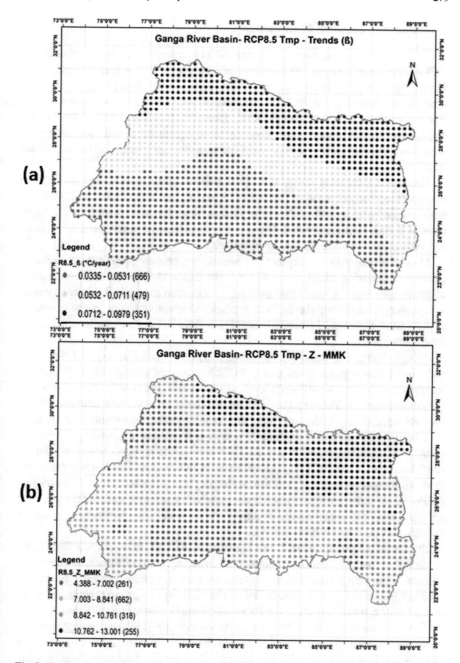

Fig. 3 Twenty-first century projections of annual average temperature in GRB considering RCP8.5 **a** trends, **b** MMK test statistic Z_{MMK} for significance of trends

- **Annual and Seasonal Precipitation**

Trends and their significance in annual 21st-century projections considering RCP4.5 and RCP8.5 emission scenarios are shown in Figs. 4 and 5, respectively. Increasing trends in the 21st-century precipitation were observed in most of the locations considering the RCP4.5 scenario. Most of the increasing trends were found significant at 1% significant level. However, trends in some of the Eastern and Northern Himalayan regions were not found significant. Small increasing trend were found in some eastern regions, but these were not significant at a 10% significant level.

Significant increasing trends at a 5% significance level considering the RCP8.5 scenario were found in most of the regions. Some regions at the foothills of Himalaya were indicating an increase in the precipitation; however, these were found insignificant at a 10% significant level. Overall annual precipitation pattern may increase significantly in the GRB during the twenty-first century.

Monsoon season (June–September) twenty-first century projected precipitation patterns considering RCP4.5 and RCP8.5 are shown in Figs. 6 and 7. A majority of the regions show a significant increase in monsoon precipitation in the twenty-first century, considering the RCP4.5 scenario. Decreasing trends in some western regions were observed; however, these trends were not found significant at a10% significance level.

A high, increasing trend (up to 5 mm/year) were found in some of the costal and Himalayan regions considering RCP8.5 projected precipitation. Significant increasing trends at a significance level, 1% were also observed in the high-altitude Himalayan regions. These results indicate that the upper and lower reaches of GRB may get significantly higher monsoon precipitation during the twenty-first century.

Winter season (January–February) twenty-first century projected precipitation patterns considering RCP4.5 and RCP8.5 are shown in Figs. 8 and 9. Significant decreasing trends at 1% significance level in Western Himalayan regions and increasing trends in some of the Eastern Himalayan regions were found considering RCP4.5 21st-century projections. So, the dominant glacier regions in the GRB may get significantly lower precipitation during winter, which may further deteriorate the condition of glaciers in the Himalayas.

Similar to the precipitation trends considering RCP45, RCP8.5 projected winter precipitation patterns indicate a significant lowering of precipitation in the initial reaches of the Ganga river and in Western and coastal regions. However, the rate of decrease is much higher than the RCP4.5. The majority of western regions and some of the lower reaches indicate a significant decrease in winter precipitation.

The previous studies in the Himalayan region indicate little to no change in annual precipitation, while a significant decrease in the monsoon rainfall has been reported (Kumar and Jain 2011; Bera 2017). While the 21st-century projection indicates that monsoon precipitation may increase significantly, and the winter precipitation may decrease. This may further speed up the melting of Himalayan glaciers, which, along with the higher amount of rainfall in monsoon months, may cause heavy flooding in the low-lying area in the basin.

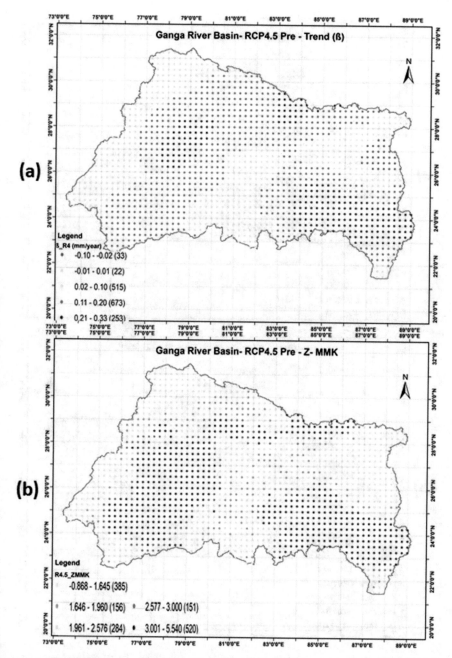

Fig. 4 Twenty-first century projections of annual precipitation in GRB considering RCP4.5 **a** trends, **b** MMK test statistic Z_{MMK} for significance of trends

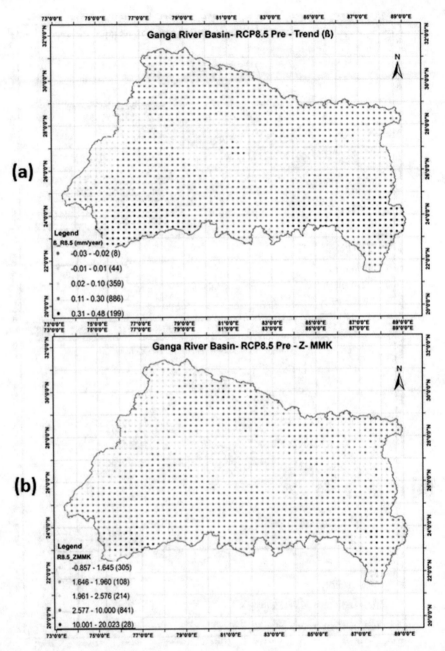

Fig. 5 Twenty-first century projections of annual precipitation in GRB considering RCP8.5 **a** trends, **b** MMK test statistic Z_{MMK} for significance of trends

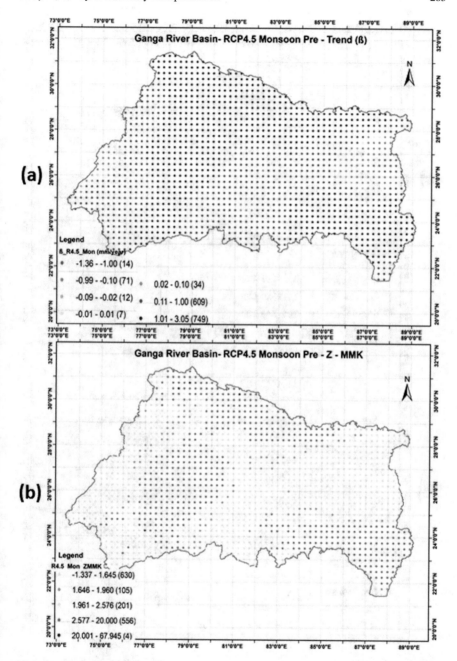

Fig. 6 Twenty-first century projections of monsoon precipitation in GRB considering RCP4.5 **a** trends, **b** MMK test statistic Z$_{MMK}$ for the significance of trends

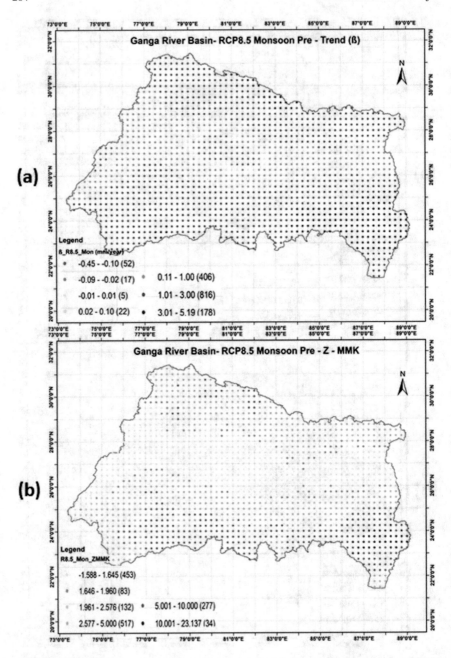

Fig. 7 Twenty-first century projections of monsoon precipitation in GRB considering RCP8.5 **a** trends, **b** MMK test statistic Z_{MMK} for significance of trends

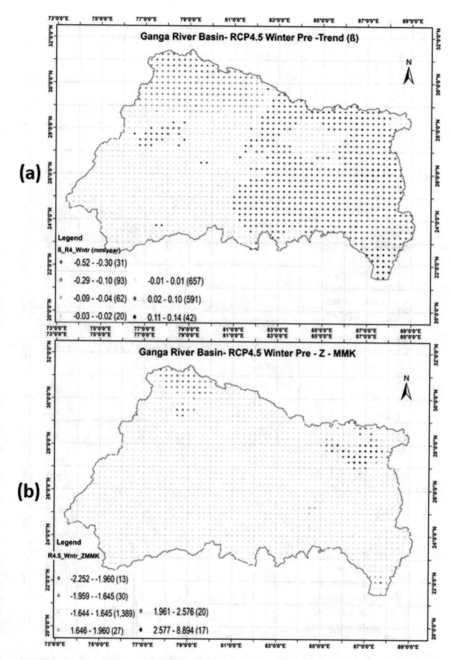

Fig. 8 Twenty-first century projections of winter precipitation in GRB considering RCP4.5 **a** trends, **b** MMK test statistic Z_{MMK} for the significance of trends

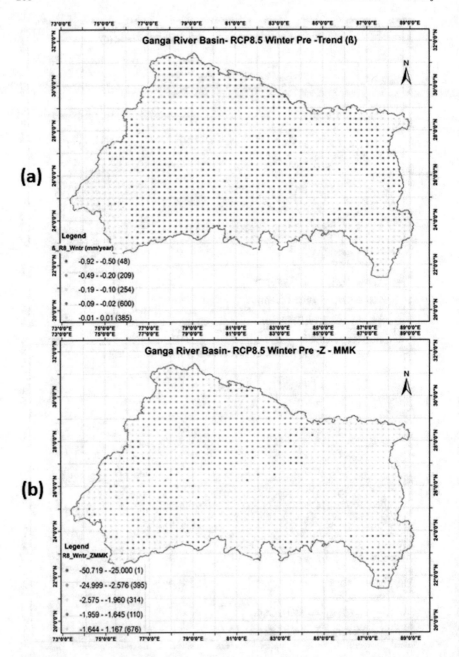

Fig. 9 Twenty-first century projections of winter precipitation in GRB considering RCP8.5 **a** trends, **b** MMK test statistic Z_{MMK} for a significance of trends

5   Summary and Conclusions

Trends in the twenty-first century projected precipitation and temperature patterns considering RCP4.5 and RCP8.5 scenarios were analyzed. The 21st-century projections indicate that the highest increase in temperature will be in the Himalayan regions while in the plains, it will be lowest. Future projections of RCP4.5 and RCP8.5 show only a small increase in the annual total amount of precipitation in most of the low and high-altitude regions. No significant change in total annual precipitation was observed in 21st century projections; however, a decrease in winter precipitation and an increase in monsoon precipitation indicates an increase in the extreme precipitation events in the twenty-first century. The annual average temperature in some locations in GRB may increase by 9.8 °C in the twenty-first century, which may lead to the melting of glaciers in the early years of the century, which may lead to floods and the winters might be drier in the absence of glaciers. These conditions may cause a serious threat to the living being, which are highly dependent on the waters of the Ganga river.

References

Abe CA, Lobo FDL, Dibike YB, Costa MPF, Dos Santos V, Novo EMLM (2018) Modelling the effects of historical and future land cover changes on the hydrology of an Amazonian basin. Water (Switzerland) 932:1–19. https://doi.org/10.3390/w10070932

Ajaaj AA, Mishra AK, Khan AA (2016) Comparison of BIAS correction techniques for GPCC rainfall data in semi-arid climate. Stoch Environ Res Risk Assess. 30:1659–1675. https://doi.org/10.1007/s00477-015-1155-9

Akinsanola AA, Ogunjobi KO (2017) Recent homogeneity analysis and long-term spatio-temporal rainfall trends in Nigeria. Theor Appl Climatol 128:275–289. https://doi.org/10.1007/s00704-015-1701-x

Ali S, Eum HI, Jaepil C, Dan L, Khan F, Dairaku K, Shrestha ML, Hwang S, Nazim W, Khan IA, Fahad S (2019) Assessment of climate extremes in future projections downscaled by multiple statistical downscaling methods over Pakistan. Atmos Res 222:114–133. https://doi.org/10.1016/J.ATMOSRES.2019.02.009

Arora H, Ojha CSP, Buytaert W, Kaushika GS, Sharma C (2017) Spatio-temporal trends in observed and downscaled precipitation over Ganga Basin. Hydrol Earth Syst Sci Discuss https://doi.org/10.5194/hess-2017-388

Barnett TP, Adam JC, Lettenmaier DP (2005) Potential impacts of a warming climate on water availability in snow-dominated regions. Nature 438:303–309. https://doi.org/10.1038/nature04141

Barnett TP, Piere DW, Hidalgo HG, Bonfils C, Santer BD, Das T, Bala G, Wood AW, Mirin TA, Cayan DR, Dettinger MD (2008) Human-induced changes in the hydrology of the western United States. Science 319:1080–1083. https://doi.org/10.1126/science.1152538

Bera S (2017) Trend analysis of rainfall in Ganga Basin, India during 1901–2000. Am J Clim Chang 06:116–131. https://doi.org/10.4236/ajcc.2017.61007

Bettolli ML, Penalba OC (2018) Statistical downscaling of daily precipitation and temperatures in southern La Plata Basin. Int J Climatol 38:3705–3722. https://doi.org/10.1002/joc.5531

Caloiero T, Coscarelli R, Ferrari E, Mancini M (2011) Trend detection of annual and seasonal rainfall in Calabria (Southern Italy). Int J Climatol 31:44–56. https://doi.org/10.1002/joc.2055

Flanner MG, Zender CS, Hess PG, Mahowald NM, Painter TH, Ramanathan V, Rasch PJ (2009) Springtime warming and reduced snow cover from carbonaceous particles. Atmos Chem Phys 9:2481–2497. https://doi.org/10.5194/acp-9-2481-2009

Garreaud RD (2007) Precipitation and circulation covariability in the extratropics. J Clim 20:4789–4797. https://doi.org/10.1175/JCLI4257.1

Gelfan A, Gustafsson D, Motovilov Y, Arheimer B, Kalugin A, Krylenko I, Lavrenov A (2017) Climate change impact on the water regime of two great Arctic rivers: modeling and uncertainty issues. Climatic Change 141:499–515. https://doi.org/10.1007/s10584-016-1710-5

Gocic M, Trajkovic S (2013) Analysis of changes in meteorological variables using Mann-Kendall and Sen's slope estimator statistical tests in Serbia. Glob Planet Change 100:172–182. https://doi.org/10.1016/j.gloplacha.2012.10.014

Goyal MK, Ojha CSP (2012) Downscaling of precipitation on a lake basin: evaluation of rule and decision tree induction algorithms. Hydrol Res. 43:215–230. https://doi.org/10.2166/nh.2012.040

Hamed KH, Ramachandra RA (1998) A modified Mann-Kendall trend test for autocorrelated data. J Hydrol 204:182–196. https://doi.org/10.1016/S0022-1694(97)00125-X

IPCC (2013) Summary for policymakers. In: Climate Change 2013: the physical science basis. Contribution of Working Group I to the Fifth Assessment Report of the Intergovernmental Panel on Climate Change (Stocker, T.F., D. Qin, G.-K. Plattner, M. Tignor, S.K. Allen, J. Boschung, A. Nauels, Y. Xia). Cambridge, United Kingdom and New York, NY, USA

Jain SK, Kumar V (2012) Trend analysis of rainfall and temperature data for India. Curr Sci 102:37–49. https://doi.org/10.2307/24080385

Johnson F, Sharma A (2009) Measurement of GCM skill in predicting variables relevant for hydroclimatological assessments. J Clim 22:4373–4382. https://doi.org/10.1175/2009JCLI2681.1

Joseph J, Ghosh S, Pathak A, Sahai AK (2018) Hydrologic impacts of climate change: comparisons between hydrological parameter uncertainty and climate model uncertainty. J Hydrol 566:1–22. https://doi.org/10.1016/j.jhydrol.2018.08.080

Kalnay E, Kanamitsu M, Kistler R, Collins W, Deaven D, Gandin L, Iredell S, Saha S, White G, Zhu Y, Leetmaa A, Reynolds R, Chelliah M, Ebisuzaki W, Higgins W, Janowiak J, Mo K, Ropelewski C, Wang J, Jenne R, Joseph D (1996) The NCEP/NCAR 40-year reanalysis project. Bull Am Meteorol Soc 77:437–471

Kendall MG (1975) Rank correlation methods, 4th edn. Charless Griffin, London (0195208374)

Khaliq MN, Ouarda TBMJ, Gachon P, Sushama l, Hilaire AS (2009) Identification of hydrological trends in the presence of serial and cross correlations: a review of selected methods and their application to annual flow regimes of Canadian rivers. J Hydrol 368:117–130. https://doi.org/10.1016/j.jhydrol.2009.01.035

Krysanova V, Vetter T, Eisner S, Huang S, Pechlivanidis I, Strauch M, Gelfan A, Kumar R, Aich V, Arheimer B, Chamorro A, Van Griensven A, Kundu D, Lobanova A, Mishra V, Plötner S, Reinhardt J, Seidou O, Wang X, Wortmann M, Zeng X, Hattermann FF et al (2017) Intercomparison of regional-scale hydrological models and climate change impacts projected for 12 large river basins worldwide—a synthesis. Environ Res Lett 12:105002. https://doi.org/10.1088/1748-9326/aa8359

Kundu A, Dwivedi S, Chandra V (2014) Precipitation trend analysis ovel Eastern Region of India using Cmip5 based climatic models. ISPRS Int Arch Photogramm Remote Sens Spat Inf Sci XL 8:1437–1442. https://doi.org/10.5194/isprsarchives-XL-8-1437-2014

Kundzewicz ZW, Robson AJ (2004) Change detection in hydrological records—a review of the methodology. Hydrol Sci J 49:7–20. https://doi.org/10.1623/hysj.49.1.7.53993

Li H, Sheffield J, Wood EF (2010) Bias correction of monthly precipitation and temperature fields from Intergovernmental Panel on Climate Change AR4 models using equidistant quantile matching. J Geophys Res Atmos 115:1–20. https://doi.org/10.1029/2009JD012882

Liu W, Wang L, Sun F, Li Z, Wang H, Liu J, Yang T, Zhou J, Qoi J (2018) Snow hydrology in the upper Yellow River basin under climate change: a land-surface modeling perspective. J Geophys Res Atmos 123:676–691. https://doi.org/10.1029/2018JD028984

Lobanova A, Liersch S, Nunes JP, Didovets I, Stagl J, Huang S, Koch H, López MDR, Rocío M, Cathrine F, Hattermann F, Krysanova V (2018) Hydrological impacts of moderate and high-end climate change across European river basins. J Hydrol Reg Stud 18:15–30. https://doi.org/10.1016/j.ejrh.2018.05.003

Mann HB (1945) Nonparametric tests against trend. Econometrica 13:245–259

Mariotti A, Pan Y, Zeng N, Alessandri A (2015) Long-term climate change in the Mediterranean region in the midst of decadal variability. Clim Dyn 44:1437–1456. https://doi.org/10.1007/s00382-015-2487-3

Modarres R, Sarhadi A (2009) Rainfall trends analysis of Iran in the last half of the twentieth century. J Geophys Res Atmos 114:1–9. https://doi.org/10.1029/2008JD010707

Mondal A, Mujumdar PP (2012) On the basin-scale detection and attribution of human-induced climate change in monsoon precipitation and streamflow. Water Resour Res 48:1–18. https://doi.org/10.1029/2011WR011468

Mullick MRA, Nur MRM, Alam MJ, Islam KMMA (2019) Observed trends in temperature and rainfall in Bangladesh using pre-whitening approach. Glob Planet Change 172:104–113. https://doi.org/10.1016/j.gloplacha.2018.10.001

Nguyen H, Mehrotra R, Sharma A (2016) Correcting for systematic biases in GCM simulations in the frequency domain. J Hydrol 538:117–126. https://doi.org/10.1016/j.jhydrol.2016.04.018

Pan Z, Christensen JH, Arritt RW, Gutowski WJ, Takle ES, Otieno F (2001) Evaluation of uncertainties in regional climate change simulations. J Geophys Res Atmos 106:17735–17751. https://doi.org/10.1029/2001JD900193

Reeves J, Chen J, Wang XL, Lund R, Lu Q (2007) A review and comparison of changepoint detection techniques for climate data. J Appl Meteorol Climatol 46:900–915. https://doi.org/10.1175/JAM2493.1

Schumacher M, Forootan E, Vandijk AIJM, Schmied HM, Crosbie RS, Kusche J, Döll P (2018) Improving drought simulations within the Murray-Darling Basin by combined calibration/assimilation of GRACE data into the WaterGAP Global Hydrology Model. Remote Sens Environ 204:212–228. https://doi.org/10.1016/j.rse.2017.10.029

Sen PK (1968) Estimates of the regression coefficient based on Kendall's Tau. J Am Stat Assoc 57:269–306

Sharma C, Ojha CSP (2021) Climate change detection in upper Ganga river basin. In: Pandey A, Mishra S, Kansal M, Singh R, Singh V (eds) Climate impacts on water resources in India. Water science and technology library. vol 95. Springer, Cham. https://doi.org/10.1007/978-3-030-51427-3_24

Sharma C, Ojha CSP (2020) Statistical parameters of hydrometeorological variables: standard deviation, SNR, Skewness and Kurtosis. In: AlKhaddar R, Singh R, Dutta S, Kumari M (eds) Advances in water resources engineering and management. Lecture Notes in Civil Engineering, vol 39. Springer, Singapore. https://doi.org/10.1007/978-981-13-8181-2_5

Sharma C, Ojha CSP (2019) Changes of annual precipitation and probability distributions for different climate types of the world. Water (Switzerland) 11:1–22. https://doi.org/10.3390/w11102092

Sharma C, Ojha CSP, Shukla AK, Pham QB, Linh NTT, Fai CM, Loc HH, Dung TD (2019) Modified approach to reduce GCM bias in downscaled precipitation: a study in Ganga River Basin. Water (Switzerland) 11:1–33. https://doi.org/10.3390/w11102097

Siderius C, Biemans H, Wiltshire A, Rao S, Franssen WHP, Kumar P, Gosain AK, Vanvliet MTH (2013) Snowmelt contributions to discharge of the Ganges. Sci Total Environ 468–469:S93–S101. https://doi.org/10.1016/j.scitotenv.2013.05.084

Collins DN (2013) Snowmelt contributions to discharge of the Ganges. Sci Total Environ 468:S93–S101. https://doi.org/10.1016/j.scitotenv.2013.05.084

Sonali P, Kumar ND (2013) Review of trend detection methods and their application to detect temperature changes in India. J Hydrol 476:212–227. https://doi.org/10.1016/j.jhydrol.2012.10.034

Tomozeiu R, Pasqui M, Quaresima S (2018) Future changes of air temperature over Italian agricultural areas: a statistical downscaling technique applied to 2021–2050 and 2071–2100 periods. Meteorol Atmos Phys 130:543–563. https://doi.org/10.1007/s00703-017-0536-7

Verma N, Devrani R, Singh V (2014) Is Ganga the longest river in the Ganga Basin, India? Curr Sci 107:2018–2022

Westra S, Alexander LV, Zwiers FW (2013) Global increasing trends in annual maximum daily precipitation. J Clim 26:3904–3918. https://doi.org/10.1175/JCLI-D-12-00502.1

Whitehead PG, Jin L, Macadam I, Janes T, Sarkar S, Rodda HJE, Sinha R, Nicholls RJ, Janes T (2018) Corrigendum to "Modelling impacts of climate change and socio-economic change on the Ganga, Brahmaputra, Meghna, Hooghly and Mahanadi river systems in India and Bangladesh". Sci Total Environ 636(15):1362–1372. https://doi.org/10.1016/j.scitotenv.2018.07.180

Yue S, Pilon P, Phinney B, Cavadias G (2002) The influence of autocorrelation on the ability to detect trend in hydrological series. Hydrol Process 16:1807–1816. https://doi.org/10.1002/hyp.1095

Characteristics of Soil Moisture Droughts in Ganga River Basin During 1948–2015

Lalit Pal and C. S. P. Ojha

Abstract Recent drought events in India are found to affect a major portion of Ganga River basin. Considering the importance of the basin in terms of residing population and contribution to agricultural sector, a rigorous understanding of agricultural droughts is crucial for ensuring food security and economic equilibrium in the country. The present study investigates the response of soil moisture to spatiotemporal pattern in meteorological variables through identifying trends in rainfall, mean temperature and soil moisture over Ganga River basin. Long-term trends in hydrologic and climatic annual time series are analyzed using Mann-Kendall test for the period 1948–2015. In the results, significant negative trend is observed in annual soil moisture over northwest and southwest region of the basin which are attributable to significant negative trend in rainfall and significant increasing trend in temperature in the respective regions. Significant increasing trend is also observed in area affected with meteorological and agricultural drought with MK-Z value of 2.63 and 2.91, respectively. Severe soil moisture drought events in Ganga River basin are identified during 1950–1955, 1965–1966, 1979–1980, 1991–1993, 2001–2003, 2008–2009 and 2012–2015. Despite of increasing trend in soil moisture droughts, basin average NDVI is found to be following a positive trend highlighting improved vegetation health over the period. The contrasting behavior of vegetation may be attributable to the advancement in agricultural practices after green revolution and considerable shift in fresh water consumption for irrigation from surface water sources to groundwater.

Keywords Soil moisture · Droughts · Trend · SPI · SMDS · Ganga river basin

L. Pal (✉) · C. S. P. Ojha
Department of Civil Engineering, Indian Institute of Technology Roorkee, Roorkee,
Uttarakhand 247667, India
e-mail: lalitpl4@gmail.com

C. S. P. Ojha
e-mail: cspojha@gmail.com

M. S. Chauhan and C. S. P. Ojha (eds.), *The Ganga River Basin: A Hydrometeorological Approach*, Society of Earth Scientists Series,
https://doi.org/10.1007/978-3-030-60869-9_19

1 Introduction

Drought is an insidious disaster among extreme climate events which affects society, economy, social life and agricultural sector (Mishra and Singh 2010). Although, the consequences of droughts may be devastating, identification of its onset is difficult to locate in time. Based on deficit in one or the other hydro-meteorological variable, droughts may be broadly classified into four categories (Dai 2011): (1) meteorological drought – occurs due to deficient precipitation; (2) hydrological drought–caused by decrease in surface runoff and depletion of groundwater level; (3) agricultural drought–characterized with prolonged precipitation deficit resulting in depletion of soil moisture in the root zone; (4) socioeconomic drought–severe economic and social losses due to combined effect of the above mentioned drought types (Heim 2002). In India, agricultural sector contributes to about 20% of the gross domestic production (GDP), thus, prolonged agricultural droughts (soil moisture deficit) may have severe implications on economic equilibrium, food security and other environmental aspects of the country.

India has a long history of severe droughts where these event of water deficit were accompanied by devastating famines during the era of British rule (1765–1947). Few of these famines were caused by widespread drought and agricultural failure due to summer monsoon failure while others resulted from policy failure of the British government (Mishra et al. 2019). However, modern India has not faced any death due to famines after 1900. Mishra et al. (2019) identified seven major drought periods (1876–1882, 1895–1900, 1908–1924, 1937–1945, 1982–1990, 1997–2004, and 2011–2015) from severity-area-duration analysis of reconstructed soil moisture over India. In addition, the analysis of spatial distribution of past severe drought event reveals a major shift in the geographic location of their occurrence. The zone of occurrence for past drought events (1918 and 1920) was mainly confined within central and southern parts of the country, however, recent droughts have occurred in the central and northern region (Mishra et al. 2019). Among the droughts between 1901–2015, the drought of 2015 is found to be most severe that has caused tremendous damage to the agricultural production and other social aspects of the region (Mishra et al. 2018). Geographically, the drought of 2015 had affected the northern and north-eastern parts, majority of which falls within the Ganga River basin.

The susceptibility of Ganga River basin to droughts in the recent past is a cause of concern for food security, and social and economic well-being of the country. The Ganga River basin is considered to be the food bowl of the country nurturing about 600 million lives which is close to half of the country's population (MoWR 2014). Agriculture is the main source of livelihood for majority of rural population in Ganga River basin. Approximately, 50% of rural population in the basin is dependent on agriculture for their livelihood. A prolonged soil moisture deficit based drought in the region may have catastrophic implications over not only the economically weak rural population of the basin but also on the socio-economic equilibrium of the nation. Therefore, rigorous analysis of spatiotemporal characteristics of drought is of paramount importance for efficient water resources management in Ganga River basin and ensuring food security of the country.

Soil moisture controls the exchange of water and energy between land surface and the atmosphere. Deficit of soil moisture in the root zone can alter the agricultural growth, resulting in reduction of crop yield (Wang et al. 2011). Therefore, soil moisture based drought indices are often used as agricultural drought indicators in the literature (Andreadis et al. 2005; Wang et al. 2011; Long et al. 2013; Thomas et al. 2014). Few commonly used agricultural drought indices are Palmer Drought Severity Index (PDSI) (Palmer 1965), Crop Moisture Index (CMI) (Palmer 1968), Water Deficit Index (WDI) (Moran et al. 1994), Reconnaissance Drought Index (RDI) (Tsakiris and Vangelis 2005) and Standardized Precipitation Evapotranspiration Index (SPEI) (Vicente-Serrano et al. 2010). Since, soil moisture plays a crucial role in crop growth and crop yield, in the present study, agricultural droughts are defined using an index that computes the empirical probability distribution of soil moisture namely Soil Moisture Drought Severity (SMDS) index (Andreadis et al. 2005). In addition, Standardized Precipitation Index (SPI) (McKee et al. 1993; Guttman 1998, 1999) is a frequently used drought index which represent meteorological drought condition in a region.

Despite the importance, long-term observations of soil moisture are essentially unavailable in India, if available, the time span is too short to draw affirmative conclusions. The alternate sources of soil moisture observations such as remote sensing based products, land surface model (LSM) simulated products and reanalysis products, etc., provide with an opportunity to study droughts in different data scarce regions of the world. Among these alternative, the Noah model simulated soil moisture dataset provided by Global Land Data Assimilation System (GLDAS) is used in various past soil moisture drought based studies, especially in India (Mishra et al. 2014; Shah and Mishra 2015; Sathyanadh et al. 2016; Mishra et al. 2018, 2019). In addition, GLDAS Noah LSM simulated soil moisture dataset is available for a longer time period (1948-present) and at fine spatial resolution (0.25°–0.25°). Subsequently, monthly soil moisture of GLDAS Noah simulations is used in the present study to compute agricultural drought index (SMDS) and identifying spatiotemporal patterns in soil moisture over Ganga River basin. The understanding of soil moisture droughts is essential for planning and management of crop production and ensuring food securing of the country, yet, limited efforts are made on the subject, especially for Ganga River basin in the existing literature. Therefore, present study aims to: (1) characterize spatial and temporal variability of major meteorological and agricultural drought events in the region; (2) understand the change in attributes of drought events over the period 1948–2015; (3) examine the impact of past drought events on the vegetation growth in the basin.

2   Study Area and Data

Ganga River basin is the largest river basin in India extending over an area of about 0.86 Mkm^2 within its administrative boundaries. It originates in Gangotri glacier in the western Himalayas and traverse across major portion of the country covering the distance of about 2758 km (Verma et al. 2014) to finally meet the Bay of Bengal. The basin covers the latitudes from $22° 30'$ to $31° 30'$ North and the longitudes from $73° 30'$ and $89° 01'$ East. The total basin area in India is shared among eleven major states viz. Himachal Pradesh, Uttarakhand, Uttar Pradesh, Madhya Pradesh, Chhattisgarh, Bihar, Jharkhand, Punjab, Haryana, Rajasthan, West Bengal and the Union Territory of Delhi. Rainfall and snow melt from the Himalayas are the major sources of fresh water in the basin. The Ganga River basin is characterized with diverse topography consisting of hilly terrains of Himalayascovered with dense forest, sparsely forested Shiwalik hills and vast fertile Indo-Gangetic Plains. The typical soil types found in the basin are sand, loam, clay and their combinations. Due to adequate availability of fresh water, majority of land area in the basin is used for agricultural practices. About 29.5% of total cultivable area of the country falls within the basin covering land area of about 58.0 MHa (GFCC 2009). The basin experiences a gradient in precipitation along east to west with eastern parts receiving higher precipitation than the western parts of the basin. A major fraction of total rainfall is received during monsoon months (June–September) with annual rainfall values ranging between 300 and 2000 mm. The temperature during winter season ranges from 2 to 15 °C, while that during summer season varies from 25 to 45 °C. The annual ET values lies in the range from 236 to 1271 mm. The geographical location and other details of the study area are given in Fig. 1.

The meteorological drought characteristics of the basin are explained by analyzing the space-time patterns of a widely used drought index namely Standardized Precipitation Index (SPI). Precipitation dataset required to compute SPI was acquired from India Meteorological Department available at a resolution of $0.25° \times 0.25°$ at daily time step. Soil moisture data used in the present study is acquired from NASA Global Land Cover Data Assimilation System (GLDAS) Noah land surface model (LSM) simulated variables. The GLDAS version 2.0 (GLDAS_NOAH025_M.2.0) and 2.1 (GLDAS_NOAH025_M.2.1) Noah simulated values of soil moisture at a resolution $0.25° \times 0.25°$ at monthly time step for the period $1948 - 2015$ are used for various analyses. GLDAS provides the simulated soil moisture data at a depth 0–10 cm and 10–40 cm from the surface. In this study, soil moisture values for the top 10 cm layer is used to compute soil moisture drought characteristics of the basin. The dataset can be downloaded from the following link: https://disc.sci.gsfc.nasa.gov/datasets?key words=GLDAS. The response of vegetation health to the meteorological and hydrological drought is analyzed using Normalized Difference Vegetative Index (NDVI) data. The GIMMS AVHRR NDVI dataset (version: 3 g V1.0) provided by NASA has been used to directly obtain NDVI value for the basin. The dataset is available at two-weekly time step from 1982 to 2015 at a spatial resolution of 1/12° (~ 8 km).

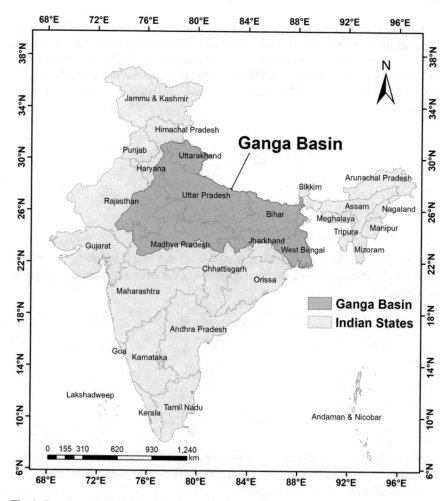

Fig. 1 Location of study area: Ganga River basin

It can be downloaded from the following link: https://ecocast.arc.nasa.gov/data/pub/gimms/3g.v1/.

3 Methodology

3.1 Standardized Precipitation Index (SPI)

Standardized Precipitation Index (SPI) (McKee et al. 1993) is a widely accepted indicator used for meteorological drought evaluation in a region. In the computation, cumulative probability is computed by fitting an appropriate probability distribution to a long-term time series of precipitation. Later, SPI values are obtained by applying inverse normal function to the computed cumulative probability values (Guttman 1998, 1999). Let x be the accumulated monthly precipitation in a given time scale (1-, 3-, 6-, 12-month, etc.). The cumulative probability function for Gamma function can be expressed as:

$$G(x) = \frac{1}{\beta^\alpha \Gamma(x)} \int_0^x x^{\alpha-1} e^{-x/\beta} dx \tag{1}$$

$$\Gamma(x) = \int_0^\infty x^{\alpha-1} e^{-x} dx \tag{2}$$

where $\Gamma(x)$ is the Gamma function, α and β are the shape and scale parameter, respectively. The maximum likelihood estimates of the function parameters are given as follows:

$$\alpha = \frac{1 + \sqrt{1 + \frac{4A}{3}}}{4A} \tag{3}$$

$$\beta = \frac{x}{\alpha} \tag{4}$$

$$A = \ln(\bar{x}) - \frac{\sum \ln(x)}{n} \tag{5}$$

where n is the number of data points in time series. The zero values of precipitation are not considered in Eq. (1), therefore, the modified cumulative probability considering zero precipitation values is computed as:

$$H(x) = q + (1 - q)G(x) \tag{6}$$

where q is the probability of $x = 0$ computed as number of events when with zero precipitation divided by total number of data points. The value of SPI is computed by transforming the value of $H(x)$ into standard normal random variable Z with mean zero and standard deviation of one.

$$SPI = Z = F^{-1}(H(x)) \tag{7}$$

3.2 Soil Moisture Drought Severity (SMDS) Index

Soil Moisture Drought Severity (SMDS) index is a probability-based index used to examine the propensity of agricultural droughts in a region. Long term time series (preferably more than 20 years) of monthly soil moisture is used to obtain estimates of SMDS index. The probability of soil moisture for a given month over n number of years may be computed using empirical cumulative probability function as:

$$P = \frac{m}{n+1} \times 100 \qquad (8)$$

where P is the soil moisture percentile (SMP) at a grid cell for a given month, m is rank of yearly data of a given month arranged in ascending order. A smaller value of P indicates more severe drought. The SMDS index is expressed as (Qin et al. 2015):

$$SMDS = \frac{100 - P}{100} \qquad (9)$$

Evidently, the value of SMDS index lies in the range from 0 to 1.

Various categories of drought as defined by Climate Prediction Center (CPC) of NOAA based on values of SMP are: moderate drought (11–20%), severe drought (6–10%), extreme drought (3–5%), and exceptional drought (0–2%). Accordingly, a drought event at a given cell is considered when the value of SMDS is greater than or equal to 0.8 (corresponding to SMP $\leq$ 20%). WMO (2012) has suggested 3–monthly and 6–monthly SPI to be comparable with soil moisture drought index (SMDS). Further, the value of standardized normally distributed quantile for 20% probability is −0.85 which is used as threshold for SPI in order to identify a drought event.

3.3 Normalized Difference Vegetation Index (NDVI)

The Normalized Difference Vegetative Index (NDVI) is widely used for characterization of vegetation in a region. Higher the NDVI value, denser and healthier the vegetation coverage in a basin. In the present study, the impact of drought events on vegetation health is examined by analyzing the spatial and temporal patterns in the Anomaly of Normalized Difference Vegetation Index (A-NDVI). Here, A-NDVI is computed as the deviation of NDVI values from the linear trend line fitted to the monthly time series. It is defined as:

$$A - NDVI(x) = NDVI(x) - T - NDVI(x) \qquad (10)$$

where, NDVI(x) is basin average annual mean NDVI in the year x and T-NDVI is the NDVI values of trend line corresponding to year x. Positive value of A-NDVI indicates improved growth of vegetation in a given year.

4   Results and Discussion

4.1   Soil Moisture Response to Space-Time Patterns in Rainfall and Temperature

The Ganga River basin is characterized with diverse climatic and topographical features with snow covered Himalayan ranges in the north to semi-humid and humid planes in the south. Accordingly, hydro-meteorological variables in the region represent varying distribution in space and time. Rainfall over Indian sub-continent is highly seasonal where majority of it (~ 70–80%) is received in monsoon season (June to September). A monthly break-up of basin average rainfall is shown in Fig. 2a.

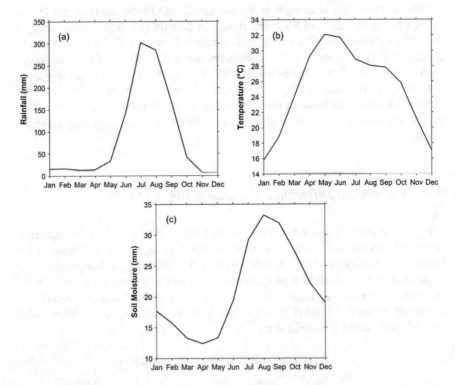

Fig. 2 Monthly distribution of basin average **a** rainfall, **b** temperature and **c** soil moisture (0–10 cm) in Ganga River basin

Evidently, monsoon months (JJAS) constitute major portion of total annual rainfall followed by a nearly dry non-monsoon period. Figure 2b represents the monthly distribution of mean temperature averaged over the basin. Mean temperature during winter season (January and February) ranges between 16 and 18 °C whereas hottest period is experienced in May month where temperature reaches as high as 35 °C approximately.

Soil moisture, on the other hand, represents a slightly delayed response to rainfall and temperature as penetration and movement of water within the soil profile take few days to months depending upon the soil composition and antecedent moisture content. A lag of nearly a month is evident in soil moisture of top 10 cm layer in Ganga River basin (Fig. 2c). A rise in soil moisture is triggered from the beginning of monsoon period leading to the peak in August month unlike rainfall which experiences peak in July month. In post-monsoon period, the recede of soil moisture is comparatively slow due to several factors including deep penetration of water, moisture holding capacity of soil, etc. Following a long dry spell of about seven to eight months, minimum soil moisture is reached in pre-monsoon season (March to May).

Owing to uneven topography and seasonal propagation of moisture bearing winds during monsoon, spatial distribution of rainfall is not homogeneous across the basin. Moisture laden winds enter the Indian subcontinent from the Bay of Bengal along southwest direction during monsoon season (JJAS) called the Indian summer monsoon (ISM). As monsoon rainfall constitutes major portion of total annual rainfall, spatial patterns in monsoon season are replicated in the annual rainfall. The spatial distribution of annual rainfall in Ganga River basin is shown in Fig. 3a. As can be seen, the southeast parts receive higher rainfall as ISM first enters in this region. A negative gradient can be observed in rainfall along south to north as moist winds get relatively drier as they propagate towards northwest. The northern parts of the basin are located in the Himalayas which provides an orographic barrier to the ISM resulting in higher rainfall in northern Himalayan belt. The annual rainfall in the basin is as high as about 3500 mm in some parts (northeastern and northwestern region) to as low as about 490 mm in others (northern region).

Figure 3c represents the spatial distribution of mean temperature in Ganga River basin. Evidently, the region with flatter topography experiences more or less same temperature of about 25–27 °C, approximately. On the other hand, parts of basin located at higher elevation in Himalayan region are comparatively colder with the mean temperature falling in the range 19–21 °C. Both rainfall and temperature are the crucial drivers of soil moisture dynamics in which rainfall defines the availability of moisture in the soil profile and temperature controls the loss of moisture from soil into the atmosphere. The spatial distribution of soil moisture in the basin is in uniformity with the spatial patterns of rainfall, substantially. The southeast and northwest regions have higher moisture availability in the top layer of soil profile similar to the patterns observed in rainfall. A positive gradient can be observed in soil moisture moving from south to north. However, higher soil moisture is also experienced in small parts located in southwest and southcentral region of the basin. It may possibly be due to presence of intense irrigation practice in the region resulting

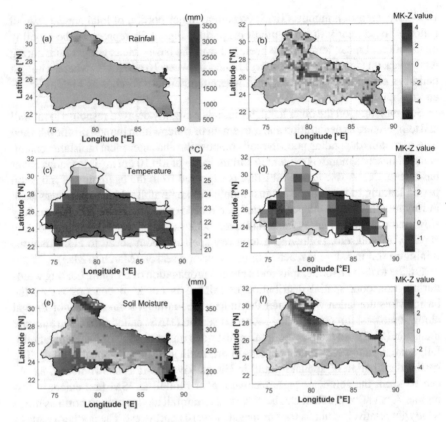

Fig. 3 **a**, **c**, **e** Spatial distribution of mean annual precipitation, temperature and soil moisture and **b**, **d**, **f** spatial distribution of MK-Z value (representing trend) of precipitation, temperature and soil moisture, respectively

in overall higher availability of moisture in soil even when magnitude of rainfall and temperature are same as that of nearby region.

A better understanding of the response of soil moisture to space-time change in meteorological variables can be developed by analyzing the trend followed by these variables. Figure 3b represents the trend in rainfall time series at grid points across Ganga River basin over the period 1948–2015. In the figure, statistically significant decreasing trend is evident in rainfall over central and northwest parts of the basin. The foothills of Himalayas receive comparatively higher rainfall in the northwest region, however, long term trend in rainfall is observed to be negative in the region. In some parts of the southeast region, significant increasing trends are evident in annual rainfall. These results show a scarcity of water in terms rainfall in north and central region, whereas, an abundance of water along southeast parts of the basin. The results of trend test applied on annual time series of mean temperature are shown in Fig. 3d. Mean temperature is observed to be following significant positive trend over southwest region of Ganga River basin. In remaining parts of the basin, temperature is found to be following decreasing trend insignificant at 5% significance level. The response of soil moisture to temporal patterns in rainfall and temperature is presented in Fig. 3f. As can be seen, significant decreasing trend is observed in soil moisture over northwest and northcentral parts of the basin. These falling trends are in uniformity with the decreasing trend observed in rainfall over the same region. Similarly, significant negative trend in soil moisture are also observed in southwest region of the basin. The decrease in soil moisture over southwest region may fairly be attributable to the increasing trend observed in temperature as no significant trend is observed in rainfall. Subsequently, annual rainfall has remained more or less constant over time whereas a significant increase in temperature may be causing a decrease in soil moisture over southwest parts of Ganga River basin. Over the eastern parts of the basin, soil moisture is observed to be following a significant increasing trend which is possibly resulting from increasing rainfall and decreasing temperature. Overall, northwest and southeast parts of the basin are highly affected from soil moisture deficit, thus, needing greater consideration from basin managers and water resources planners.

4.2 Droughts in Ganga River Basin

Several drought events are experienced in India over past one and a half century that are found responsible for devastating famines and huge loss of human life and economy of the country. Among these events, the most recent severe droughts have affected the northern parts of country, covering a major portion of Ganga River basin. In response, an improved understanding of soil moisture based drought events, especially in Gang River basin, becomes important for efficient management of water resources and basin planning. In the first step, temporal changes in drought affected area are computed in terms of meteorological (SPI) and agricultural (SMDS) drought over the period 1948–2015.

Drought affected area in each month is calculated based on thresholds (SPI-3 < − 0.85 and SMDS > 0.80) defined in the methodology section. Annual time series of affected area is computed by taking average of the monthly values and temporal patterns are identified using Mann-Kendall test (Mann 1945; Kendall 1975). Figure 4 shows the inter-annual variation in area affected by drought based on SPI and SMDS. The annual series of drought affected area based on both SPI and SMDS follow an increasing trend significant at 5% level of significance with MK-Z value of 2.63 and 2.91, respectively. The results show that the spatial extent of drought post-2000 has been consistently larger than the years before 2000. It indicates that increasing fraction of Ganga River basin is becoming susceptible to meteorological and agricultural drought in the recent times. In addition, spatial extent of drought defined on soil moisture is higher than that on precipitation for most of the years. Figure 4 further highlights that the SPI-based drought prone area before 1970 was lesser than that for SMDS-based drought. However, the number of years with SPI-based drought area larger than SMDS-based drought are observed to be increased post-1970. Droughts

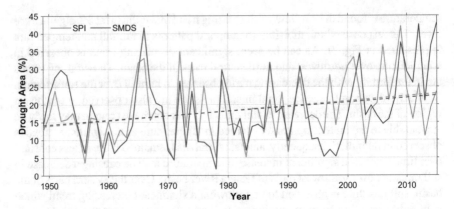

Fig. 4 Inter-annual variability of drought affected area between 1948–2015 based on SPI and SMDS in the Ganga River basin

have effected a considerable area of the basin from the beginning of analysis period. The decreasing trend in soil moisture further aggravates the deficit causing an increase in spatial extent of both meteorological and hydrological droughts.

Figure 5 represents 13-month moving average values of drought affected area derived from monthly values of SPI-3 and SMDS. A threshold value of 20% area is selected for identifying extreme drought events occurred in the past over the study period. Previous studies in India identified three major drought events (meteorological and hydrological) in the year 1987, 2002 and 2015 during the period 1948–2015 (Mishra et al. 2018, 2019). The drought events in year 1987 and 2002 affected mainly the southeast and northwest region of the country. However, 2015 drought event was concentrated in the Indo-Gangetic plain that caused severe water crisis in the region (Mishra et al. 2018, 2019). In the present study, meteorological and agricultural drought characteristics are studied distinctly over Ganga River basin considering its importance to national economy and food security of the country. Accordingly, eight and six such extreme drought events are identified in the basin based on SPI and SMDS, respectively.

The months corresponding to peak drought affected area for each event are shown in Fig. 5. Severe drought events with spatial coverage over 20% of the basin area are observed during 1950–1955, 1965–1966, 1979–1980, 1991–1993, 2001–2003, 2008–2009 and 2012–2015. Further, the frequency as well as spatial extent of both meteorological and agricultural drought events have increased after 1960. The most severe drought conditions in the basin is observed in the recent past where peak affected area had reached about 65% of the basin based on SPI and SMDS. The soil moisture droughts in the basin are accompanied by significant precipitation deficit computed using SPI values. It shows that the decrease in precipitation have major implications over soil moisture deficit observed in Ganga River basin over the period 1948–2015.

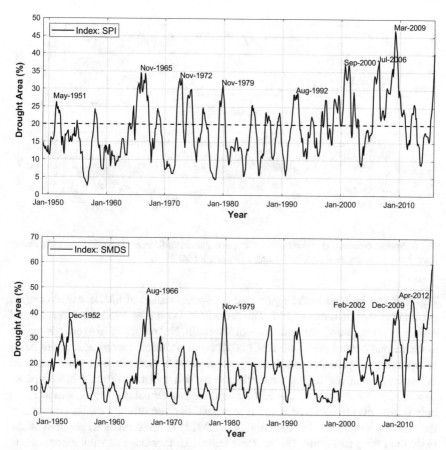

Fig. 5 Temporal variability of drought affected area between 1948–2015 computed as 13-month moving average values from monthly data based on **a** SPI and **b** SMDS. (Months corresponding to peak affected area are marked for severe drought events)

4.3 Spatial Pattern in Extreme Drought Events

The spatial distribution of three major drought events identified based on affected area is shown in Fig. 6. The three drought events selected based on SPI-3 are in the year 1965, 2000 and 2009 whereas based on SMDS are in 1952, 1979 and 2012. In case of meteorological droughts, the drought event in November 1965 occurred in the western and central parts of the basin whereas September 2000 event was concentrated mainly in central and eastern parts of the basin. The drought of March 2009 had the largest spatial extent which covered the northern, central and eastern parts of the basin. These results reveal that the location of drought affected areas has shifted over the period from western to eastern parts of the basin. Whereas, the central region of the basin had consistently been susceptible to droughts over the period 1948–2015. The spatial features of SMDS based droughts are also shown in

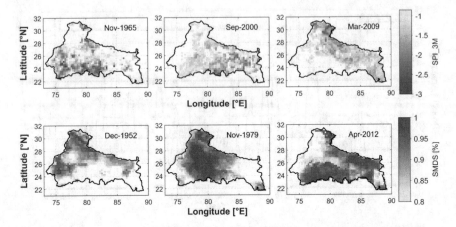

Fig. 6 Spatial distribution of three major drought events identified based on affected area in Ganga River basin between 1948 and 2015 using **a** SPI and **b** SMDS

Fig. 6. It can be seen in the figure that the spatial extent of SMDS based drought events is higher than that of SPI. The drought in December 1952 was concentrated mainly in western and northern parts of the basin, November 1979 event covered in northern and central parts and April 2012 event occurred over almost entire basin except for some parts in western region. It may also be noted here that parts of basin falling in Himalayan region were affected by soil moisture drought in almost all past events. However, in case of SPI-based droughts, Himalayan region was affected only during events occurred in the recent past. Further, drought events occurred in the recent past (December 2009 and April 2012) covered much larger area of the basin than the past events. The possible reason for increase in spatial extent of soil moisture based drought could be the continuous depletion of ground water levels and precipitation deficit in the basin.

4.4 Impact of Drought on Vegetation

The vegetation characteristics of the basin are examined using spatial and temporal patterns in NDVI value obtained GIMMS NOAA/AVHRR NDVI dataset. The spatial distribution of annual mean NDVI is shown in Fig. 7a. The parts of basin falling in the lesser Himalayan region are covered with dense forest thus representing higher NDVI value ranging between 0.6–0.9. On the other hand, the northern most parts of the basin falling in greater Himalayan region are covered with snow and glacier, thus, the value of NDVI is less than zero for these areas. The basin average value of annual NDVI lies in the range 0.4–0.45 which shows a fair coverage of vegetation in the basin. The temporal variation in annual mean NDVI value is shown in Fig. 7b. Annual NDVI in the basin is observed to be following an increasing trend over the period

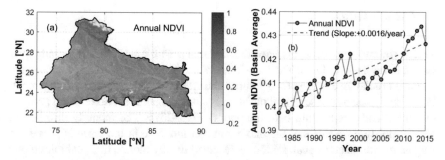

Fig. 7 **a** Spatial distribution of annual mean NDVI between 1982–2015. **b** Temporal variation and trend in basin average annual mean NDVI in Ganga River basin

1982–2015 at the rate of 0.0016/year. Increasing pattern in NDVI with increase in severity of both meteorological and agricultural droughts indicates that the effect of these drought events is not reflected over the vegetation characteristics of the basin. Development of canal irrigation projects and advancement in agricultural practices may be the possible reason for such contrasting behavior.

The inter-annual variation of A-NDVI with SPI and SMPS is shown in Fig. 8. Evidently, a poor correlation exists between A-NDVI values and both SPI and SMDS with R^2 value being 0.135 and 0.10, respectively. Though the inter-annual variation in A-NDVI and SPI are in uniformity before year 2000, considerable deviations are evident between the two values after 2000. On the other hand, the temporal variations in A-NDVI and SMDS are not uniform over the entire period. While severity of droughts has increased with time, increasing trend are also evident in NDVI which shows that the vegetation health in the basin is not effected by the past drought events. However, continued intensification of extreme drought events in the near future is expected to effect the vegetation health of the basin.

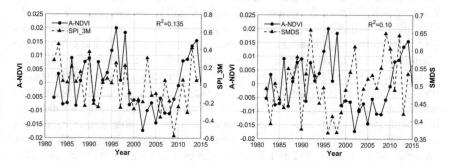

Fig. 8 Inter-annual variation of SPI and SMDS with A-NDVI between 1982 and 2015

5 Conclusions

Recent soil moisture droughts have resulted in a severe water crisis especially over Ganga River basin in India. The knowledge of characteristics of past droughts in the region is crucial for ensuring food security and efficient water resources management. In this study, the response of soil moisture is studied against spatiotemporal patterns in rainfall and temperature over the period 1948–2015. The monthly distribution of soil moisture reveals that the maximum soil moisture in the basin is present in August month whereas peak rainfall is received in July month, thus, exhibiting one month lagged response. In the results of trend test, significant negative trend is observed in annual soil moisture over northwest and southwest region of the basin. The decreasing trends in soil moisture over northwest parts are possibly due to significant negative trend in rainfall. On the other hand, significant increasing trend in temperature are found responsible for decreasing trend observed in soil moisture over southwest region of the basin. The characteristics of meteorological and agricultural droughts in Ganga River basin are examined using 3-monthly SPI and monthly SMDS indices, respectively. Temporal changes in spatial extent of drought affected area are examined for past SPI and SMDS based drought events. The area affected with both meteorological and agricultural drought is observed to be following significant increasing trend with MK-Z value of 2.63 and 2.91, respectively.

Severe soil moisture drought events in Ganga River basin are identified during 1950–1955, 1965–1966, 1979–1980, 1991–1993, 2001–2003, 2008–2009 and 2012–2015. Almost all soil moisture droughts were accompanied by severe deficit in rainfall highlighting dominance of rainfall in defining soil moisture response of the region. Among the past events, recent drought of 2012–2015 covered almost entire Ganga River basin. The recent droughts (meteorological and agricultural) have affected the eastern parts of the basin whereas past events were mainly concentrated in western region. The parts of Ganga River basin falling in Himalayan range are found consistently susceptible to soil moisture drought in almost all past event during 1948–2015. Frequent soil moisture drought events are postulated to have degrading effect over vegetation health of a region. The vegetation characteristics of the basin are examined through identifying temporal and spatial patterns in NDVI. The Ganga River basin is found to be characterized with fairly healthy vegetation cover with basin average NDVI value falling between 0.4–0.45. An increasing trend (0.0016/year) is observed in annual NDVI over the period 1982–2015. The contrasting behavior of vegetation in the basin against soil moisture deficit is attributable to the advancement in agricultural practices after green revolution and considerable shift in fresh water consumption for irrigation from surface water sources to groundwater. Subsequently, inter-annual variation in SPI and SMDS reflect poor correlation with annual NDVI with R^2 value being 0.135 and 0.10, respectively. In the coming future, continued soil moisture deficit may have serious implication over crop yield and food security of the region. The present study investigates the characteristics of past meteorological and agricultural drought and their impact over vegetation of Ganga River basin. As more number of farmers in basin have adopted groundwater as a primary source of water for irrigation practices, the impact of deficit in rainfall and soil moisture should also be studied on quantity and quality of groundwater reserves.

Acknowledgements The authors are grateful to the anonymous reviewers for their useful comments and suggestions.

References

Andreadis KM, Clark EA, Wood AW, Hamlet AF, Lettenmaier DP (2005) Twentieth-century drought in the conterminous United States. J Hydrometeorology 6(6):985–1001

Dai A (2011) Drought under global warming: a review. Wiley Interdisci Rev Clim Change 2(1):45–65

GFCC (2009) Ganga basin, ganga flood control commission, Patna, Ministry of Water Resource, Govt. of India

Guttman NB (1998) Comparing the palmer drought index and the standardized precipitation index 1. JAWRA J Am Water Res Assoc 34(1):113–121

Guttman NB (1999) Accepting the standardized precipitation index: a calculation algorithm 1. JAWRA J Am Water Res Assoc 35(2):311–322

Heim RR Jr (2002) A review of twentieth-century drought indices used in the United States. Bull Am Meteorol Soc 83(8):1149–1166

Kendall MG (1975) Rank correlation methods 4th ed. Charles Griffin, p 202

Long D, Scanlon BR, Longuevergne L, Sun AY, Fernando DN, Save H (2013) GRACE satellite monitoring of large depletion in water storage in response to the 2011 drought in Texas. Geophys Res Lett 40(13):3395–3401

Mann HB (1945) Nonparametric tests against trend. Econometrica: J Econometric Soc 13(3):245–259

McKee TB, Doesken NJ, Kleist J (1993) The relationship of drought frequency and duration to time scales. In: Proceedings of the 8th conference on applied climatology. vol 17. no 22. American Meteorological Society, Boston, MA, pp. 179–183

Mishra AK, Singh VP (2010) A review of drought concepts. J Hydrol 391(1–2):202–216

Mishra V, Shah R, Thrasher B (2014) Soil moisture droughts under the retrospective and projected climate in India. J Hydrometeorology 15(6):2267–2292

Mishra V, Shah R, Azhar S, Shah H, Modi P, Kumar R (2018) Reconstruction of droughts in India using multiple land-surface models (1951–2015). Hydrol Earth Syst Sci 22(4):2269–2284

Mishra V, Tiwari AD, Aadhar S, Shah R, Xiao M, Pai DS, Lettenmaier D (2019) Drought and famine in India, 1870–2016. Geophys Res Lett 46(4):2075–2083

Moran MS, Clarke TR, Inoue Y, Vidal A (1994) Estimating crop water deficit using the relation between surface-air temperature and spectral vegetation index. Remote Sens Environ 49(3):246–263

MoWR (Min. of Water Resources, Govt. of India) (2014) Ganga Basin—Version 2.0

Palmer WC (1965) Meteorological droughts. US Department of Commerce, Weather Bureau, Research Paper No. 45, pp 58

Palmer WC (1968) Keeping track of crop moisture conditions, nationwide: the new crop moisture index. Weatherwise 21:156–161

Qin Y, Yang D, Lei H, Xu K, Xu X (2015) Comparative analysis of drought based on precipitation and soil moisture indices in Haihe basin of North China during the period of 1960–2010. J Hydrol 526:55–67

Sathyanadh A, Karipot A, Ranalkar M, Prabhakaran T (2016) Evaluation of soil moisture data products over Indian region and analysis of spatio-temporal characteristics with respect to monsoon rainfall. J Hydrol 542:47–62

Shah RD, Mishra V (2015) Development of an experimental near-real-time drought monitor for India. J Hydrometeorology 16(1):327–345

Thomas AC, Reager JT, Famiglietti JS, Rodell M (2014) A GRACE-based water storage deficit approach for hydrological drought characterization. Geophys Res Lett 41(5):1537–1545

Tsakiris G, Vangelis HJEW (2005) Establishing a drought index incorporating evapotranspiration. Eur Water 9(10):3–11

Verma N, Devrani R, Singh V (2014) Is ganga the longest river in the ganga basin, India? Curr Sci 107(12):2018–2022

Vicente-Serrano SM, Beguería S, López-Moreno JI (2010) A multiscalar drought index sensitive to global warming: the standardized precipitation evapotranspiration index. J Clim 23(7):1696–1718

Wang A, Lettenmaier DP, Sheffield J (2011) Soil moisture drought in China, 1950–2006. J Clim 24(13):3257–3271

World Meteorological Organization (WMO) (2012) Standardized precipitation index user guide (WMO-No.1090), Geneva

Strategic Analysis of Water Resources in the Ganga Basin, India

Jyoti P. Patil, Suman Gurjar, C. A. Bons, and M. K. Goel

Abstract GangaWIS is a comprehensive tool that integrates various hydrological components of the Ganga River basin and supports the policymakers in analyzing the impact of future developments and climate change scenarios in combination with multiple interventions. It describes the functioning of the water system of the Ganga basin within India concerning rainfall-runoff, surface water and groundwater flow, storage and diversion of water for various purposes, water quality, and ecology. The hydrology and the rainfall-runoff process has been divided into two different models: SPHY and WFlow. They are both fully distributed models working on a grid of square cells. SPHY is used to describe the hydrological process in the mountainous areas in the Himalaya. The rainfall-runoff processes for the non-mountainous part of the Ganga Basin are simulated with the WFlow model. The river discharges calculated by the SPHY model for the Himalayas are used as upstream boundaries for the WFlow model. The water resources model RIBASIM describes the management and use of water. Its hydrological input is derived from the river discharges calculated by WFlow. RIBASIM uses a schematization of links and nodes to describe the flow of water in the rivers, the storage in reservoirs, the diversion into canals, and the use and return flow by different functions. Furthermore, return flows can be divided over rivers, canals, and Groundwater. The information on discharges and water levels calculated by RIBASIM are used by the groundwater model to describe the interaction between surface and Groundwater. Groundwater movement is simulated with iMOD, the Deltares extension of the well-known MODFLOW

J. P. Patil (✉) · S. Gurjar · M. K. Goel
National Institute of Hydrology (NIH), Roorkee, India
e-mail: jyoti.nihr@gov.in

S. Gurjar
e-mail: sumangurjar25@gmail.com

M. K. Goel
e-mail: goel.mk1@gmail.com

C. A. Bons
Deltares, Delft, The Netherlands
e-mail: Kees.Bons@deltares.nl

© The Author(s), under exclusive license to Springer Nature Switzerland AG 2021   309
M. S. Chauhan and C. S. P. Ojha (eds.), *The Ganga River Basin: A Hydrometeorological Approach*, Society of Earth Scientists Series,
https://doi.org/10.1007/978-3-030-60869-9_20

code for solving the groundwater flow equation. iMOD uses the same calculation grid as Wflow but is only applied to the alluvial fraction of the basin. The Ganga river basin model is capable of assessing the impacts of future developments/climate change scenarios and various interventions/measures at basin scale by comparison of simulation results. Using this model, different scenarios are developed: present, pristine, and 2040 with three different possible climate change developments: no climate change, climate change following the RCP4.5 scenario, and climate change following the RCP8.5 scenario. The scenarios used model parameters such as precipitation, temperature, land use, infrastructure, population, industry, and agriculture to develop a corresponding model output of the river flow, water quality, and groundwater levels. The dashboard of GangaWIS depicts the various indicators to assess the impact of the different scenarios, such as state of groundwater development, lowest discharge, volume of water stored in reservoirs, agricultural crop production, deficit irrigation, and drinking water, surface water quality index, the amount of Groundwater extracted and ecological, hydrological and socio-economic status. GangaWIS can be used to analyze and visualize various data (temporal/spatial) and model results, and it can provide relevant measured and modeled information to multiple users such as data managers, modelers, policymakers, and decisionmakers.

Keywords Ganga basin · iMOD · Ribasim · SPHY · Strategic planning · Water quality · WFlow

1 Introduction

Water is a supporter of natural systems and lives on earth. For centuries, it has been observed that societies are evolved on riverbanks, lakes, or available water resources. The water streams or rivers provide life to habitat, including people and the ecosystem. It also supports the spiritual, cultural, and recreational activities of the society. The ever-growing population, industrial pollution, inefficient irrigation practices, encroachment, flood plain alteration, and many other activities put pressure on available water resources and make their planning difficult. Therefore, river basin planning plays an important role in supporting the decisionmakers for managing available water resources of the area. However, basin planning should be strategic, considering social and environmental benefits as well as the overall economic development of the basin by setting the best possible water management goals with the help of different stakeholders and government bodies.

Ganga river basin is the largest river basin in India, extending over the states of Uttarakhand, Uttar Pradesh, Haryana, Himachal Pradesh, Delhi, Bihar, Jharkhand, Rajasthan, Madhya Pradesh, Chhattisgarh, and West Bengal. It lies between East longitudes 73° 30 and 89° 0 and North latitudes of 22° 30 and 31° 30, covering an area of 1,086,000 km^2, extending over India, Nepal and Bangladesh (Fig. 1). It has a catchment area of 861,404 km^2 in India, constituting 26% of the country's landmass and supporting about 43% of the population. The annual average rainfall

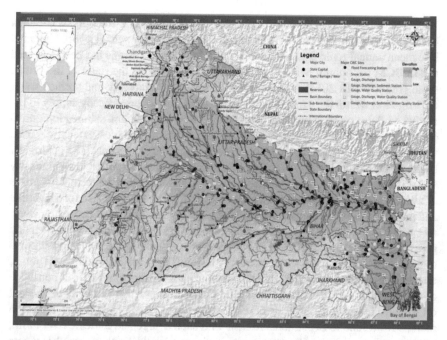

Fig. 1 Ganga basin with sub-basin, drainage, and major water resources structure (*Source* India-WRIS 2014)

in the basin varies between 39 cm, 200 cm, with an average of 110 cm. 80% of the rainfall occurs during the monsoon months, i.e., between June and October. Because of large temporal variations in precipitation over the year, there is wide fluctuation in the flow characteristics of the river. Rainfall, subsurface flows, and snowmelt from glaciers are the main sources of water in river Ganga (NMCG 2017).

More than 60% of the water flowing into the Ganga basin comes from the Himalayan streams joining the Ganga from the north. The Peninsular streams combine to contribute only 40% of the water, even though the catchment area of the peninsular streams extends well over 60% of the entire Ganga basin. The reported average Water Resources Potential of Ganga basin is 525 BCM. According to the assessment, the total utilizable surface water resource in the Ganga is 250 BCM. The tributaries which contribute the largest amount of water per annum are, the Ghaghara including the Gomti (113.5 BCM), followed by Kosi-Mahananda (81.85 BCM), the Gandak-Burhi Gandak together (58.96 BCM), Yamuna (57.2 BCM), Sone-East of Sone (44.14 BCM), the Chambal (32.55 BCM) and Ramganga (17.79 BCM) (NRCD 2009).

The Ganga basin consists of about 276,947 surface water bodies. The majority of water bodies that account for 98.9% of total water bodies having a size range of 0–25 ha. There are 23 major water bodies larger than 2500 ha. The Ganga average water resource potential is 525,020 MCM, out of which 250,000 MCM is potentially utilizable surface water. The total live storage capacity of the Ganga basin is 42,060

MCM, with an extra amount of 18,600 MCM still under construction. The Ganga basin has a vast reservoir of Groundwater, replenished every year at a very high rate. The mean annual replenishable Groundwater in India as a whole has been assessed at 433 BCM per annum, with about 202.5 BCM (46.8%) in the states of the Ganga basin (India-WRIS 2014). The conjunctive use of groundwater and surface water for irrigation within the canal command areas not only ensures steady supply to the cultivated fields on time but also helps reduce waterlogging and salinization due to consequent downward movement of subsurface moisture. The most extensively used water sources for irrigation in the basin are the groundwater wells. The ecological health of the Ganga River and some of its tributaries have deteriorated significantly as a result of high pollution loads (from the point and non-point sources), high levels of water abstraction for consumptive use (mostly for irrigation but also for municipal and industrial uses), and other flow regime and river modifications caused by water resources infrastructure (dams and barrages for diverting and regulating the river and generating hydropower). The major water resources challenge in the Ganga basin is to judiciously manage the water resources to fulfill the ever-increasing agriculture, domestic, and industrial water demands without harming the eco-system. About 77% of the population in the basin is engaged in agriculture, which is mostly dependent on irrigation, as almost 85% of rainfall in the basin takes place in four monsoon months from June to September. To deal with such a large and intricate river basin, in cooperation with the Government of India, the World Bank assigned Deltares and its partners AECOM India and FutureWater to develop a strategic planning tool 'GangaWIS' to deal with such a large and intricate river basin. GangaWIS (Ganga Water Information System), is based on Deltares' platform FEWS (Bons and Vander Vat 2016).

The objective of this article is to give an overview of GangaWIS, its framework components, and its application to develop a set of scenarios for the development of the Ganga basin, to build a strong and accessible knowledge base to support strategic basin planning in river Ganga.

2   Material and Methods

The preparation of a river basin model and a water information system for the Ganga basin, jointly called GangaWIS, was an important component. The Ganga river basin model aims to support strategic basin planning by assessing the impact of different scenarios. The water information system serves to store and disseminate all relevant information for planning, i.e., maps, measurements, and input and output of the river basin model. The Ganga river basin model has a wide scope that allows an integrated assessment of impacts related to hydrology, geohydrology, water resources management, water quality, and ecology. Although the level of detail is limited to keep the model manageable and the complexity understandable, the model contains sufficient detail for meaningful assessment of strategies and scenarios. The model area covers the Ganga river basin within India. Upstream parts of the basin in Nepal and China have been included in calculating flows to the Indian part of the basin (Bons 2018).

2.1 GangaWIS Framework

The main purpose of GangaWIS was to build and apply a comprehensive rainfall-runoff and river system model, including water quality and ecological health module for the entire Ganga river basin. Therefore, a robust modeling framework was developed that can assess the consequences of alternative strategies or development scenarios on the environment and for different water use sectors (Vat et al. 2016). The GangaWIS framework is based on the following individual sub-models:

1. Distributed catchment hydrological models for the simulation of rainfall-runoff and subsurface flows on a daily time step and fine spatial resolution (WFLOW—SPHY);
2. 3D geo-hydrological model to simulate the groundwater dynamics, the interaction with the surface water such as the Ganga River and the impact of groundwater extraction (iMOD-MODFLOW);
3. A water management tool to simulate the Ganga River Basin, the operation of its water resources infrastructure, water demand and allocation (RIBASIM);
4. A catchment water quality (DWAQ) and ecology model to quantify catchment loads and pollutant concentrations as well as the impact on the ecology of the river and its floodplains and the ecosystem services;
5. The Ganga Water Information System (WIS) and dashboard to store and manage all data and model results in a geodatabase and to allow dedicated querying of model results; and
6. Delft-FEWS as a connector between the models and the information system.

A schematic overview of the complete river basin modeling framework and individual submodels is presented in Fig. 2 (Van der Vat 2018). Data from the storage layer of GangaWIS feed the models, and the model output is stored in the system for future dissemination. The workflow routines are managed by Delft-FEWS.

The description of the models, integrated with GangaWIS, is given below.

2.1.1 SPHY

SPHY is the hydrological model, which is used in the Himalayan parts of the Ganga basin. The SPHY model calculates the discharge (m^3/s) at the outflow locations of the model, to be used as input to the Wflow model. The SPHY model also calculates snow and glacier melt processes.

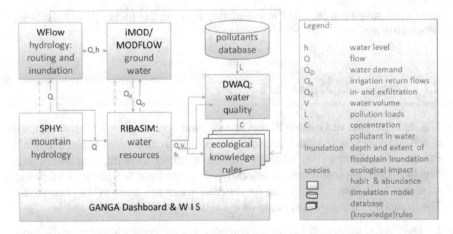

Fig. 2 Components of the Ganga basin model and their interaction

2.1.2   Wflow

Wflow calculates the runoff from the meteorological inputs, precipitation, temperature, and evaporation. For each cell, Wflow calculates the horizontal processes of the runoff, the direction of the flow, following the local drainage direction derived from the elevation model, and the vertical processes, how much water will infiltrate, and how much will be a direct runoff. The water that infiltrates is the recharge. The recharge calculated by Wflow for the non-irrigated areas in the basin is one of the boundary conditions of the iMOD model.

2.1.3   Ribasim

Ribasim is the water management software to assess the demands from domestic, industrial, and agricultural uses and to then divert water to those users. It calculates all non-natural fluxes in the water system, e.g., water demand and return flow from irrigated areas (Krogt 2008). As input, Ribasim uses the natural discharge calculated by Wflow for each sub-basin. The sub-basins are derived from the Ribasim schematization, making sure that for each node in the Ribasim schematization, the input can be generated from the Wflow model.

Ribasim calculates the return flow from irrigated areas to the Groundwater (recharge) for each irrigation node in the schematization. The scalar time series from each node are spread over the corresponding area of the irrigated zones in iMOD.

Abstractions from the Groundwater are calculated in Ribasim. This includes abstractions for irrigation, domestic and industrial usage. The abstractions are taken from the Groundwater and thus must be connected to the iMOD model as well. The abstractions are grouped per district and are sent to the iMOD model.

Ribasim then calculates the discharge in the rivers and main canals. There is always leakage from these rivers and canals to the Groundwater; this leakage can be simulated with the iMOD model. For this simulation, water levels in the rivers and canals are required. To calculate the water levels, the discharge from Ribasim is translated to a water level using Q-h relations for each section. The calculated water levels then input to the iMOD model.

2.1.4 iMOD

iMOD is used to simulate groundwater levels and flows. The input to the iMOD model is described in the previous sections. As an output, iMOD calculates the water levels in the basin and the exchange of water with the surface water. This flux is used as an input of the next iteration run of the Ribasim model to calculate the total water balance of the system per district and for the whole basin.

2.1.5 Dwaq

Dwaq is the Water Quality model used in GangaWIS. It is based on a rich library from which relevant substances and processes can be selected.

2.1.6 Ecological Knowledge Rules

The ecological module allows a combination of quantitative and qualitative information in a semi-quantitative evaluation framework based on location-specific knowledge rules. The definition of these knowledge rules is a collaborative effort of experts and stakeholders.

2.2 Modeling in GangaWIS

The workflow in GangaWIS starts with the generation of meteorological input outside the system and importing this data into the database. Once the meteorological data is in the database, the SPHY model is run. The SPHY model generates input for the Wflow model. Together, Wflow and SPHY calculate the hydrological inputs for the Ribasim model. All water use and water management data are entered into the Ribasim schematization by the modeler. The schematization is included in GangaWIS.

The Ribasim model and the iMOD model influence each other. For example, the exchange of water between the rivers and the Groundwater is simulated by iMOD based on water levels provided by RIBASIM. This again influences the amount of

Table 1 Concepts employed in GangaWIS

Scenario	Describes developments exogenous to the decision under consideration, i.e., developments that cannot be controlled by the decision 1. Includes a static climate, based on rainfall and temperature projections, and a level of socio-economic developments
Strategy	Consists of a logical combination of measures that may provide a solution to a given water resources system problem
Case	It is a unique combination of a strategy and a scenario
Indicator	An indicator describes the state of the water resources system Indicators are defined at different aggregation levels in time and space; they are provided by GangaWIS components through overview graphs, maps, and scorecards. The indicators are presented on the dashboard
Case management	The state-specific measures and strategies can be configured as a case But, calculations for any specific case will always be performed for the entire Ganga River basin

water in the rivers. Therefore, the exchange flux between the rivers and the Groundwater, calculated by iMOD, must be fed back to the Ribasim model. Because the models are not coupled directly, i.e., on a time step basis, Ribasim must be run a second time after the iMOD run before a realistic and stable result can be achieved. Finally, the results are displayed on the dashboard with the help of indicators.

Following concepts are employed in GangaWIS (Table 1):

3 Application of the Integrated Model

Water management in the Ganga Basin faces significant challenges to cope with present population pressures and climate change impacts and uncertainties surrounding future developments. The GangaWIS model framework was built up with available historical and present data to stimulate the processes in the basin and reflect changes as a result of human or natural interventions in past years. The model was calibrated and validated with the observed data at different locations in the Ganga basin, and then it was used for analyzing the behavior of the system under different conditions. The strategic basin planning includes analysis of the present situation, the scope of future development, and identify the factors which will be impacted with upcoming development (Bons 2018).

3.1 Scenarios

Five scenarios are considered in the assessment using the GangaWIS: Present scenario, pristine scenario, 2040 with present climate scenario, 2040 with moderate future climate change (RCP 4.5), and 2040 with extreme future climate change (RCP

Table 2 Scenarios applied to run GangaWIS

Parameters	Present scenario	Pristine scenario	2040 scenario
Land use	LULC (NMCG 2013)	The natural vegetation of climatic zones	LULC (NMCG 2013)
Infrastructure	Main barrages, dams, canals, treatment plants, and drains are included	None	Main barrages, dams, canals, treatment plants, and drains are included
Population	2011 census data	None	Projected autonomous growth to 2040
Industry	CPCB data	None	Projected autonomous growth to 2040
Irrigated agriculture	Based on the Ministry of Agriculture and Farmers Welfare data	None	Projected autonomous growth to 2040
Temperature	Based on EUWATCH and FWDEI data	Based on EUWATCH and FWDEI data	Based on EUWATCH and FWDEI data and RCP 4.5 or 8.5 scenario of greenhouse gas concentration
Precipitation	Based on IMD observations for India	Based on IMD observations for India	Based on IMD observations for India and RCP 4.5 or 8.5 scenario of greenhouse gas concentration

8.5). The parameters and data sources used for the different scenarios are summarized in Table 2.

The scenarios settings were used to develop model inputs parameters like land use, infrastructure, population, industry, and agriculture settings as well as precipitation and temperature, which approximate the expected situation, so that the model output provides a simulation of the water uses, river flow, water quality, and groundwater levels.

The 'pristine scenario' describes the water resources situation before human interventions changed the Ganga basin. It is a reference case for comparison of the present and future simulated scenarios and strategies.

The present scenario described the situation in the basin around 2015, for which the most recent data is available.

**2040 was selected as a suitable year to develop predictions of future conditions that would make sense in planning. Apart from a scenario without a change of climate, two scenarios of climate change were generated based on RCP 4.5 and 8.5 scenarios of IPCC.

Table 3 Strategies used in GangaWIS

Strategy	Description
Do nothing	All existing infrastructure, diversions, irrigated areas and wastewater treatment systems as included and assumed to continue to operate as they do at present
Approved infrastructure	Presently approved infrastructure developments including irrigation projects, expansions of irrigation area, new storage, diversions, or conveyance capacity of infrastructure
Inter Basin Transfer Links (IBTL)	Upstream of Farakka and The Yamuna–Rajasthan links which export water out of the Ganga basin
NMCG planned treatment	Additional treatment as planned by NMCG
Improved treatment	A significant higher increase in treatment capacity than planned by NMCG WWTP
Increased irrigation efficiency with 20%	20% higher efficiency in all agricultural areas
Conjunctive use	Groundwater abstraction capacity is reduced by half for all over-extracted nodes
E-flow	Reduction of diversions to realize "Moderate" environmental flow status (between 40 and 60% deviation from the pristine reference situation)

3.2 Strategies

The various strategies used to run the GangaWIS are summarized in Table 3.

3.3 Indicators

The results of the planning models and assessment studies are more effective if presented as indicators on which decision-makers base their decisions. The following indicators were used in GangaWIS to show the results in the Dashboard (Table 4).

4 Results and Discussion

The GangaWIS model was run for the different scenarios and strategies, as illustrated in earlier sections, and the results were analyzed on the dashboard, comparing the indicator scores for various scenarios. All scenarios described have impacts on the water resources situation in the basin, primarily when no additional interventions are implemented.

Table 4 List of indicators used in dashboard of GangaWIS

Indicator	Description
State of groundwater development	Percentage of the area where the simulated groundwater abstraction amounts to 90% or more of the simulated recharge
Lowest discharge	Lowest monthly simulated discharge at the end of basin or state
The volume of water stored in reservoirs	The total sum of simulated water stored in the main basin reservoirs at the end of the monsoon period, October, for a 1/10 dry year
Agricultural production	The ratio between the actual and potential harvested area at basin/state level
A deficit in irrigation water	Difference between simulated irrigation water supply and simulated demand as a percentage of the simulated demand for a 1/10 dry year
A deficit in drinking water	Difference between simulated drinking water supply and simulated demand as a percentage of the simulated demand for a 1/10 dry year
The volume of groundwater extracted	The total simulated volume of Groundwater, in billion cubic meters, abstracted for public water supply and irrigation during the 1/10 driest hydrological year, the year with the 1/10 highest abstraction
Surface water quality index	CPCB classification in five categories (A to E) based on criteria for designated best use using simulated concentrations
Environmental flow	Three main environmental flow indicators are calculated within the Ganga river basin model: hydrological changes, species habitat suitability, and ecosystem service availability

1. The impacts are most visible in the hydrological indicators when results of present scenarios were compared with 2040 (RCP 4.5 and RCP 8.5) in the 'Do Nothing' strategy. The percentage of areas with critical groundwater use increases significantly, and the lowest discharge flow in the river in dry years reduces significantly.

2. The scenario assessments indicate a significant decrease in future water availability, water quality, and ecological status in the event no additional interventions are made. Future changes are mainly determined by socioeconomic factors, much less by climate change.

3. The intervention that has the most positive impact is the improvement of municipal wastewater treatment. Whether centralized or decentralized, whether high or low technology, reduction in pollution loads gives a return on investment both in the availability of clean water for downstream uses, including ecosystem services, as well as a drastic reduction in water-related illnesses and deaths.

4. The next intervention, which has a beneficial impact is the increase in efficiency of all water uses: irrigation, domestic, and industrial water use. However, it can be expected that farmers will increase their cropped areas in tune with the increased efficiency resulting in higher production, but not fewer abstractions from surface or Groundwater.

5. There is no single intervention that solves all problems. Combinations of different interventions are required. However, the set of currently considered far-reaching interventions, requiring huge investments and facing significant technical challenges, is insufficient to deal with future difficulties regarding water availability, water quality, and ecology, let alone restoring the system to present conditions.

6. All users must realize that water availability will be insufficient to meet all the rising demands, and there are no easy and simple technical solutions.

7. Robust strategies need to be implemented, aiming at a reduction in demands in all sectors, but at the same time, trade-offs need to be made between different sectors. The agricultural sector will have to adapt to lower water availability in terms of choice of crops, planting season, and water efficiency. Farmers will need to develop a flexible approach: depending on the monsoon, they may have to select irrigated or non-irrigated crops even when irrigated crops are already of high efficiency.

8. The consequences of these conclusions are far-reaching and involve departments and ministries outside the traditional water resources realm. Non-technical interventions such as incentives to change cropping patterns and practices to reduce water demand are needed.

9. Impact of strategies: The strategy that includes the approved projects will have a beneficial effect on agricultural crop failures, achieving a two percent reduction. Groundwater usage will decrease slightly, accompanied by an increase in reservoir volumes at the end of the monsoon season. When the approved infrastructure and the identified IBTL are assumed to be operational, agricultural crop production is increased significantly. However, most resources that are made available in the IBTL are consumed in the additional irrigated areas along the canals, and little water is transferred to the destination river. The strategies with planned treatment and improved treatment do impact basin water quality, but as the indicator is determined by the lowest value, this is not visible in the water quality indicator scores at the basin level. However, the ecological status and socio-economic, environmental flow indicators that include fish increase, primarily because of improved oxygen levels in the rivers. The environmental flow strategy that reduces abstractions scores significantly better on the hydrological and socio-economic environmental flow indicators as it revives some of the river dynamics. However, as water will be less available for irrigation and other purposes, the groundwater extraction will increase, resulting in more over-extracted areas and a significantly higher deficit in irrigation and drinking water (Bons 2018).

10. Impact of combination of strategies: When a number of the suggested strategies are combined, it is expected that this powerful approach will achieve significant benefits. For this assessment, the interventions on approved infrastructure, IBTL, 20% efficiency increase in agriculture, reduced groundwater use in over-extracted areas, and improved wastewater treatment were combined in one run. The main conclusion from the results is that the combination of interventions does improve the situation compared to the 2040_RCP4.5 scenarios somewhat, primarily because it will produce more agricultural output with almost the same indicator values. However, the strategy is nowhere near sufficient to achieve the present conditions. In other words, the conditions in the basin will significantly deteriorate between now and 2040, even if all these interventions are implemented. Results would be better in terms of irrigation and drinking water deficits when groundwater abstractions are not limited in over-extracted areas. However, that would be a very unsustainable solution as Groundwater would be severely depleted, leading to the same shortages at a later stage (Bons 2018).

5 Conclusion

The content presented here is the strategic analysis of water resources in the Ganga basin using an integrated tool developed by Deltares and its partners AECOM India and FutureWater, in cooperation with the Government of India. The Ganga water information system (GangaWIS), which combines models with a database and tools, presents the input data, and the simulation results in graphical and map format. The system is operational for impact assessment of socio-economic and climates changes scenarios and management strategies.

All users of water resources in the Ganga Basin must realize there will not be enough water to meet all the rising demands, and there are no simple and one technical solution. Ambitious strategies need to be implemented that reduce requirements in all sectors; at the same time, trade-offs need to be made between different sectors. The agricultural sector will have to adapt to lower water availability in terms of crop choice, planting season, and water efficiency. Farmers will need to develop a flexible approach; depending on the monsoon, they may have to select irrigated or non-irrigated crops even when irrigated crops are already of high efficiency. Domestic and industrial demands for the year could be 'reserved' in reservoirs so that domestic and industrial supply can always be met in the downstream regions, allocating only the non-reserved volume to agriculture. The findings of the scenarios and strategies envisaged in GangaWIS will be helpful to a decisionmaker and all the stakeholders of the Ganga basin, India.

Acknowledgements This project was funded by the South Asia Water Initiative under World Bank contract number 8005347. The funding support from the governments of Australia, Norway, and the United Kingdom are gratefully acknowledged. Authors would like to acknowledge The World Bank and Government of India for carrying out the work in the Ganga basin and for associating the authors to this development. The content of this chapter is mainly from the reports published for the work on 'Strategic Basin Planning for Ganga River Basin in India.' The permission given by The World Bank for publishing this work is duly acknowledged by the authors.

References

Bons CA (2018) Strategic basin planning for Ganga river basin in India—Ganga river basin planning assessment report. Main volume and appendices. Deltares with AECOM and FutureWater for the World Bank and the Government of India, Report 1220123-002-ZWS-0003

Bons CA, Van der Vat M (2016) Analytical work and technical assistance to support strategic basin planning for Ganga river basin in India. Inception report. World Bank Project 1220123-000, Reference 1220123-000-ZWS-0013

India-WRIS (2014) Ganga basin report. India-WRIS project, RRSC-WEST, NRSC, ISRO, Jodhpur, India

Krogt VD (2008) RIBASIM version 7.00. Technical reference manual. WNM Deltares, Delft

NMCG (2013) Ganga river basin management plan. Interim report. Prepared by Consortium of 7 Indian Institute of Technologies (IITs). Ministry of Environment and Forests (MoEF), GOI, New Delhi

NMCG (2017) Namami Gange. Projects along river Ganga available under private funding. National Mission for Clean Ganga (NMCG). CSR e-book. Available at https://nmcg.nic.in/csr/csrebook/csrebook.html#p=7

NRCD (2009) Status paper on River Ganga. State of environment and water quality. National River Conservation Directorate, Ministry of Environment and Forests, Government of India

Van der Vat M (2018) Ganga river basin model, and information system. Report and documentation. Deltares, Delft, The Netherlands. Available online: https://cwc.gov.in/sites/default/files/ganga-river-basin-modeland-wis-report-and-documentation.pdf. Accessed on 10 Oct 2019

Van der Vat M, Lutz A, Mark H, Boderie P, Roelofsen F, Fernando MM, Langenberg V, Bons K (2016) Conceptualization of river basin model, surface water groundwater interaction analysis, and environmental flow assessment source of Sarda Sahay canal command information

Printed in the United States
by Baker & Taylor Publisher Services